T0189342

Lecture Notes in Computer Science 14721

The series Lecture Notes in Computer Science (LNCS), including its subseries Lecture Notes in Artificial Intelligence (LNAI) and Lecture Notes in Bioinformatics (LNBI), has established itself as a medium for the publication of new developments in computer science and information technology research, teaching, and education.

LNCS enjoys close cooperation with the computer science R & D community, the series counts many renowned academics among its volume editors and paper authors, and collaborates with prestigious societies. Its mission is to serve this international community by providing an invaluable service, mainly focused on the publication of conference and workshop proceedings and postproceedings. LNCS commenced publication in 1973.

Fiona Fui-Hoon Nah · Keng Leng Siau
Editors

HCI in Business, Government and Organizations

11th International Conference, HCIBGO 2024
Held as Part of the 26th HCI International Conference, HCII 2024
Washington, DC, USA, June 29 – July 4, 2024
Proceedings, Part II

Editors
Fiona Fui-Hoon Nah
Singapore Management University
Singapore, Singapore

Keng Leng Siau
Singapore Management University
Singapore, Singapore

ISSN 0302-9743 ISSN 1611-3349 (electronic)
Lecture Notes in Computer Science
ISBN 978-3-031-61317-3 ISBN 978-3-031-61318-0 (eBook)
https://doi.org/10.1007/978-3-031-61318-0

This Springer imprint is published by the registered company Springer Nature Switzerland AG
The registered company address is: Gewerbestrasse 11, 6330 Cham, Switzerland

Foreword

This year we celebrate 40 years since the establishment of the HCI International (HCII) Conference, which has been a hub for presenting groundbreaking research and novel ideas and collaboration for people from all over the world.

The HCII conference was founded in 1984 by Prof. Gavriel Salvendy (Purdue University, USA, Tsinghua University, P.R. China, and University of Central Florida, USA) and the first event of the series, "1st USA-Japan Conference on Human-Computer Interaction", was held in Honolulu, Hawaii, USA, 18–20 August. Since then, HCI International is held jointly with several Thematic Areas and Affiliated Conferences, with each one under the auspices of a distinguished international Program Board and under one management and one registration. Twenty-six HCI International Conferences have been organized so far (every two years until 2013, and annually thereafter).

Over the years, this conference has served as a platform for scholars, researchers, industry experts and students to exchange ideas, connect, and address challenges in the ever-evolving HCI field. Throughout these 40 years, the conference has evolved itself, adapting to new technologies and emerging trends, while staying committed to its core mission of advancing knowledge and driving change.

As we celebrate this milestone anniversary, we reflect on the contributions of its founding members and appreciate the commitment of its current and past Affiliated Conference Program Board Chairs and members. We are also thankful to all past conference attendees who have shaped this community into what it is today.

The 26th International Conference on Human-Computer Interaction, HCI International 2024 (HCII 2024), was held as a 'hybrid' event at the Washington Hilton Hotel, Washington, DC, USA, during 29 June – 4 July 2024. It incorporated the 21 thematic areas and affiliated conferences listed below.

A total of 5108 individuals from academia, research institutes, industry, and government agencies from 85 countries submitted contributions, and 1271 papers and 309 posters were included in the volumes of the proceedings that were published just before the start of the conference, these are listed below. The contributions thoroughly cover the entire field of human-computer interaction, addressing major advances in knowledge and effective use of computers in a variety of application areas. These papers provide academics, researchers, engineers, scientists, practitioners and students with state-of-the-art information on the most recent advances in HCI.

The HCI International (HCII) conference also offers the option of presenting 'Late Breaking Work', and this applies both for papers and posters, with corresponding volumes of proceedings that will be published after the conference. Full papers will be included in the 'HCII 2024 - Late Breaking Papers' volumes of the proceedings to be published in the Springer LNCS series, while 'Poster Extended Abstracts' will be included as short research papers in the 'HCII 2024 - Late Breaking Posters' volumes to be published in the Springer CCIS series.

I would like to thank the Program Board Chairs and the members of the Program Boards of all thematic areas and affiliated conferences for their contribution towards the high scientific quality and overall success of the HCI International 2024 conference. Their manifold support in terms of paper reviewing (single-blind review process, with a minimum of two reviews per submission), session organization and their willingness to act as goodwill ambassadors for the conference is most highly appreciated.

This conference would not have been possible without the continuous and unwavering support and advice of Gavriel Salvendy, founder, General Chair Emeritus, and Scientific Advisor. For his outstanding efforts, I would like to express my sincere appreciation to Abbas Moallem, Communications Chair and Editor of HCI International News.

July 2024 Constantine Stephanidis

HCI International 2024 Thematic Areas and Affiliated Conferences

- HCI: Human-Computer Interaction Thematic Area
- HIMI: Human Interface and the Management of Information Thematic Area
- EPCE: 21st International Conference on Engineering Psychology and Cognitive Ergonomics
- AC: 18th International Conference on Augmented Cognition
- UAHCI: 18th International Conference on Universal Access in Human-Computer Interaction
- CCD: 16th International Conference on Cross-Cultural Design
- SCSM: 16th International Conference on Social Computing and Social Media
- VAMR: 16th International Conference on Virtual, Augmented and Mixed Reality
- DHM: 15th International Conference on Digital Human Modeling & Applications in Health, Safety, Ergonomics & Risk Management
- DUXU: 13th International Conference on Design, User Experience and Usability
- C&C: 12th International Conference on Culture and Computing
- DAPI: 12th International Conference on Distributed, Ambient and Pervasive Interactions
- HCIBGO: 11th International Conference on HCI in Business, Government and Organizations
- LCT: 11th International Conference on Learning and Collaboration Technologies
- ITAP: 10th International Conference on Human Aspects of IT for the Aged Population
- AIS: 6th International Conference on Adaptive Instructional Systems
- HCI-CPT: 6th International Conference on HCI for Cybersecurity, Privacy and Trust
- HCI-Games: 6th International Conference on HCI in Games
- MobiTAS: 6th International Conference on HCI in Mobility, Transport and Automotive Systems
- AI-HCI: 5th International Conference on Artificial Intelligence in HCI
- MOBILE: 5th International Conference on Human-Centered Design, Operation and Evaluation of Mobile Communications

List of Conference Proceedings Volumes Appearing Before the Conference

1. LNCS 14684, Human-Computer Interaction: Part I, edited by Masaaki Kurosu and Ayako Hashizume
2. LNCS 14685, Human-Computer Interaction: Part II, edited by Masaaki Kurosu and Ayako Hashizume
3. LNCS 14686, Human-Computer Interaction: Part III, edited by Masaaki Kurosu and Ayako Hashizume
4. LNCS 14687, Human-Computer Interaction: Part IV, edited by Masaaki Kurosu and Ayako Hashizume
5. LNCS 14688, Human-Computer Interaction: Part V, edited by Masaaki Kurosu and Ayako Hashizume
6. LNCS 14689, Human Interface and the Management of Information: Part I, edited by Hirohiko Mori and Yumi Asahi
7. LNCS 14690, Human Interface and the Management of Information: Part II, edited by Hirohiko Mori and Yumi Asahi
8. LNCS 14691, Human Interface and the Management of Information: Part III, edited by Hirohiko Mori and Yumi Asahi
9. LNAI 14692, Engineering Psychology and Cognitive Ergonomics: Part I, edited by Don Harris and Wen-Chin Li
10. LNAI 14693, Engineering Psychology and Cognitive Ergonomics: Part II, edited by Don Harris and Wen-Chin Li
11. LNAI 14694, Augmented Cognition, Part I, edited by Dylan D. Schmorrow and Cali M. Fidopiastis
12. LNAI 14695, Augmented Cognition, Part II, edited by Dylan D. Schmorrow and Cali M. Fidopiastis
13. LNCS 14696, Universal Access in Human-Computer Interaction: Part I, edited by Margherita Antona and Constantine Stephanidis
14. LNCS 14697, Universal Access in Human-Computer Interaction: Part II, edited by Margherita Antona and Constantine Stephanidis
15. LNCS 14698, Universal Access in Human-Computer Interaction: Part III, edited by Margherita Antona and Constantine Stephanidis
16. LNCS 14699, Cross-Cultural Design: Part I, edited by Pei-Luen Patrick Rau
17. LNCS 14700, Cross-Cultural Design: Part II, edited by Pei-Luen Patrick Rau
18. LNCS 14701, Cross-Cultural Design: Part III, edited by Pei-Luen Patrick Rau
19. LNCS 14702, Cross-Cultural Design: Part IV, edited by Pei-Luen Patrick Rau
20. LNCS 14703, Social Computing and Social Media: Part I, edited by Adela Coman and Simona Vasilache
21. LNCS 14704, Social Computing and Social Media: Part II, edited by Adela Coman and Simona Vasilache
22. LNCS 14705, Social Computing and Social Media: Part III, edited by Adela Coman and Simona Vasilache

23. LNCS 14706, Virtual, Augmented and Mixed Reality: Part I, edited by Jessie Y. C. Chen and Gino Fragomeni
24. LNCS 14707, Virtual, Augmented and Mixed Reality: Part II, edited by Jessie Y. C. Chen and Gino Fragomeni
25. LNCS 14708, Virtual, Augmented and Mixed Reality: Part III, edited by Jessie Y. C. Chen and Gino Fragomeni
26. LNCS 14709, Digital Human Modeling and Applications in Health, Safety, Ergonomics and Risk Management: Part I, edited by Vincent G. Duffy
27. LNCS 14710, Digital Human Modeling and Applications in Health, Safety, Ergonomics and Risk Management: Part II, edited by Vincent G. Duffy
28. LNCS 14711, Digital Human Modeling and Applications in Health, Safety, Ergonomics and Risk Management: Part III, edited by Vincent G. Duffy
29. LNCS 14712, Design, User Experience, and Usability: Part I, edited by Aaron Marcus, Elizabeth Rosenzweig and Marcelo M. Soares
30. LNCS 14713, Design, User Experience, and Usability: Part II, edited by Aaron Marcus, Elizabeth Rosenzweig and Marcelo M. Soares
31. LNCS 14714, Design, User Experience, and Usability: Part III, edited by Aaron Marcus, Elizabeth Rosenzweig and Marcelo M. Soares
32. LNCS 14715, Design, User Experience, and Usability: Part IV, edited by Aaron Marcus, Elizabeth Rosenzweig and Marcelo M. Soares
33. LNCS 14716, Design, User Experience, and Usability: Part V, edited by Aaron Marcus, Elizabeth Rosenzweig and Marcelo M. Soares
34. LNCS 14717, Culture and Computing, edited by Matthias Rauterberg
35. LNCS 14718, Distributed, Ambient and Pervasive Interactions: Part I, edited by Norbert A. Streitz and Shin'ichi Konomi
36. LNCS 14719, Distributed, Ambient and Pervasive Interactions: Part II, edited by Norbert A. Streitz and Shin'ichi Konomi
37. LNCS 14720, HCI in Business, Government and Organizations: Part I, edited by Fiona Fui-Hoon Nah and Keng Leng Siau
38. LNCS 14721, HCI in Business, Government and Organizations: Part II, edited by Fiona Fui-Hoon Nah and Keng Leng Siau
39. LNCS 14722, Learning and Collaboration Technologies: Part I, edited by Panayiotis Zaphiris and Andri Ioannou
40. LNCS 14723, Learning and Collaboration Technologies: Part II, edited by Panayiotis Zaphiris and Andri Ioannou
41. LNCS 14724, Learning and Collaboration Technologies: Part III, edited by Panayiotis Zaphiris and Andri Ioannou
42. LNCS 14725, Human Aspects of IT for the Aged Population: Part I, edited by Qin Gao and Jia Zhou
43. LNCS 14726, Human Aspects of IT for the Aged Population: Part II, edited by Qin Gao and Jia Zhou
44. LNCS 14727, Adaptive Instructional System, edited by Robert A. Sottilare and Jessica Schwarz
45. LNCS 14728, HCI for Cybersecurity, Privacy and Trust: Part I, edited by Abbas Moallem
46. LNCS 14729, HCI for Cybersecurity, Privacy and Trust: Part II, edited by Abbas Moallem

47. LNCS 14730, HCI in Games: Part I, edited by Xiaowen Fang
48. LNCS 14731, HCI in Games: Part II, edited by Xiaowen Fang
49. LNCS 14732, HCI in Mobility, Transport and Automotive Systems: Part I, edited by Heidi Krömker
50. LNCS 14733, HCI in Mobility, Transport and Automotive Systems: Part II, edited by Heidi Krömker
51. LNAI 14734, Artificial Intelligence in HCI: Part I, edited by Helmut Degen and Stavroula Ntoa
52. LNAI 14735, Artificial Intelligence in HCI: Part II, edited by Helmut Degen and Stavroula Ntoa
53. LNAI 14736, Artificial Intelligence in HCI: Part III, edited by Helmut Degen and Stavroula Ntoa
54. LNCS 14737, Design, Operation and Evaluation of Mobile Communications: Part I, edited by June Wei and George Margetis
55. LNCS 14738, Design, Operation and Evaluation of Mobile Communications: Part II, edited by June Wei and George Margetis
56. CCIS 2114, HCI International 2024 Posters - Part I, edited by Constantine Stephanidis, Margherita Antona, Stavroula Ntoa and Gavriel Salvendy
57. CCIS 2115, HCI International 2024 Posters - Part II, edited by Constantine Stephanidis, Margherita Antona, Stavroula Ntoa and Gavriel Salvendy
58. CCIS 2116, HCI International 2024 Posters - Part III, edited by Constantine Stephanidis, Margherita Antona, Stavroula Ntoa and Gavriel Salvendy
59. CCIS 2117, HCI International 2024 Posters - Part IV, edited by Constantine Stephanidis, Margherita Antona, Stavroula Ntoa and Gavriel Salvendy
60. CCIS 2118, HCI International 2024 Posters - Part V, edited by Constantine Stephanidis, Margherita Antona, Stavroula Ntoa and Gavriel Salvendy
61. CCIS 2119, HCI International 2024 Posters - Part VI, edited by Constantine Stephanidis, Margherita Antona, Stavroula Ntoa and Gavriel Salvendy
62. CCIS 2120, HCI International 2024 Posters - Part VII, edited by Constantine Stephanidis, Margherita Antona, Stavroula Ntoa and Gavriel Salvendy

Preface

The use and role of technology in the business and organizational context have always been at the heart of human-computer interaction (HCI) since the start of management information systems. In general, HCI research in such a context is concerned with the ways humans interact with information, technologies, and tasks in the business, managerial, and organizational contexts. Hence, the focus lies in understanding the relationships and interactions between people (e.g., management, users, implementers, designers, developers, senior executives, and vendors), tasks, contexts, information, and technology. Today, with the explosion of the metaverse, social media, big data, and the Internet of Things, new pathways are opening in this direction, which need to be investigated and exploited.

The 11th International Conference on HCI in Business, Government and Organizations (HCIBGO 2024), an affiliated conference of the HCI International (HCII) conference, promoted and supported multidisciplinary dialogue, cross-fertilization of ideas, and greater synergies between research, academia, and stakeholders in the business, managerial, and organizational domain.

HCI in business, government, and organizations ranges across a broad spectrum of topics from digital transformation to customer engagement. The HCIBGO conference facilitates the advancement of HCI research and practice for individuals, groups, enterprises, and society at large. A considerable number of papers accepted to HCIBGO 2024 encompass recent developments in the area of digital commerce and marketing, expanding on virtual influencer strategies and impact, e-commerce trends, consumer attitudes, gamification, and dark patterns' effects. Moreover, motivated by the digital transformation and reshaping of business operations, submissions have explored productivity, gamification strategies in work environments, and professional digital well-being, as well as teleworking and virtual collaboration. Particular emphasis was given by conference authors to the impact of Artificial Intelligence (AI) in business, addressing generative AI, automated content creation, predictive analysis, fraud detection, as well as transparency and explainability in public services. Finally, a considerable number of submissions were devoted to user experience and service efficiency, focusing on service quality assessment and improvement, consumer requirements, and user experience studies in businesses and e-government services.

Two volumes of the HCII 2024 proceedings are dedicated to this year's edition of the HCIBGO conference. The first covers topics related to Digital Commerce and Marketing, Artificial Intelligence in Business, and Workplace, Well-being and Productivity. The second focuses on topics related to Teleworking and Virtual Collaboration, and Improving User Experience and Service Efficiency.

The papers of these volumes were accepted for publication after a minimum of two single-blind reviews from the members of the HCIBGO Program Board or, in some cases, from members of the Program Boards of other affiliated conferences. We would like to thank all of them for their invaluable contribution, support, and efforts.

July 2024

Fiona Fui-Hoon Nah
Keng Leng Siau

11th International Conference on HCI in Business, Government and Organizations (HCIBGO 2024)

Program Board Chairs: **Fiona Fui-Hoon Nah** and **Keng Leng Siau,** *Singapore Management University, Singapore, Singapore*

- Kaveh Abhari, *San Diego State University, USA*
- Andreas Auinger, *University of Applied Sciences Upper Austria, Austria*
- Denise Baker, *National Institute of Occupational Health and Safety (CDC-NIOSH), Mine Safety and Research, USA*
- Gaurav Bansal, *University of Wisconsin - Green Bay, USA*
- Valerie Bartelt, *University of Tampa, USA*
- Kaveh Bazargan, *Allameh Tabataba'i University, France*
- Xiaofang Cai, *Southern University of Science and Technology, P.R. China*
- Langtao Chen, *University of Tulsa, USA*
- Constantinos K. Coursaris, *HEC Montreal, Canada*
- Brenda Eschenbrenner, *University of Nebraska at Kearney, USA*
- Ann Fruhling, *University of Nebraska at Omaha, USA*
- Jie Mein Goh, *Simon Fraser University, Canada*
- Netta Iivari, *University of Oulu, Finland*
- Qiqi Jiang, *Copenhagen Business School, Denmark*
- Yi-Cheng Ku, *Fu Jen Catholic University, Taiwan*
- Samuli Laato, *Tampere University, Finland*
- Murad Moqbel, *University of Texas Rio Grande Valley, USA*
- Xixian Peng, *Zhejiang University, P.R. China*
- Martin Stabauer, *Johannes Kepler University, Austria*
- Werner Wetzlinger, *University of Applied Sciences Upper Austria, Austria*
- I-Chin Wu, *National Taiwan Normal University, Taiwan*
- Dezhi Wu, *University of South Carolina, USA*
- Nannan Xi, *Tampere University, Finland*
- Haifeng Xu, *Shanghai Jiao Tong University, P.R. China*
- Yu-chen Yang, *National Sun Yat-sen University, Taiwan*
- Xuhong Ye, *Zhejiang University of Technology, P.R. China*
- Jie Yu, *University of Nottingham Ningbo China, P.R. China*

The full list with the Program Board Chairs and the members of the Program Boards of all thematic areas and affiliated conferences of HCII 2024 is available online at:

http://www.hci.international/board-members-2024.php

HCI International 2025 Conference

The 27th International Conference on Human-Computer Interaction, HCI International 2025, will be held jointly with the affiliated conferences at the Swedish Exhibition & Congress Centre and Gothia Towers Hotel, Gothenburg, Sweden, June 22–27, 2025. It will cover a broad spectrum of themes related to Human-Computer Interaction, including theoretical issues, methods, tools, processes, and case studies in HCI design, as well as novel interaction techniques, interfaces, and applications. The proceedings will be published by Springer. More information will become available on the conference website: https://2025.hci.international/.

General Chair
Prof. Constantine Stephanidis
University of Crete and ICS-FORTH
Heraklion, Crete, Greece
Email: general_chair@2025.hci.international

https://2025.hci.international/

Contents – Part II

Teleworking and Virtual Collaboration

Improving User Experience and Service Efficiency

Contents – Part I

Artificial Intelligence in Business

Workplace, Wellbeing and Productivity

Teleworking and Virtual Collaboration

A Bibliometric Approach to Existing Literature on Teleworking

Jorge Cruz-Cárdenas[1,2](✉), Carlos Ramos-Galarza[3], Ekaterina Zabelina[4], Olga Deyneka[5], and Andrés Palacio-Fierro[1,2]

[1] Research Center in Business, Society, and Technology, ESTec, Universidad Indoamérica, Quito, Ecuador
{jorgecruz,andrespalacio}@uti.edu.ec
[2] Facultad de Administración y Negocios, Universidad Indoamérica, Quito, Ecuador
[3] Centro de Investigación MIST, Facultad de Psicología, Universidad Indoamérica, Quito, Ecuador
caramos@puce.edu.ec
[4] Department of Psychology, Chelyabinsk State University, Chelyabinsk, Russia
[5] Department of Political Psychology, St. Petersburg State University, St. Petersburg, Russia

Abstract. Teleworking or telecommuting implies that employees perform their duties mainly from home or outside the facilities of the organizations to which they belong. The growth of academic and scientific interest in teleworking has accelerated markedly since the Covid-19 pandemic. The present study contributes to the academic research area of teleworking by organizing and structuring existing knowledge using a bibliometric approach. This study has selected Scopus as a database, and the applied search strategy has generated 8,692 documents. The descriptive analyses (using performance measures) conducted on this body of literature have confirmed the rapid increase in academic and scientific publications. The results indicate that the research on teleworking can be characterized as a multidisciplinary field, where the institutions and universities of developed countries play a leading role. Science mapping establishes the following four areas of research: 1) technological aspects that support teleworking, 2) advantages and disadvantages of teleworking, 3) human aspects of teleworking, and 4) Covid-19, teleworking, and home. The analysis of overlay visualization of terms helps identify the historical evolution of the topics of interest. This evolution starts from issues focused on technology and then addresses the issue of the human being as the center of interest. In recent years, the interest has focused on the Covid-19 pandemic. Finally, the present work discusses these research results and provides guidelines for future studies.

Keywords: Teleworking · Telecommuting · Bibliometric Analysis · Covid-19 · Human Resources

1 Introduction

Teleworking means that employees perform their duties outside the facilities of their companies or organizations; often, their work locations are their homes [1]. Teleworkers maintain contact with their organizations, based on technological equipment and services

F. F.-H. Nah and K. L. Siau (Eds.): HCII 2024, LNCS 14721, pp. 3–12, 2024.
https://doi.org/10.1007/978-3-031-61318-0_1

(mainly information and communication technologies [ICTs]) [2], such as computers, the internet, databases, smartphones, and cloud storage. Additionally, teleworkers have the support of other tools, such as specific software. A study has found that in developed countries, up to 37% of jobs can be adequately performed through teleworking, but this percentage decreases in low-income countries [3].

Teleworking was conceived and the tools for its application were created decades ago, but the Covid-19 pandemic expanded the use of teleworking [2]. Covid-19 led many countries to adopt quarantine and social distancing measures from 2020 to 2022. Although Covid-19 created the need for teleworking, the advanced state of technological development, which in the organizational world is called digital transformation [4], enabled its large-scale implementation.

For teleworking to be successful, in addition to the necessary infrastructure and equipment, teleworkers must possess certain characteristics, such as self-discipline, self-motivation, and the ability to manage time [5]. Teleworking offers various advantages, of which the most important are work–life balance, job satisfaction, commitment to the organization, and savings in travel time (between the home and the workplace) [5, 6]. However, teleworking also has disadvantages and problems (especially when it is not designed appropriately), such as psychological tension, professional and social isolation, perceived threats to professional and career advancement, intensification of the work pace, and distractions at home [1, 5, 6].

Teleworking and other scenarios about work in the future have attracted great interest among academics and professionals, as reflected in the huge increase in the number of relevant publications, which skyrocketed with the emergence of the Covid-19 pandemic [7–9]. In such situations of rapid growth, knowledge can accumulate in a disorderly manner and even with duplication of efforts [10]. Therefore, there is a need for efforts aimed at establishing the structure of existing knowledge and generating guidelines for future work. In this way, the present work seeks to make its contribution in that direction using a bibliometric approach. The general objective of this study is to conduct a bibliometric analysis of the state of knowledge about teleworking.

2 Method

Given the large amount of accumulated knowledge about teleworking, particularly since the Covid-19 pandemic, the bibliometric approach would be the most appropriate for this study's objective [11]. Once the method to be used was defined, it was necessary to establish the stages for its deployment. The authors decided to follow three stages that are widely used in a systematic literature review or a bibliometric analysis [12]: 1) Identify the topic to be investigated. 2) Establish the search strategy. 3) Analyze the documents.

Regarding stage 1, the topic to be investigated, as already defined and justified in the *Introduction* section, refers to the existing literature on teleworking. As for stage 2, the search strategy is developed in this section. Finally, in stage 3, the analysis is presented in the *Bibliometric Analysis* section.

The first decision on the search strategy concerned the database to be used. The Scopus database was chosen, and the selection criterion was to obtain good journal coverage without sacrificing the quality of the included content [4]. Once the database was defined, the next step was to determine the search string. For this purpose, an exploratory investigation of keywords and key terms was conducted. First, "teleworking" and related terms used in the documents were searched. The definitive search string was "teleworking" OR "telecommuting" OR "telework" OR "home office" OR "home-office" OR "remote work"; this was applied to the titles, abstracts, and keywords of the documents. The definitive search was carried out in September 2023. Two delimitations were also applied. The first was to include only articles, conference papers, and book chapters to ensure that the documents had gone through a review process. The second delimitation was to include only the documents published in English in order to perform all the proposed analyses. In this way, 8,692 documents were obtained, and the database was downloaded in a CSV file.

3 Bibliometric Analysis

This section is divided into two groups of very common bibliometric analyses: performance analysis and science mapping [11, 13]. In the performance analysis, the authors used descriptive statistics on the annual rhythm of publications, their disciplinary areas, and the leading institutions and countries. Regarding science mapping, two types of analyses were carried out, one of co-occurrence of words and the other of overlay visualization of terms.

As indicated, the first group of analysis was aimed at measuring performance in the field of teleworking research. The first approach to the body of literature obtained was that of the historical evolution of the annual volume of publications. Figure 1 presents this evolution, where it is easy to distinguish two stages. The first stage covered a long period characterized by slow growth in the number of publications, starting with 118 in 2000 and reaching 175 in 2018. The second stage showed rapid growth, beginning with 236 published documents in 2019 and continued to increase per year. In 2022 alone (the last full year from which the data were obtained), scientific production rose to 1,478 documents. This explosive growth in the number of annual publications can be attributed to the emergence of the Covid-19 pandemic, which caused the massive adoption of teleworking during the years that the health crisis lasted [7–9].

The second analysis developed in this study was aimed at identifying the main areas of knowledge related to research on teleworking. The results of this analysis are presented in Table 1, showing the disciplines that had more than 1,000 associated documents.

Table 1 highlights the five disciplines or areas of knowledge that contributed to or allowed the understanding of teleworking: social science; computer science; engineering; business, management, and accounting; and medicine. This shows the multidisciplinary nature of teleworking studies, which can be approached from various perspectives. Thus, teleworking can be tackled according to its social, organizational, and health implications or the technological tools used. It is also important to clarify that the same study could be associated with more than one area of knowledge in the Scopus database. It should be noted that other studies oriented to the analysis of topics on technologies and organization also found this multidisciplinary nature of the research [14].

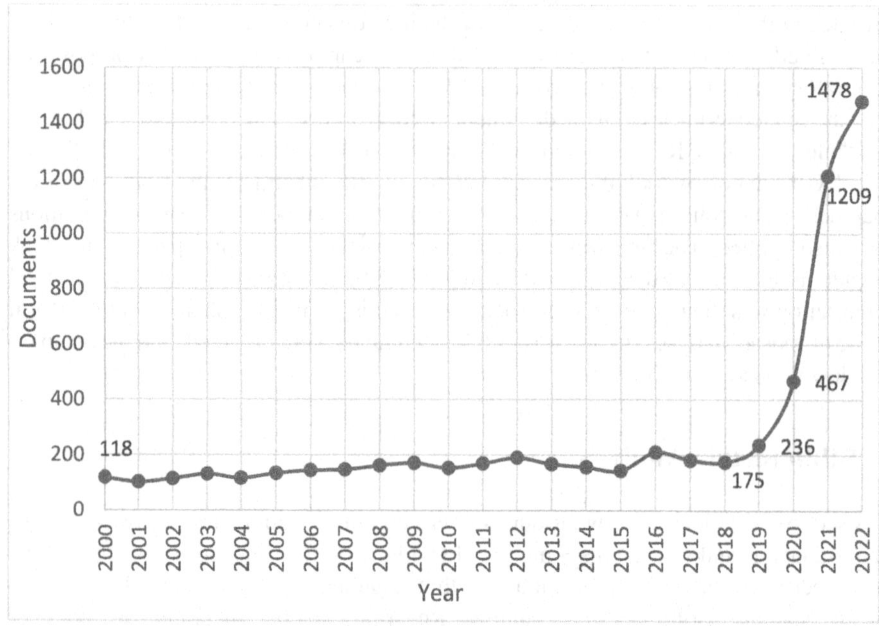

Fig. 1. Evolution of the annual publication of documents (source: Scopus)

Table 1. Disciplines associated with research on teleworking

Subject area	Documents	%
Social science	2,858	32.9
Computer science	2,273	26.2
Engineering	2,092	24.1
Business, management, and accounting	1,621	18.6
Medicine	1,422	16.4

Another topic of interest involved the types of documents included in the present study. As shown in Table 2, journal articles comprised the predominant type of publication on teleworking (68.4%). Conference papers (23.6%) and book chapters (8.0%) followed at a good distance.

The next topic of interest in the analysis of the literature on teleworking concerned the institutions with which the publications' authors were affiliated. Table 3 lists eight institutions or universities with 40 or more associated documents, which stood out from the others for their production. Interestingly, all of these institutions are located in developed countries such as Great Britain, the United States, Canada, and Portugal.

A complementary overview is offered in Table 4, listing the countries from which teleworking research comes. As a cutoff point, it was established that to be included, the country must have 300 or more documents associated with it. As can be seen in

Table 2. Main publication formats

Document type	Number of documents	%
Journal article	5,945	68.4
Conference paper	2,051	23.6
Book chapter	696	8.0
Total	8,692	100.0

Table 3. Universities and institutions that research teleworking the most

University/Institution	Associated documents
Home Office Central Research Establishment	71
University of California, Davis	60
King's College London	50
University College London	50
University of Toronto	45
Universidade de Lisboa	44
University of Oxford	41
University of Manchester	40

Table 4, five countries stood out: United States, United Kingdom, Germany, Japan, and Canada. Both Table 3 (institutions) and Table 4 (countries) show the developed countries' significant weight in research on teleworking and the absence of emerging and developing countries. This dominant role of developed countries has been observed in other bibliometric studies on human behavior in technological environments [4]. Additionally, Table 4 reveals the clear dominance of the United States and the United Kingdom, with 25.0% and 16.2%, respectively. It is necessary to clarify that since the same document can have several co-authors, it can be associated with several institutions and countries.

Table 4. Countries that research teleworking the most

Country	Documents	%
United States	2,177	25.0
United Kingdom	1,408	16.2
Germany	383	4.4
Japan	382	4.4
Canada	344	4.0

Once the different performance analyses of the scientific field related to teleworking had been completed, two types of science mapping analyses were developed in this study. The first was the analysis of co-occurrence of terms. Through this analysis, the authors sought to establish the relevant thematic clusters [15, 16], which are approximations of the areas or lines of research on the topic of interest [4]. VOSviewer software version 1.6.20 [15] was used for this analysis.

The co-occurrence analysis of words/terms was applied to the documents' keywords, which totaled 33,370. The software was instructed to only use frequent keywords, defined as those occurring 30 or more times. In addition, a thesaurus was structured, where similar words or terms were grouped, and terms that did not contribute to the analysis were eliminated. For example, terms such as "covid," "covid-19," and "coronavirus" were grouped under the term "covid." After this filtering, the final analysis of co-occurrence of words was conducted, with 217 words and terms. The results of the analysis are presented in Fig. 2.

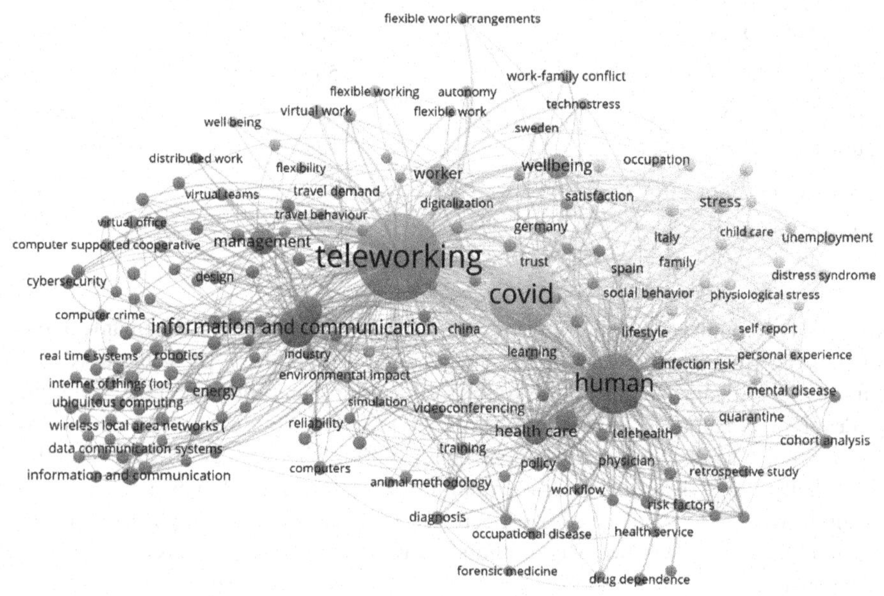

Fig. 2. Analysis of co-occurrence of words/terms (Color figure online)

As shown in Fig. 2, the results of the co-occurrence analysis of words/terms are four thematic clusters associated with the research areas. In Fig. 2, each cluster is represented with a different color. Additionally, each word or term is represented with a node, the size of the node being proportional to the frequency with which the analyzed documents used each word [15].

Cluster 1 (93 items, red color) was the most numerous and had the term "information and communication" as its central node. Additional words and terms in this cluster included "data communication systems," "ubiquitous computing," "computers," "distributed work," among others. The presence of all these terms related Cluster 1 to technological aspects in the deployment of teleworking.

Cluster 2 (43 items, green color) had as its central node the word "teleworking." Other words and terms associated with this cluster were "worker," "wellbeing," "travel demand," "flexibility," "flexible work," "work family conflict," and "technostress." This composition of the cluster associated it with both the advantages and negative impacts of teleworking.

In Cluster 3 (42 items, blue color), the central node was "human." Additional words and terms included "health care," "training," "policy," "workflow," "learning," "risk factors," "diagnosis," "drug dependence," among others. This case indicated that Cluster 3 focused on the human dimension of teleworking.

Finally, Cluster 4 (39 items, yellow color) had "covid" as its central term. This cluster's members included words and terms such as "family," "social behavior," "child care," "physiological stress," "lifestyle," and "quarantine." This integration of terms associated Cluster 4 with teleworking in the context of Covid-19 and the closest environment of the teleworker that was his or her home.

The second science mapping analysis, the so-called overlay visualization of terms, allows the evolution of the research interest to be visually displayed over the years [15]. This analysis is presented in Fig. 3, where the keywords and key terms that were most recently used by the publications' authors are shown with yellow nodes. The words and terms that were most used several decades ago are presented with purple nodes. Finally, words and phrases with green nodes are between these two extremes. The size of each node is proportional to the frequency of use of the word or term.

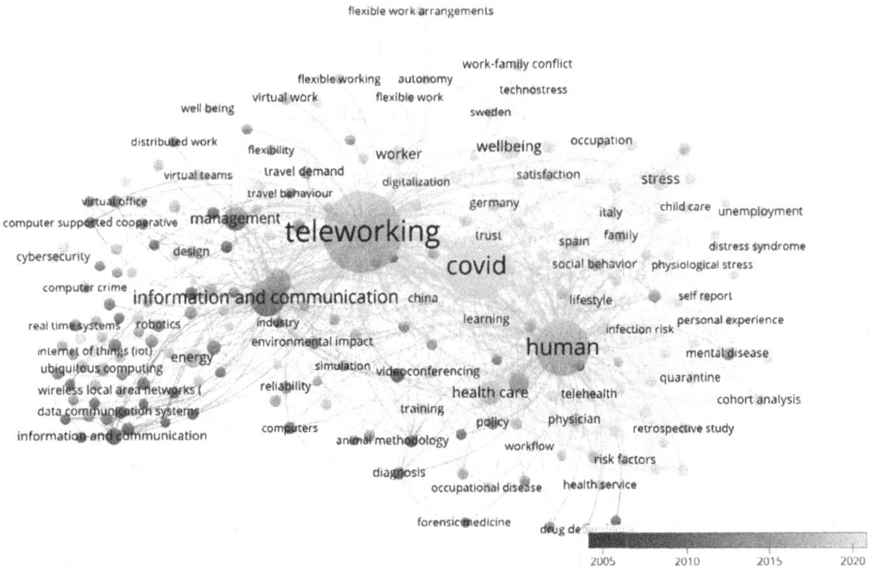

Fig. 3. Overlay visualization of terms (Color figure online)

As shown in Fig. 3, the word that stood out among the most current terms was "covid," followed by others, such as "worker," "wellbeing," and "stress." This made sense since (as indicated in a previous section) Covid-19 was one of the causes of the growth of teleworking studios [7–9]. On the other extreme, among the oldest outstanding phrases were "information and communication," "data communication systems," "video conferencing," among others, that is, terms related to technologies and technological tools. Finally, intermediate in time were terms such as "teleworking," "human," among others. In short, the beginnings of research on teleworking were focused on technological aspects; eventually, it moved to social and human aspects. Finally, the impacts of Covid-19 on the adoption of teleworking dominated, an adoption that was often improvised and produced negative effects [8].

4 Discussion and Implications

This study examined the state of knowledge about teleworking using a bibliometric approach. With this method, 8,692 relevant documents were analyzed, with the aims of establishing the performance of scientific production in the research area and mapping science.

The first result of this study is the identification of the explosive growth of annual publications on teleworking. This expansion begins with the Covid-19 outbreak and determines that much of the research occurs in the context of the pandemic. This context is characterized by the urgent and sometimes improvised implementation of teleworking [8]. Therefore, more studies on teleworking in conditions other than emergency implementation are recommended. Once the worst effects of the pandemic have passed, it will be informative to study how organizations and people assimilated the massive experience of technology adoption and experimentation with teleworking [4].

The second important conclusion reached by this study points to the multidisciplinary nature of research on teleworking. In this way, disciplines such as social sciences, computer sciences, engineering, business administration, and medical sciences contribute to teleworking research. This feature of multidisciplinary research is shared by other areas of study focused on organizational and human issues in technological environments [14]. Although future studies can continue to investigate teleworking in any of the mentioned disciplines, a wide field for interdisciplinary research also opens.

The third conclusion discloses the concentration of existing studies in developed countries and the little contribution from developing countries. This characteristic is also present in other studies on human behavior in technological environments [4]. Therefore, the present study recommends more research on teleworking in developing countries, where most of the world's population lives. Although in these countries, the percentage of jobs that can be converted to teleworking is lower than in developed countries [3], the rapid growth rate of many of the economies in developing countries must be considered.

This study carried out several analyses using the bibliometric approach of science mapping. An interesting result is the co-occurrence of terms, revealing the areas of research on the analyzed topic [4, 15]. These research areas are fundamental so that future work can be conceptually situated. In this way, the present study identifies four thematic areas: 1) technological aspects that support teleworking, 2) advantages and disadvantages of teleworking, 3) the human dimension of teleworking, and 4) Covid-19, teleworking, and home. In this sense, it is necessary to highlight again that given the large number of studies on teleworking in the context of Covid-19, this pandemic and its effects on teleworking may be overstated [8]. Therefore, companies and organizations must consider that in the current post-pandemic era, the implementation of teleworking can be more planned and lead to greater benefits and fewer disadvantages.

The next result of science mapping is the so-called overlay visualization of terms. This method makes it possible to verify that decades ago, there was great interest in the issues of technologies applied to teleworking. The next stage was the emphasis on topics that placed human beings at the center of attention. Finally, with the Covid-19 outbreak, most of the focus shifted to studying teleworking within the framework of the pandemic. As a useful reflection for future research, once the worst effects of the pandemic have passed, the topics of interest are expected to change. It can be foreseen that given the rapid pace of technological advancement, the issue of technologies, particularly artificial intelligence, will be strongly taken up. Additionally, it is anticipated that the interest in studying teleworking with human beings at its center will flourish again, evidently in the post-pandemic state.

Finally, it is necessary to recognize the limitations of the present study. It adopted the search strategies of selecting the Scopus database for its balance between coverage and quality [4] and only including documents in English. These decisions surely left out a significant number of publications. However, the authors of the present study consider these decisions necessary for quality criteria (which led to including only documents indexed in Scopus) and for the feasibility of science mapping analyses (which led to including only documents in English).

References

1. Bentley, T.A., Teo, S.T.T., McLeod, L., Tan, F., Bosua, R., Gloet, M.: The role of organisational support in teleworker wellbeing: a socio-technical systems approach. Appl. Ergon. **52**, 207–215 (2016)
2. Belzunegui-Eraso, A., Erro-Garcés, A.: Teleworking in the context of the Covid-19 crisis. Sustainability **12**, 3662 (2020)
3. Dingel, J.I., Neiman, B.: How many jobs can be done at home? J. Public Econ. **189**, 104235 (2020)
4. Cruz-Cárdenas, J., Zabelina, E., Guadalupe-Lanas, J., Palacio-Fierro, A., Ramos-Galarza, C.: COVID-19, consumer behavior, technology, and society: a literature review and bibliometric analysis. Technol. Forecast. Soc. Change **173**, 121179 (2021)
5. Charalampous, M., Grant, C.A., Tramontano, C., Michailidis, E.: Systematically reviewing remote e-workers' well-being at work: multi-dimensional approach. Eur. J. Work Organ. Psychol. **28**(1), 51–73 (2019)
6. Felstead, A., Henseke, G.: Assessing the growth of remote working and its consequences for effort, well-being and work-life balance. New Technol. Work Employ. **32**(3), 195–212 (2017)
7. Kraus, S., Ferraris, A., Bertello, A.: The future of work: how innovation and digitalization re-shape the workplace. J. Innov. Knowl. **8**(4), 1004380 (2023)
8. Wang, B., Liu, Y., Qian, J., Parker, S.K.: Achieving effective remote working during the Covid-19 pandemic: a work design perspective. Appl. Psychol. **70**(1), 16–59 (2021)

9. Xiong, A., Xia, S., He, Q., Ameen, N., Yan, J., Jones, P.: When will employees accept remote working? The impact of gender and internet skills. J. Innov. Knowl. **8**(3), 100402 (2023)
10. Cruz-Cárdenas, J., Zabelina, E., Deyneka, O., Palacio-Fierro, A., Guadalupe-Lanas, J., Ramos-Galarza, C.: Smartphones and higher education: mapping the field. In: Salvendy, G., Wei, J. (eds.) HCII 2023. LNCS, vol. 14052, pp. 253–261. Springer, Cham (2023). https://doi.org/10.1007/978-3-031-35921-7_17
11. Donthu, N., Kumar, S., Mukherjee, D., Pandey, N., Lim, W.N.: How to conduct a bibliometric analysis: an overview and guidelines. J. Bus. Res. **133**, 285–296 (2021)
12. Linnenluecke, M.K., Marrone, M., Singh, A.K.: Conducting systematic literature reviews and bibliometric analyses. Aust. J. Manag. **45**(2), 175–194 (2020)
13. Noyons, E.C.M., Moed, H.F., Van Raan, A.F.J.: Integrating research performance analysis and science mapping. Scientometrics **46**, 591–604 (1999)
14. Cruz-Cárdenas, J., Ramos-Galarza, C., Guadalupe-Lanas, J., Palacio-Fierro, A., Galarraga-Carvajal, M.: Bibliometric analysis of existing knowledge on digital transformation in higher education. In: Meiselwitz, G., et al. (eds.) HCII 2022. LNCS, vol. 13517, pp. 231–240. Springer, Cham (2022). https://doi.org/10.1007/978-3-031-22131-6_17
15. Van Eck, N.J., Waltman, L.: Manual for VOSviewer version 1.6.20 (2023). https://www.vosviewer.com/documentation/Manual_VOSviewer_1.6.15.pdf
16. Van Eck, N.J., Waltman, L.: Software survey: VOSviewer, a computer program for bibliometric mapping. Scientometrics **84**(2), 523–538 (2010)

Configurational Perspectives in Social Media Research: A Systematic Literature Review

Kailing Deng and Langtao Chen(✉)

University of Tulsa, Tulsa, OK 74104, USA
{kailing-deng,langtao-chen}@utulsa.edu

Abstract. Social media sites collectively foster a rich social structure that facilitates relationship-building between users with diverse backgrounds at different levels of analysis. Given the complexity and richness of social media research, traditional approaches and methods may fall short of representing the interconnectedness of the studied context. We argue that adopting configurational perspectives is a promising remedy. This literature review highlights the findings of a systematic investigation of 56 studies on social media and social networking sites (SNSs) with a configurational lens. Specifically, we find that prior studies have focused on the behavioral aspect of social media with positive and negative effects. Other research interests include content creation, collaboration, firm performance, software or platform performance, or a combination of multiple contexts. Furthermore, we identify commonalities and gaps in the literature and set potential directions to shape future research in this emerging research area.

Keywords: Configurational Perspectives · Social Media · Literature Review · fsQCA · Information Systems

1 Introduction

Social media has been widely adopted by individuals and organizations for various purposes, including personal experience sharing, information and knowledge exchange, product/service promotion, election, and emergency response. With its unique capability of facilitating the establishment of interpersonal connections among individuals with diverse backgrounds, social media has contributed to the formation of resilient social frameworks. Prominent social media platforms such as Facebook, X (formerly Twitter), TikTok, Instagram, WhatsApp, YouTube, LinkedIn, Stack Exchange, and Reddit have transformed the way people communicate, share information, interact, and collaborate. Those platforms are predominantly shaped by user-generated content and wield substantial influence across various domains, encompassing consumer behavior, entrepreneurship, political discourse, and venture capitalism [1]. According to Pew Research, around seven-in-ten Americans used social media to connect with others, get news content, share information, and find amusement in 2021.[1] Unsurprisingly, there is substantial

[1] https://www.pewresearch.org/internet/fact-sheet/social-media/ (accessed on November 25th, 2023).

F. F.-H. Nah and K. L. Siau (Eds.): HCII 2024, LNCS 14721, pp. 13–26, 2024.
https://doi.org/10.1007/978-3-031-61318-0_2

literature addressing various aspects of social media, such as social media marketing [2], knowledge and experience sharing in social media [3–6], and disaster social media [7].

Specifically, scholars have investigated the multifaceted aspects of social media across a diverse range of analyses. At the individual level, social media platforms have become deeply ingrained in our daily routines, serving as indispensable tools for a myriad of purposes. These routines include staying up to date on daily news and crucial events, seeking entertainment, fostering connections with family and friends, accessing reviews and recommendations for products/services and locations, addressing informational and emotional supports [8–10], seeking and contributing professional knowledge to solve complicated problems [4, 5, 11, 12], and even facilitating workplace management. At the group or community level, online forums or communities have proven efficacious in uniting individuals who share common interests and objectives, thereby propelling the prominence of content generation and influencers. While the predominant content on social media platforms comprises personal statuses or updates on contemporary events, a subset of user-generated content is dedicated to soliciting support, reflecting instances where individuals actively seek assistance and guidance. Organizations also actively participate in social media channels, primarily with the objective of soliciting feedback from stakeholders [13]. Such engagement serves as a valuable resource for revenue generation, strategic adjustment, and the enhancement of overall organizational performance.

Due to the wide range of social media research topics and the complexity of social media mechanisms, various theoretical perspectives have been utilized to guide the exploration of social media phenomena. For example, a variety of personal behavior and social behavior theories have been used to study the causes and effects of social media adoption and usage [14]. Among the theoretical perspectives relevant to social media are the configurational perspectives originally suggested for organizational analysis [15]. Configurations refer to combinations of conceptually distinct characteristics that frequently occur together [15]. The tenet of configurational perspective is that "the whole is best understood from a systemic perspective and should be viewed as a constellation of interconnected elements" [16, p. 2]. Given configurational perspectives' potential to deeply understand intricate nonlinear relationships and complex interdependencies at individual, group, and organizational levels, we argue that configurational perspectives serve as a promising theoretical foundation that helps researchers to reveal complicated mechanisms embedded in various social media contexts. Furthermore, with the gradual adoption of fuzzy set qualitative comparative analysis (fsQCA) as a novel quantitative method for analyzing configurational patterns in Information Systems (IS) and relevant fields [16, 17], we expect the plethora of social media research to adopt configurational perspectives in years to come.

With the rapid growth of research interest in both social media and configurational perspectives, a systematic review of the extant literature is needed. To the best of our knowledge, no prior research has systematically reviewed social media configurations. Therefore, this study offers a significant contribution to the literature by comprehensively assessing the status quo of social media research that includes a variety of configurations.

Based on the deepened understanding of the current literature on social media configurations, we aim to provide guidance for future research efforts to unpack the complicated mechanisms of configurations within the context of social media more thoroughly.

This paper is organized as follows. The next section explains the literature review method, including identification of relevant literature and data analysis, followed by preliminary findings presented in Sect. 3. Section 4 concludes the literature review.

2 Research Method

We searched existing research papers across multiple disciplines to create the final sample of review. The literature searches were conducted in the Scopus database and the Association for Information Systems (AIS) e-Library. The two databases were comprehensive since they also index other potentially relevant databases such as ACM, IEEE, and Springer. Evidence suggests that conducting the searches in a few comprehensive databases is preferable in terms of maintaining rigor and clarity [18].

We followed a systematic literature review procedure for searching and screening articles, as outlined in Fig. 1. Titles and abstracts of English-language articles published in peer-reviewed journals and conference proceedings were searched. The search fields were defined in Scopus as abstract and subject area, and in AIS e-Library as abstract only. These search parameters were used to find relevant articles by refining key search terms. Specifically, in Scopus, we were interested in including (1) research that examined configurations theoretically or methodologically; and (2) empirical studies explicitly utilizing fsQCA, a popular method for accessing the effect of configurations of variables leading to an outcome in a nonlinear fashion. In addition, we focused our search on pertinent articles within three subject areas, including "*Business, Management and Accounting*," "*Decision Science*," and "Social Science." In addition, we searched relevant literature in the AIS e-Library by using key terms such as "fsQCA", "QCA", "social media", and "social networking." To ensure maximum inclusion, we also employed additional searches in variation.

The literature search resulted in 268 papers from the Scopus database and 20 from the AIS e-Library database. The initial sample was compiled into a reference management software, where the references were organized and categorized. The review process was primarily conducted by the first author and then reviewed by the second author. Unclear cases or categorizations were discussed among the research team before we reached agreement.

The initial 288 papers collected from the literature search underwent additional analyses to eliminate duplicate studies, false positives, and non-English papers. Articles were screened based on a review of titles, abstracts, and keywords, with authors reading the full text when necessary. To be included in the final sample, the study must: (1) use fsQCA at least as part of the empirical analysis; or (2) examine a configuration of factors without using a configurational approach but could benefit from using one. A study was excluded if it: (1) did not clearly define or describe a configurational perspective; (2) was atheoretical; (3) utilized a configurational approach in contexts other than social media; or (4) only focused on algorithm design. Inevitably, our search strategy led to hits where the key terms (i.e., configuration, QCA, social media, or social networking) were not actually relevant to the resulting articles. For instance, some irrelevant articles mentioned one of our key search terms in the abstract as a future research direction. These articles were included in the initial review but excluded in the final analysis.

Upon further examination, we removed 3 papers that duplicate in the databases. Among the remaining papers, 4 studies were identified as false hits where at least one of the key search terms included in the concerned papers showed no relevance to this review. Additionally, we excluded one non-English article from further analysis. We iteratively applied inclusion and exclusion criteria to further assess eligibility. After conducting a backward search, we added 3 more studies to the final article pool. As a result, the final sample contains 56 empirical articles, with a discernible majority (47 articles) published in journals and a comparatively small number of studies (9 articles) appearing in conference proceedings.

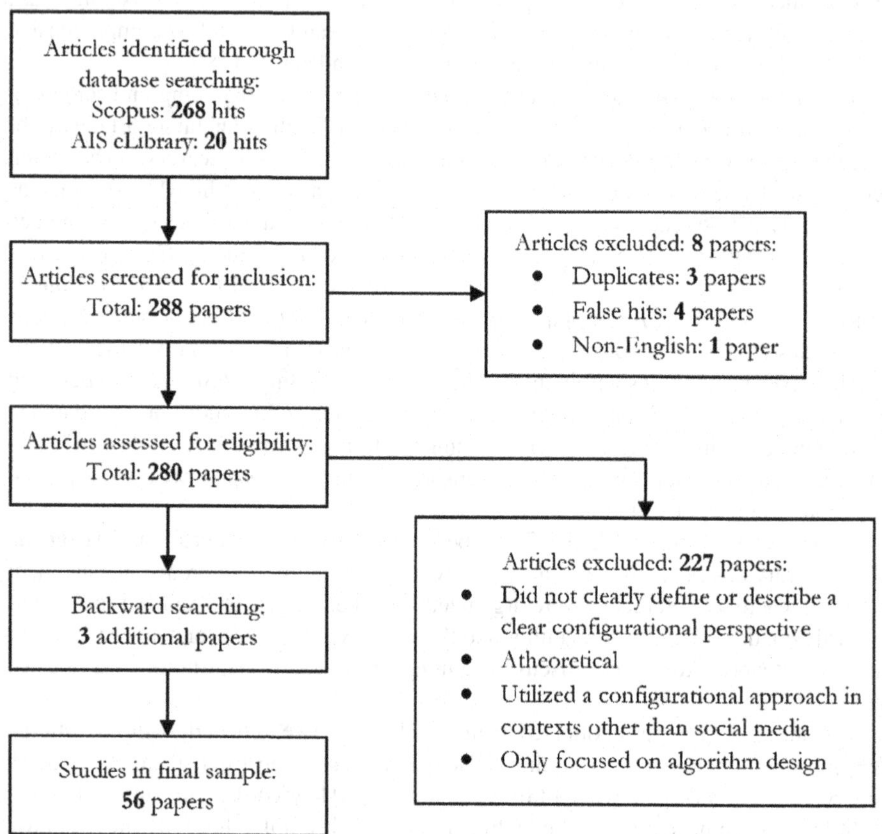

Fig. 1. Sampling procedure.

3 Preliminary Findings and Discussions

3.1 An Overview of Configurative Perspectives in Social Media Research

In the analysis of articles included in the final sample, we systematically categorized studies based on several key thematic dimensions, including social media context, empirical methods employed, predominant use of fsQCA, unit of analysis, and domains of independent and dependent variables. Table 1 summarizes all studies included in the final sample of this review.

Table 1. Summary of studies in three periods.

	2013–2015		2016–2019		2020–2023		Total
	IS	Non-IS	IS	Non-IS	IS	Non-IS	
	0	4	7	8	13	24	56
Publication type							
Journal	0	3	3	8	11	22	47
Conference	0	1	4	0	2	2	9
Method used to analyze configurations							
fsQCA	0	3	4	8	12	15	42
Not fsQCA	0	1	3	0	1	9	14

We found that extant empirical investigations encompassed a broad range in terms of the scope of data collection and analysis, spanning from extensive large-scale or multi-year experiments to exploratory studies involving only a few participants. Notably, there has been an increasing trend in employing a configurational approach, both in quantitative and qualitative dimensions. Specifically, the volume of academic literature adopting fsQCA either as the principal method or as a substantively significant component of empirical investigations has consistently risen in recent years (see details in Table 1).

We also found that scholars in the IS discipline encounter challenges of theorizing and empirically examining the intricate nature of digital phenomena. Consequently, the dynamics of theoretical multiplicity—signifying the applicability of multiple theoretical perspectives—and configurational multiplicity—indicating the existence of multiple configurations of pertinent factors within a given theoretical framework—have garnered considerable research attention [19]. Given the recognized capacity of fsQCA to amalgamate qualitative inquiry with quantitative exploration by means of configural analysis for explicating intricate phenomena [20], IS researchers have undertaken endeavors to provide systematic methodological directives concerning the application of fsQCA in the field [19, 21, 22].

Further, we observed that a noticeable portion of the papers did not anchor on specific theoretical foundations. Instead, they relied primarily on ad hoc evidence from relevant literature in an attempt to establish theoretical support. Among the limited number of theoretically grounded studies, the affordance theory has maintained popularity. The widespread adoption of social media technologies, particularly the prevalence of social network sites, has led to an increased application of the affordance theory in social media research [4]. Scholars have increasingly employed the concept of affordances as a valuable theoretical lens to investigate the impact of new media on society. For example, using both structural equation modelling (SEM) and fsQCA, Hossain et al. [23] assessed the extent to which social media fake news contributes to supply chain disruption and identified various affordances within social media that contribute to the dissemination of fake news. Similarly, in the context of tourism, Zhao et al. [24] utilized a mixed-method approach and found that the presence of deindividuation affordances combined with the absence of tournament affordances is associated with identity-based attachment. Additionally, the presence of personalization affordances in conjunction with the absence of tournament affordances forms an essential configuration that reinforces bond-based attachment.

While the theoretical stances of configurational perspectives in social media research were not explicitly mapped in this review, our coding revealed a pattern that some studies only passively addressed the theoretical foundation via definitions or simply citing prior literature, such as general mentioning of the literature on business innovation, gamification, social media marketing, online customer engagement, and online platforms. As a result, insufficient elaboration on theoretical foundations leads to findings that lack substantive grounding.

In addition, our analysis suggested an increasing trend of using configuration perspectives and fsQCA in the literature. This aligns with the general view that configuration theory provides an alternative paradigmatic lens for gaining a deeper understanding of the intricacies within digital ecodynamics [25]. Nonetheless, we argue that besides being methodologically thorough, research in this area has an urgent need to be theoretically rigorous.

3.2 Social Media Context

Table 2 summarizes all studies in terms of their social media context. As shown in Table 2, a considerable number of studies (24 articles) did not specify a particular social media context, such as a platform or social networking site. Instead, those studies were set in a general social media context. For example, to investigate antecedents of virtual satisfaction in a cross-cultural context, a group of researchers distributed questionnaires among college students, prompting them to reflect on their experiences using social media networks without confining the inquiry to one or a few specific sites [26]. The remaining work, consisting of 32 articles, was anchored on a specific social media context. Among them, the most popular platforms include Facebook (8 articles), E-commerce sites (3 articles), Weibo (3 articles), and online forums (3 articles), with 7 articles investigating multiple leading social network sites.

Table 2. Summary of social media context.

Category	Social Media Context	Number of Studies
General context	General social media sites	24
Specific context	Facebook	8
	Multiple SNSs	7
	E-commerce sites	3
	Weibo	3
	Online Forums	3
	Instagram	2
	X/Twitter	2
	WeChat	2
	TikTok	1
	Dazhong Dianping	1
	Total	56

Being an emergent economic entity, China has seen its prominent social media platforms, such as Weibo and WeChat, attracting considerable attention over the years. Weibo, the Chinese counterpart to X (formerly Twitter), has evolved into the foremost micro-blogging platform in China since its launch in 2009. WeChat is a popular mobile messaging app in China, operationally analogous to WhatsApp. Scholars focusing on the examination of Weibo and WeChat have recognized the intricate and dynamic nature inherent in these expansive social media platforms. They have deliberately employed configurational research design methods, such as fsQCA or mixed method design with fsQCA.

3.3 Units of Analysis

Our results show that the landscape of social media research has undergone rapid evolution. Regarding the spectrum of research topics, a substantial number of scholars have delved into the behavioral aspects of social media. Remarkably, some researchers have uncovered factors that impede users from sustaining their engagement [27]. Among the 56 studies, individual-level research (comprising 40 articles) emerges as a predominant trend, primarily clustering around the behavioral dimensions of social media use across diverse contexts, such as online shopping and purchase intention [28–30], digital marketing and advertising [30, 31], online satisfaction [26, 32], and other pertinent domains.

Another emerging theme involves the exploration of the content within individual posts or microblogs. Specifically, 7 articles underscored the importance of scrutinizing post-level characteristics, employing methods such as text mining [33], content analysis [34], or a mixed-methods design [35–37]. For example, using machine learning and fsQCA, Li and colleagues [35] examined COVID-19-related posts and revealed the configurational patterns combating online infodemics.

There were 5 studies addressing varied organizational contexts. Collectively, researchers have shown interest in investigating the effects of using social media sites on entrepreneurial behaviors. Some studies investigated the employee side of social media use in areas such as innovativeness [38, 39], while other studies explored the association between social enterprise systems and organizational performance [40]. In essence, studies in this category focused on incorporating social media into organizational work roles. The proficient management and utilization of social media were shown to stimulate employee engagement, subsequently contributing to enhanced innovativeness, retention, and motivation among employees. Furthermore, studies advocated for an elevated social media engagement strategy encompassing diversity, thereby fostering a positive organizational image [41].

Lastly, researchers exhibited interest in comprehending the country-level ramifications of social media use, with studies delving into technological implications and ensuing social changes. For instance, Capatina et al. [42], employing machine learning and fsQCA, discerned country-level disparities in the digital marketing landscape while studying the effects of an artificial intelligence-based software on user expectations. Grounded in the global context of the COVID-19 pandemic, Orlandi et al. [43] argued that configurations encompassing the personal background and characteristics of social media users, combined with the presence or absence of social media penetration, explained country-level gaps in vaccine coverage.

3.4 Application of fsQCA

In terms of methodology, we found that studies not employing an explicit configuration method such as fsQCA predominantly utilized qualitative and mixed-method approaches, though with a pronounced focus on unraveling patterns or configurations [33, 44–46]. Research adopting qualitative methods primarily aimed to investigate the mechanisms and motivations underlying individuals' engagement with social media, with a meaning attached to their experiences with these platforms. For instance, Steward and Schultze [46] conducted interviews with members of a social movement activist group to elucidate and identify configurations related to the generation of virtual solidarity, ultimately contributing to collective actions such as protests.

Similarly, mixed-methods design has shown potential in social media research by collecting both quantitative (closed-ended) and qualitative (open-ended) data, integrating these datasets, and deriving interpretations to comprehend research problems. As an illustration, Lee and Hallak [47] utilized a mixed-methods design to explore the impacts of offline and online social capital on tourism SME (small to medium-sized enterprises) performance, supplementing their cluster analysis results with subsequent interviews conducted with SME owners.

IS researchers, like their counterparts in other management and social science fields, predominantly utilize regression-based analysis. This involves employing multiple measurement indicators in questionnaire-based surveys to collect empirical data on users' perceptions of various IS-related attributes, aiming to explain diverse IS use phenomena, such as continuous intention to use and actual use. A notable bias in employing regression analysis, especially structural equation modeling (SEM), lies in the assumption of symmetric relationships between variables. In reality, the relationships between variables are frequently asymmetric. Hence, a viable substitute for regression analysis is essential, and one promising alternative is the application of configuration analysis.

Configuration theory posits that different combinations of initial conditions can result in the same outcome. Consequently, the association between an outcome and its preconditions is typically asymmetric rather than symmetric [48, 49]. For example, in the context of technology use, meeting one condition may not necessarily lead to end users' intention to use, even if another condition is fulfilled. The asymmetric relationships and combined complexities in such scenarios defy modeling by conventional regression-based methods.

Configuration methods outperform traditional qualitative and econometric analyses by providing additional evidence of the interconnections among pivotal factors. These methods excel at revealing patterns and configurations that might be overlooked or dismissed as statistically insignificant results in conventional analyses. Therefore, we posit that adopting a configurational approach can provide additional insights, especially in situations where relationships are dynamic and asymmetric, particularly concerning IS user behaviors [16, 17].

Specifically, we recommend employing a set-theoretical configurational analysis technique, namely fuzzy-set qualitative comparative analysis (fsQCA), as a methodological alternative to complement mainstream regression-based methods in IS research. This technique provides IS researchers with a novel data analysis tool, a fresh perspective of theorizing, and an improved understanding of IS phenomena, especially in the context of IS user behaviors.

IS behavioral research has focused primarily on detecting the causality of IS use phenomena, with regression-based models being the common approach utilized. Thereby, casual symmetry has been assumed. However, casual asymmetry is also found within IS behavioral research. Indeed, social media research frequently grapples with causal ambiguity and causal asymmetry. Determining the causal relationships between social media activities and outcomes can be challenging due to the intricate and dynamic nature of online interactions. Additionally, the asymmetric nature of these relationships, where the influence of variables may not be reciprocated in the same manner, adds complexity to the analysis. As a result, navigating the causal landscape in social media research demands careful consideration of inherent ambiguities and asymmetries.

Unlike regression-based methods that can be sensitive to outliers, fsQCA is less prone to this concern as its analysis centers on identifying subsets of the data. In fsQCA, each observation is treated as a combination of conditions, making the inclusion or exclusion of a specific data point merely impact the evaluation of that combination without affecting the overall assessment of other causal combinations. Moreover, sample representativeness is less critical for fsQCA, as it does not rely on the assumption that data are drawn from a specific probability distribution [48]. The over-representation or under-representation of a specific group of users has minimal impact on the existence of other configurations.

Yet, fsQCA has its own limitations. fsQCA depends significantly on researchers' prior knowledge for selecting conditions and outcomes to simplify configurations. Interpreting results obtained from fsQCA is labor-intensive and may pose a risk of subjective bias, especially when dealing with complex solutions. Moreover, for fsQCA to yield meaningful conclusions, the calibration of data is essential, a step not required in classic regression-based analyses. Another pertinent limitation, especially in the context of social media research, is that fsQCA does not allow for the direct utilization of latent variables. Consequently, it is recommended to integrate the benefits of a measurement model test using SEM with an asymmetric relationship test using fsQCA [49].

3.5 Expanding Firm-Level Social Media Research

Business entities are increasingly acknowledging the significance of social media as a key determinant of equity value, contributing not only to enhanced short-term performance but also yielding enduring productivity advantages. Our synthesized findings propose that the integration of social media within firms augments meta-knowledge, which in turn helps define and clarify roles and responsibilities, fostering innovative approaches to firm-level strategic management.

Thus, we propose two prospective research directions. First, research can specifically examine the financial implications of firm-level social media strategy using a configurational approach, as this kind of research is scarce in extant literature. Given that the examined articles centered their firm-level strategy on aspects such as innovation management [38, 39], there is an opportunity to make theoretical and practical implications by delving into the financial ramifications. Based on the historical complexity and debate surrounding the monetary value of IS usage, we contend that configurational methods, including fsQCA, hold the potential to unveil latent and underexplored strategic portfolio options, thereby augmenting financial returns.

Second, researchers may look into the impact of software/application/platform capabilities on performance factors. Among the 56 reviewed articles, merely two studies have endeavored to explore software-centric or general technological capabilities [41, 42]. The prevailing consensus among most studies recognizes social media primarily for its information-sharing and exchange capabilities. This trend may be attributed to the lack of adequate methodologies addressing the intricate interconnectedness of capability-related factors from a technical standpoint. Hence, we highlight the necessity to elucidate the impact of software/application/platform capabilities by unveiling configurations that encompass determining factors from diverse perspectives.

3.6 Research in Collaboration and Content Creation

There is a widespread agreement that social media platforms facilitate open online communication, content generation, and collaboration, enabled by diverse combinations of networking modes and channels. However, in comparison to other research domains and contexts examined in this review, content creation and collaboration have received relatively less attention. Only three articles, directly or indirectly, explored content creation either as an independent or dependent variable [35, 50, 51]. Additionally, another four articles delved into the factors influencing online collaboration [35–37, 52]. Thus, we propose that future research should scrutinize social media and value creation, emphasizing intricate nonlinear issues such as the impact of social and structural interplay on performance and the sustainability of software/application/platform using configurational approaches.

Our final suggestion revolves around online communities—an online environment where a shared sense of communal identity stimulates engagement, thereby augmenting satisfaction [53]. Notably, social media platforms are instrumental in fostering the emergence of virtual knowledge [4, 8, 11, 12]. Nevertheless, given the nascent stage of research in this domain, the assessment of social intricacies involved, such as the stability of networking platforms uniting community members, remains underexplored. Stanko [36] and Stanko and Allen [37] offered examples of investigating online communities through a configurational approach, wherein traditional econometric findings are complemented with fsQCA analysis. This method enabled them to establish a robust theoretical and methodological foundation, revealing dynamic patterns in which users within an innovation-focused online community collaboratively contribute to knowledge creation. Thus, we suggest that future research also consider using configurational approaches in online communities such as knowledge sharing and self-help forums.

4 Conclusion

Configurational perspectives provide a novel way of theorizing complex relationships that characterize social media phenomena. Despite ample existing social media research exploring diverse relationships in a variety of contexts, adopting configurational perspectives can reshape our understanding of many social media aspects. This research conducts a systematic review of social media research that addresses various configurations. We are optimistic that more social media research in the future will adopt configurational perspectives to investigate rich configurations that have not been fully addressed before, thereby enhancing our understanding of complex nonlinear relationships and interdependencies within social media contexts. This research offers directions for future research in this emerging and exciting research area.

References

1. Greenwood, B.N., Gopal, A.: Research note—tigerblood: newspapers, blogs, and the founding of information technology firms. Inf. Syst. Res. **26**(4), 812–828 (2015)
2. Alves, H., Fernandes, C., Raposo, M.: Social media marketing: a literature review and implications. Psychol. Mark. **33**(12), 1029–1038 (2016)
3. Ahmed, Y.A., Ahmad, M.N., Ahmad, N., Zakaria, N.H.: Social media for knowledge-sharing: a systematic literature review. Telemat. Inform. **37**, 72–112 (2019)
4. Chen, L., Baird, A., Straub, D.: Why do participants continue to contribute? Evaluation of usefulness voting and commenting motivational affordances within an online knowledge community. Decis. Support. Syst. **118**, 21–32 (2019)
5. Chen, L.: Predicting the usefulness of questions in Q&A communities: a comparison of classical machine learning and deep learning approaches. In: Fui-Hoon Nah, F., Siau, K. (eds.) HCII 2022. LNCS, vol. 13327, pp. 153–162. Springer, Cham (2022). https://doi.org/10.1007/978-3-031-05544-7_12

6. Chen, L.: What do user experience professionals discuss online? Topic modeling of a user experience Q&A community. In: Nah, F., Siau, K. (eds.) HCII 2023. LNCS, vol. 14038, pp. 365–380. Springer, Cham (2023). https://doi.org/10.1007/978-3-031-35969-9_25
7. Houston, J.B., et al.: Social media and disasters: a functional framework for social media use in disaster planning, response, and research. Disasters **39**(1), 1–22 (2015)
8. Chen, L., Baird, A., Straub, D.: Fostering participant health knowledge and attitudes: an econometric study of a chronic disease-focused online health community. J. Manag. Inf. Syst. **36**(1), 194–229 (2019)
9. Chen, L., Baird, A., Straub, D.: A linguistic signaling model of social support exchange in online health communities. Decis. Support. Syst. **130**, 113233 (2020)
10. Chen, L.: A classification framework for online social support using deep learning. In: Nah, FH., Siau, K. (eds.) HCII 2019. LNCS, vol. 11589, pp. 178–188. Springer, Cham (2019). https://doi.org/10.1007/978-3-030-22338-0_14
11. Chen, L., Baird, A., Straub, D.: The impact of hierarchical privilege levels and non-hierarchical incentives on continued contribution in online Q&A communities: a motivational model of gamification goals. Decis. Support. Syst. **153**(3), 113667 (2022)
12. Chen, L.: The effect of gamification on knowledge contribution in online Q&A communities: a perspective of social comparison. In: Nah, F.FH., Siau, K. (eds.) HCII 2021. LNCS, vol. 12783, pp. 471–481. Springer, Cham (2021). https://doi.org/10.1007/978-3-030-77750-0_30
13. Phang, C.W., Kankanhalli, A., Tan, B.C.Y.: What motivates contributors vs. Lurkers? An investigation of online feedback forums. Inf. Syst. Res. **26**(4), 773–792 (2015)
14. Ngai, E.W.T., Tao, S.S.C., Moon, K.K.L.: Social media research: theories, constructs, and conceptual frameworks. Int. J. Inf. Manag. **35**(1), 33–44 (2015)
15. Meyer, A.D., Tsui, A.S., Hinings, C.R.: Configurational approaches to organizational analysis. Acad. Manag. J. **36**(6), 1175–1195 (1993)
16. Fiss, P.C., Marx, A., Cambré, B.: Configurational theory and methods in organizational research: introduction. In: Fiss, P.C., Cambré, B., Marx, A. (eds.) Configurational Theory and Methods in Organizational Research, vol. 38, pp. 1–22. Emerald Group Publishing Limited (2013)
17. Liu, Y., Mezei, J., Kostakos, V., Li, H.: Applying configurational analysis to IS behavioural research: a methodological alternative for modelling combinatorial complexities. Inf. Syst. J. **27**(1), 59–89 (2017)
18. Paré, G., Trudel, M.-C., Jaana, M., Kitsiou, S.: Synthesizing information systems knowledge: a typology of literature reviews. Inf. Manag. **52**(2), 183–199 (2015)
19. Park, Y., Fiss, P.C., El Sawy, O.A.: Theorizing the multiplicity of digital phenomena: the ecology of configurations, causal recipes, and guidelines for applying QCA. Manag. Inf. Syst. Q. **44**, 1493–1520 (2020)
20. Kraus, S., Ribeiro-Soriano, D., Schüssler, M.: Fuzzy-set qualitative comparative analysis (fsQCA) in entrepreneurship and innovation research–the rise of a method. Int. Entrep. Manag. J. **14**, 15–33 (2018)
21. Mattke, J., Maier, C., Weitzel, T., Gerow, J.E., Thatcher, J.B.: Qualitative comparative analysis (QCA) in information systems research: status quo, guidelines, and future directions. Commun. Assoc. Inf. Syst. **50**(1), 8 (2022)
22. Pappas, I.O., Woodside, A.G.: Fuzzy-set qualitative comparative analysis (fsQCA): guidelines for research practice in information systems and marketing. Int. J. Inf. Manag. **58**, 102310 (2021)
23. Hossain, M.A., Chowdhury, M.M.H., Pappas, I.O., Metri, B., Hughes, L., Dwivedi, Y.K.: Fake news on facebook and their impact on supply chain disruption during COVID-19. Ann. Oper. Res. **327**(2), 683–711 (2023)

24. Zhao, H., Liu, F., Li, Y., Lim, E., Tan, C.-W.: A configurational view of the role of affordances in enhancing members' attachment to social networking sites. In: 41st International Conference on Information Systems, pp. 1–17 (2020)
25. El Sawy, O.A., Malhotra, A., YoungKi, P., Pavlou, P.A.: Seeking the configurations of digital ecodynamics: it takes three to tango. Inf. Syst. Res. **21**(4), 835–848 (2010)
26. Krishen, A.S., Berezan, O., Agarwal, S., Kachroo, P., Raschke, R.: The digital self and virtual satisfaction: a cross-cultural perspective. J. Bus. Res. **124**, 254–263 (2021)
27. Xie, X.Z., Tsai, N.C.: The effects of negative information-related incidents on social media discontinuance intention: evidence from SEM and fsQCA. Telemat. Inform. **56** (2021)
28. Bawack, R.E., Bonhoure, E., Kamdjoug, J.R.K., Giannakis, M.: How social media live streams affect online buyers: a uses and gratifications perspective. Int. J. Inf. Manag. **70** (2023)
29. Gunawan, D.D., Huarng, K.-H.: Viral effects of social network and media on consumers' purchase intention. J. Bus. Res. **68**(11), 2237–2241 (2015)
30. Khan, A., Rezaei, S., Valaei, N.: Social commerce advertising avoidance and shopping cart abandonment: a fs/QCA analysis of german consumers. J. Retail. Consum. Serv. **67** (2022)
31. Carlson, J., Gudergan, S.P., Gelhard, C., Rahman, M.M.: Customer engagement with brands in social media platforms: configurations, equifinality and sharing. Eur. J. Mark. **53**(9), 1733–1758 (2019)
32. Pappas, I.O., Papavlasopoulou, S., Mikalef, P., Giannakos, M.N.: Identifying the combinations of motivations and emotions for creating satisfied users in SNSs: an fsQCA approach. Int. J. Inf. Manag. **53** (2020)
33. Aguerri, J.C., Molnar, L., Miró-Llinares, F.: Old crimes reported in new bottles: the disclosure of child sexual abuse on Twitter through the case #MeTooInceste. Soc. Netw. Anal. Min. **13**(1) (2023)
34. Kwon, J., Lin, H., Deng, L., Dellicompagni, T., Kang, M.Y.: Computerized emotional content analysis: empirical findings based on charity social media advertisements. Int. J. Advert. **41**(7), 1314–1337 (2022)
35. Li, Z., Zhao, Y., Duan, T., Dai, J.: Configurational patterns for COVID-19 related social media rumor refutation effectiveness enhancement based on machine learning and fsQCA. Inf. Process. Manag. **60**(3) (2023)
36. Stanko, M.A.: Toward a theory of remixing in online innovation communities. Inf. Syst. Res. **27**(4), 773–791 (2016)
37. Stanko, M.A., Allen, B.J.: Disentangling the collective motivations for user innovation in a 3D printing community. Technovation **111** (2022)
38. Bouwman, H., Nikou, S., de Reuver, M.: Digitalization, business models, and SMEs: how do business model innovation practices improve performance of digitalizing SMEs? Telecommun. Policy **43**(9) (2019)
39. Pateli, A.G., Mikalef, P.: Configurations explaining the impact of social media on innovation performance. In: Proceedings ot the 21st Pacific Asia Conference on Information Systems: ''Societal Transformation Through IS/IT'', PACIS 2017. Association for Information Systems (2017)
40. Martini, M., Cavenago, D., Marafioti, E.: Exploring types, drivers and outcomes of social e-HRM. Empl. Relat. **43**(3), 788–806 (2021)
41. Fernandez-Lores, S., Crespo-Tejero, N., Fernández-Hernández, R.: Driving traffic to the museum: the role of the digital communication tools. Technol. Forecast. Soc. Change **174** (2022)
42. Capatina, A., Kachour, M., Lichy, J., Micu, A., Micu, A.E., Codignola, F.: Matching the future capabilities of an artificial intelligence-based software for social media marketing with potential users' expectations. Technol. Forecast. Social Change **151** (2020)

43. Orlandi, L.B., Zardini, A., Rossignoli, C., Ricciardi, F.: To do or not to do? Technological and social factors affecting vaccine coverage. Technol. Forecast. Soc. Change **174**, 121283 (2022)
44. Al-Shiridah, G., Mahdi, K., Safar, M.: Facebook feedback capacity modeling. Soc. Netw. Anal. Min. **3**, 1417–1431 (2013)
45. Gupta, M., Sharma, T.G., Thomas, V.C.: Network's reciprocity: a key determinant of information diffusion over Twitter. Behav. Inf. Technol. (2021)
46. Stewart, M., Schultze, U.: Producing solidarity in social media activism: the case of My Stealthy Freedom. Inf. Organ. **29**(3) (2019)
47. Lee, C., Hallak, R.: Investigating the effects of offline and online social capital on tourism SME performance: a mixed-methods study of New Zealand entrepreneurs. Tour. Manag. **80** (2020)
48. Fiss, P.C.: A set-theoretic approach to organizational configurations. Acad. Manag. Rev. **32**(4), 1180–1198 (2007)
49. Park, Y., Sawy, O.A.E.: Discovering the multifaceted roles of information technologies with a holistic configurational theory approach. In: 45th Hawaii International Conference on System Sciences, pp. 5204–5212 (2012)
50. Korzynski, P., Paniagua, J.: Score a tweet and post a goal: social media recipes for sports stars. Bus. Horiz. **59**(2), 185–192 (2016)
51. Krishen, A.S., Berezan, O., Agarwal, S., Robison, B.: Harnessing the waiting experience: anticipation, expectations and WOM. J. Serv. Mark. **34**(7), 1013–1024 (2020)
52. Wang, S.M., Hou, H.T.: Exploring learners' cognitive processing behavioral patterns of a collaborative creativity project using facebook to support the online discussion. In: Proceedings - IEEE 14th International Conference on Advanced Learning Technologies, ICALT 2014, pp. 505–507. Institute of Electrical and Electronics Engineers Inc. (2014)
53. Ray, S., Kim, S.S., Morris, J.G.: The central role of engagement in online communities. Inf. Syst. Res. **25**(3), 528–546 (2014)

IT Project Management Complexity Framework: Managing and Understanding Complexity in IT Projects in a Remote Working Environment

Megan Rebecca Evans and Tevin Moodley(✉)

University of Johannesburg, Kingsway Avenue, University Road, Auckland Park, Johannesburg 2092, South Africa
220021998@student.uj.ac.za, tevin@uj.ac.za

Abstract. The COVID-19 pandemic has brought about a sudden shift towards remote working in almost all industries, including the IT sector. Even after the lifting of lockdown restrictions, many IT companies have opted to continue with a remote or hybrid work situation. However, with the increase in remote working, new challenges have emerged, such as decreased effective communication among developers and between developers and customers. This decrease in communication can lead to misunderstandings about project requirements, system domain knowledge, and team standards and protocols, thereby increasing the risk of scope, cost, and time impediments. As technology continues to advance in the 21st century, IT project complexity naturally increases. However, a lack of understanding of the complexity of an IT system further increases the risk of project impediments. Therefore, it is crucial to analyze current Agile methodology practices and propose new or adjusted practices to mitigate the risk associated with complexity. This paper outlines an Agile framework incorporating these adjustments, which can help IT companies manage their projects more effectively and efficiently. However, further research into this topic is necessary to investigate the concept of project complexity. It is widely misunderstood, and focusing on complexity in IT projects can help IT companies better understand and manage the risks associated with complex projects. Additionally, research can be carried out on remote and hybrid working, specifically regarding team communication and misunderstanding. Such research can help companies overcome the challenges that arise due to remote working and ensure that they can continue to work efficiently and effectively even in a remote or hybrid work environment.

Keywords: Remote-working · Complexity · IT Project Management · Agile

1 Introduction

Complexity within an IT system poses the risk of project scope, budget, and time impediments due to the system having many interdependent parts. With

F. F.-H. Nah and K. L. Siau (Eds.): HCII 2024, LNCS 14721, pp. 27–37, 2024.
https://doi.org/10.1007/978-3-031-61318-0_3

the global transition of work to a remote setting due to the COVID-19 pandemic, a new set of challenges has arisen that project managers need to combat in order to ensure project success [12]. This paper will focus on social complexity within IT projects in a remote context, which describes the complexity caused by the diversity and the number of actors (team members) working and communicating with each other [5]. With one in five projects failing due to poor communication [1], this paper explores the effects of COVID-19 on remote working, the relationship between project complexity, remote working and communication, and how best to mitigate the risk of subsequent IT project failure.

This paper establishes a deeper understanding of the problem: Sect. 2 describes the transition from in-person working to remote working due to the COVID-19 pandemic and, furthermore, the challenges that arose from distributed working situations within teams. The most prevalent challenge in this paper is the lack of communication within teams and the subsequent lack of understanding of the problem and solution the team is working on. Next, the concept of complexity is introduced in Sect. 2.2, which details the fact that the topic is not fully understood and how increased complexity results in communication breakdowns within teams.

This paper proposes a modified Agile framework in Sect. 3, with adjustments justified in the proceeding sections. These adjustments include 3.1, which breaks down the processes that positively and negatively affect complex IT projects, and Sect. 3.2, which outlines the importance of regular meetings to ensure a shared understanding and more profound knowledge of the system within all members of the team. Next, Sect. 3.3 describes the need for documentation, contrary to some agile frameworks, followed by 3.5, which adds some practical strategies that project managers can use to further enhance their way of working, including video-calling with cameras on and an increase of pair-programming between team members to widen their knowledge areas and context of the system.

2 Problem Areas

Having a better understanding of the complexity of IT projects and the issues arising from miscommunication in distributed teams can help project managers proactively and effectively adapt their working methods. This can help ensure their teams work cohesively and towards a common goal. Effective communication is crucial for remote or hybrid teams to ensure that they have a shared understanding of the problem they are trying to solve and the solution they are creating.

2.1 Remote Working: Communication

COVID-19 is credited as being the driving factor in the adoption of a remote working system and the use of technology to deliver core business [13]. The pandemic saw a global shift from in-person teams to remote working, where processes were facilitated by virtual communication [23]. However, many remote workers felt a sense of disconnect within their teams [23].

The Information Technology (IT) industry has witnessed a considerable increase in remote jobs over the past five years. According to a study by [30], remote jobs in the IT industry grew by 300% between 2014 and 2019. However, the impact of remote working has not been the same across all industries [13]. Several organisations in the IT industry have stated that they plan to adopt remote or hybrid working as the new norm [32]. They will operate permanently, with 40% of employees working remotely and an additional 42% working in a hybrid environment [16]. Remote working allows for the globalisation of teams [7], making it possible to attract multinational employees and, in some cases, a lower-cost workforce [7,32]. However, cultural differences, varying work schedules due to time zone differences and home-life schedules, skills, experience, and background can lead to misunderstandings of project requirements [23,29].

The greatest challenge faced when working from home is the lack of face-to-face communication and eye contact [2], which is to the detriment of the team's performance [32]. Communication is a fundamental mechanism used in coordinating a project and its requirements [23]. An insufficient shared understanding is caused by a lack of communication within a team and is exacerbated by a distributed work situation [23], with 46% of project leaders struggling to identify when an employee requires support and 33% struggling to gauge whether an employee understands their tasks correctly when working remotely [11].

Effective teamwork requires a shared understanding of the problem and its potential solutions. However, such understanding can be of two types: implicit and explicit [23]. Implicit understanding is based on assumptions and unspecified facts that team members hold, while explicit understanding is derived from clear facts and specifications, such as system documentation [23]. To achieve a shared understanding of the problem and its solution, it is important for team members to communicate openly and honestly. This means actively listening to each other's perspectives and challenging assumptions where necessary. It is also important to document key decisions and agreements reached during team discussions so that everyone is on the same page. In addition, a shared understanding of the problem and its solution can be facilitated through the use of visual aids and diagrams. This can help team members to better understand complex concepts and see how different parts of the problem fit together. Fostering a shared understanding of the problem and its solution is essential for effective teamwork. It helps to ensure that team members are working towards the same goals and are able to make informed decisions based on a common understanding of the problem.

Remote work was believed to lead to reduced effective communication and a more significant amount of conflict [12], and the lack of direct communication between colleagues subsequently led to workers being more independent of each other [6]. In the context of software projects, informal communication is vital in understanding stakeholder needs and team member progress [7,23]. Remote working removes the opportunities for informal exchanges between coworkers that would result in an implicit understanding of the system and requirements and also leads to a lack of team spirit, which would otherwise encourage colleagues to share the rules of the game and participate in collaborative sessions [7,23].

Communication between workers in a remote setting is more time-consuming and requires greater precision. Project managers are less likely to allow and promote online work if it requires high interdependence (collaboration) between team members [6,15]. The transition to remote working negatively affected software engineers' productivity, specifically in large, long-running projects [27]. Inappropriate or inadequate planning of communication channels and procedures leads to delays and misunderstandings of messages communicated online [28], and in larger teams, a greater number of communication lines results in reduced effective communication within the team [29].

The idea that a diverse set of individuals working together makes communication and cohesion within a project more complex is known as social complexity [5], which is just one aspect of complexity within an IT project that could lead to project, cost or scope impediment. Understanding the risk of complexity will enable project managers to better plan their projects and processes.

2.2 Complexity

"Complexity" describes the number and interdependence of components within a system [20,24], usually entailing an ambiguous scope, long duration, high cost and many stakeholders [4]. Complexity, as a subject, is not fully understood [24], with no commonly accepted definition that can be applicable to all project cases [18], despite a large number of definitions proposed over the years [18]. In fact, more than 30 definitions have been recorded, ultimately claiming that complexity relates to information or to time and space [9]. [9] states that there is a problem with distinguishing complex from complicated in a "rational fashion".

High complexity and reduced managerial control are believed to be inversely proportional [21]. However, a project's complexity can be reduced by an increase in the project manager's and worker's experience and skill level [20]. Additionally, as project complexity increases, communication problems increase [20], therefore increasing the need for more sophisticated communication levels [20]. A project's level of complexity helps to determine the project's planning and control practices. Still, it can impede the correct identification of goals and objectives, which affects the time, cost and quality of an IT project [18].

Complexity in IT projects introduces additional management challenges [18], requiring exceptional management, with greater attention to design and architecture [25].

2.3 Poor Management

Project Management involves defining the processes and standards of a team according to a project methodology, which ultimately leads to project success

[22]. However, [22] argues that there is no single best project management framework, and the most appropriate framework depends on the context of the project. Moreover, traditional project management approaches may not be suitable for complex projects.

The "Iron Triangle" of project management refers to the relationship between cost, time, and scope and often results in a focus on project outputs to the detriment of customer satisfaction measures [3]. To improve decision-making, [31] suggest that project managers should have a better understanding of risks, as well as their own perceptions and assessments of risk. Ultimately, experience plays an important role in achieving successful project outcomes, as it helps identify risks and project goals more effectively.

3 Literature Review and Proposed Framework

The paper proposes a framework that is visually represented in Fig. 1. The framework has been modified, and the changes are highlighted in green. These modifications are supported by the literature that is discussed in the following sections: Analysis of Agile methodologies (Sect. 3.1), Knowledge sharing (Sect. 3.2), Richer Resources and Documentation (Sect. 3.3), Project Risk Analysis (Sect. 3.4), and Adjusted Agile practices (Sect. 3.5).

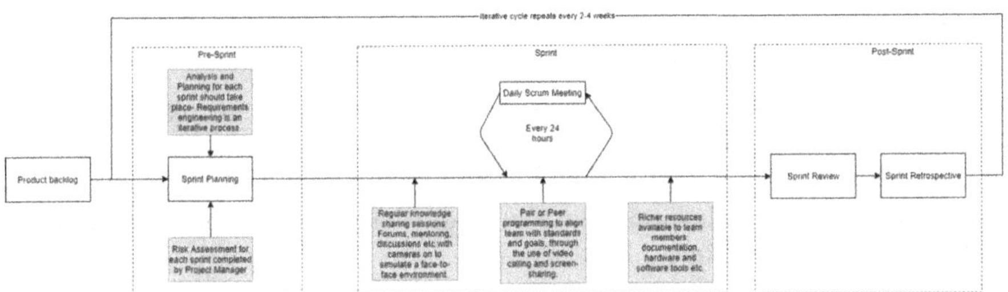

Fig. 1. Modified Scrum methodology for a Remote team working on a Complex Project

3.1 An Analysis of Agile Methodologies: SCRUM and Extreme Programming

The Agile methodology has become increasingly popular in recent years as a project management approach that enables teams to adapt and react quickly to changes in their environment, user requirements, deadlines, and more. According to a study by [8], Agile is designed to support short iterations, where continuous integration allows for faster feedback, enabling teams to adjust their approach on the fly and make faster progress towards their goals.

However, as [29] note, there are some challenges to achieving true agility in some contexts. For example, if all communication is formal due to legalities or worldwide distributed development, it may be difficult to achieve the level of collaboration and flexibility that Agile requires. In such cases, it may be necessary to find ways to work around these challenges, such as by using technology to facilitate communication or finding ways to bring team members together for more face-to-face interaction. Overall, the Agile methodology remains a powerful tool for enabling teams to work more efficiently and effectively, but it is important to recognise that it is not a one-size-fits-all solution. Teams must carefully evaluate their specific needs and challenges and adapt their approach accordingly in order to achieve the best possible outcomes. The use of traditional techniques and frameworks for complex problems is found to be "inappropriate", and when dynamic problems are treated statically, cost and time overruns are expected [5].

Agile supports the notion of short, informal stages, constant communication with the customer, re-prioritisation of requirements, frequent communication amongst the team and pair programming, which is supported by the proposed framework in Sect. 3. However, the literature also supports co-located teams, minimal documentation (with a preference for oral communication), and (explicitly) informal communication [10,21,25,29]. The approach of short iterations is maintained in the proposed framework, as the project scope may change due to the system's complexity, as discussed.

3.2 Knowledge Sharing

Regular knowledge sharing sessions can be highly beneficial for teams working within the Agile framework. These sessions allow team members to discuss their areas of expertise, share knowledge, and learn from each other. By including these sessions within each sprint, the team can continuously improve their skills and knowledge throughout the project.

Moreover, these sessions can help create a collaborative work environment where everyone feels comfortable sharing ideas and asking questions. This can lead to better communication, more effective problem-solving, and a stronger sense of teamwork. Additionally, knowledge sharing sessions can help mitigate the risk of knowledge silos, where team members hold onto information that could be useful to others.

To combat the impediment of remote working on team communication within a complex problem, project managers and team members need to actively ensure that each member has context of the system and extensive resources made available to them to minimise miscommunication. A deeper understanding of the problem, solutions, processes, and protocols within the team is achieved through knowledge-sharing guilds, frequent cross-team communication, and collaboration amongst team members, which will mitigate gaps in knowledge and understanding within and across teams in an organisation. Knowledge can be defined as information, experience, context, interpretation and reflection, meaning that knowledge depends on the context of a specific system [9].

Agile software development is defined by team collaboration [27], prioritising interactions between team members over processes [21]. An organisation's strength and success are derived from its knowledge, intellectual assets, and subsequent transferal of such [14]. A lack of knowledge sharing poses a greater risk to remote teams [26], so to combat the lack or decreased amount of communication caused by remote-work [26], increased and intentional communication needs to be prioritised. A shared understanding of the project fosters collaboration and a shared understanding, which results in team members being able to predict other's behaviour, reduce conflict within the team and increase team spirit and motivation [23].

Additionally, both explicit and implicit knowledge transfers should be facilitated by the project manager: explicit knowledge sharing being formal, involving project documentation, training and interviews, and implicit knowledge sharing being informal, including coaching, mentoring and communities of practice [26].

Knowledge-sharing guilds allow for teams to proactively build an implicit, shared understanding of a project's requirements through regularly scheduled meetings [23]. Organisations were also found to have an increase in shared understanding through accidental scope creep, rework, technical debt and disruptions of service [23], which fostered learning within the team, allowing new team members to learn about team processes by studying how previous incidents or processes took place [23]. Finally, developers improve through feedback, so regular code reviews and feedback methods increase knowledge of the domain and the team's processes within developers [23].

In a project context where team members may belong to different departments, knowledge sharing bridges the gap between members and team [14]. Change propagation describes the effect that a change to one component has on other components within a system [19], so frequent communication will foster a greater understanding of the scope and cohesive requirements within a team [14], which in turn will mitigate the risk of change and change propagation within the system [19]. Guilds allow for cross-functional communication across teams, making it easier for project stakeholders to coordinate activities and communicate throughout the duration of a project [23].

3.3 Richer Resources and Documentation

To further enable remote developers to share context, protocols, and experience and to prevent misunderstandings and gaps in knowledge, the proposed framework suggests that a set of resources needs to be made available for future reference.

Technical issues may hinder online collaboration [26], so providing the team with adequate hardware and alternative connectivity solutions may increase collaboration. Richer technological resources were shown to have a direct, positive effect on team productivity and performance [15].

Contrary to Agile techniques, a lack of documentation is believed to result in requirements not being understood or explicitly understood at the start of a project [29], leading to developers making arbitrary design choices, making it difficult to evaluate whether the system fulfils its requirements or not [25]. Additionally, poor or no documentation relies on source code as documentation, which results in knowledge (design, reasoning, etc.) existing only in the heads of developers [29]. In contrast, it was reported that clear documentation allows for appropriate testing [23].

Documentation is necessary for ensuring that a cohesive, consistent product is built [29], especially in the context of large, complex, distributed software projects due to a large set of requirements, designs, features and decisions made [29].

3.4 Project Risk Analysis Within the Agile Methodology

Despite the rapid growth of agile methodologies, risk management is often neglected due to the inherent nature of agility [21]. The proposed framework suggests that Project Managers should aim to devise a Risk Assessment for each sprint, which devises the probability and impact of each identified risk [21]. Mitigating risk within a complex project requires reducing the "unknowns." [17]

3.5 Adjusted Agile Practices for Remote-Working

A shift in remote-working habits like video calling to simulate face-to-face communication will see workers grow as individuals and teams communicate more effectively [7,8]. Regular planned discussions amongst the team allow for synchronisation where developers discuss their progress and future plans for the system [29]. Documentation should occur between these meetings, accompanied by informal communication between team members using platforms like email, online chats or video-conferencing [29].

Scrum typically involves "stand-up" meetings, but intentional pair or peer programming and collaboration need to be incorporated. 73% of candidates interviewed claimed that pair programming assists in speeding up the software development process and aligns the entire team to their shared coding standards [8]. To combat the challenge of a remote-working situation, the use of screen sharing allows for coworkers to pair-program and work with visual components remotely [23].

A limitation discussed by [8] is a lack of attention to design and architectural issues, so a deliberate effort in improving the analysis and planning of each sprint should be made- requirements engineering methods are not sequential but iterative [25], so should be performed during every iteration of the project. Many projects use prototyping as a method of communicating with their customer [25], which builds a shared understanding of the system's requirements through wireframing the system [23]. A coding standard should be set within the team, and as such, that code cannot be identifiable by someone's style [10], for example, naming conventions.

4 Conclusion

COVID-19 saw a sudden increase in remote-working within IT organisations, with the majority of whom maintaining a remote or hybrid working situation. Distributed work results in a decrease in communication, giving rise to new challenges like misunderstood requirements and inconsistent development practices. As technology advances, the complexity of IT projects is increasing, and current Agile project management frameworks need to be adapted to overcome the complexity of the projects and team dynamics, as discussed. Understanding the communication barriers of a remote working team and proactively fostering team relationships and active knowledge sharing is key in mitigating the potential of project failure when executing a complex IT project. Frequent communication and adequate resources, including hardware and documentation, allow team members to communicate requirements and challenges effectively, ultimately reducing the "unknowns" of the system and scope.

The proposed framework aims to combat the shortfalls of current agile approaches by increasing the opportunities for knowledge sharing by creating intentional sessions involving the whole team, devising risk analysis at each iteration of the project, increasing knowledge pools by enforcing that documentation is kept up to date, and that communication and collaboration are improved through video-on protocols and screen-sharing tools. These Steps should ensure that teams work cohesively as a team, with a shared understanding of the problem and a deeper knowledge of the system and processes within the team.

The shortfalls of this paper stem from the fact that complexity, as a concept, is not entirely understood, especially in the context of IT projects. Furthermore, remote, hybrid, and distributed working are still relatively new concepts, and although there has been some progress in better understanding them, they are still bound to change (and hopefully improve) as technology and the tools available improve. Future research into this topic could include case studies of organisations working remotely on complex IT projects, including perspectives from project managers, architects, technical leads and developers.

References

1. The high cost of low performance: The Essential Role of Communications. Project Management Institute (2013)
2. Al-Habaibeh, A., Watkins, M., Waried, K., Javareshk, M.B.: Challenges and opportunities of remotely working from home during Covid-19 pandemic. Global Trans. **3**, 99–108 (2021). https://doi.org/10.1016/j.glt.2021.11.001
3. Badewi, A.: The impact of project management (pm) and benefits management (bm) practices on project success: Towards developing a project benefits governance framework. Int. J. Project Manage. **34**(4), 761–778 (2016). https://doi.org/10.1016/j.ijproman.2015.05.005
4. Cerezo-Narváez, A., Pastor-Fernández, A., Otero-Mateo, M., Ballesteros-Pérez, P.: The influence of knowledge on managing risk for the success in complex construction projects: the IPMA approach. Sustainability **14**(15), 9711 (2022). https://doi.org/10.3390/su14159711

5. Cristóbal, J.S.: Complexity in project management. Proc. Comput. Sci. **121**, 762–766 (Nov 2017). https://doi.org/10.1016/j.procs.2017.11.098
6. Danielak, W., Wysocki, R.: The impact of remote work during the covid-19 pandemic on the development of competences in selected areas of project management. Annales Universitatis Mariae Curie-Skłodowska, sectio H - Oeconomia **56**(2), 7–20 (2022). https://doi.org/10.17951/h.2022.56.2.7-20
7. Delany, K.: What challenges will organisations face transitioning for the first time to the new normal of remote working? Hum. Resour. Dev. Int. **25**(5), 642–650 (2021). https://doi.org/10.1080/13678868.2021.2017391
8. Dybå, T., Dingsøyr, T.: Empirical studies of agile software development: a systematic review. Inf. Softw. Technol. **50**(9–10), 833–859 (2008). https://doi.org/10.1016/j.infsof.2008.01.006
9. Emblemsvåg, J.: Risk and complexity - on complex risk management. J. Risk Fin. **21**(1), 37–54 (2020). https://doi.org/10.1108/jrf-09-2019-0165
10. Erickson, J., Lyytinen, K., Siau, K.: Agile modeling, agile software development, and extreme programming. J. Database Manage. **16**(4), 88–100 (2005). https://doi.org/10.4018/jdm.2005100105
11. Felfe, J., et al.: Working from home: Opportunities and risks for working conditions, leadership, and health, pp. 335–341 (12 2022). https://doi.org/10.24405/14574
12. Ferrara, B., Pansini, M., De Vincenzi, C., Buonomo, I., Benevene, P.: Investigating the role of remote working on employees' performance and well-being: an evidence-based systematic review. Int. J. Environ. Res. Public Health **19**(19), 12373 (2022). https://doi.org/10.3390/ijerph191912373
13. George, G., Lakhani, K.R., Puranam, P.: What has changed? the impact of covid pandemic on the technology and innovation management research agenda. J. Manage. Stud. **57**(8), 1754–1758 (2020). https://doi.org/10.1111/joms.12634, https://onlinelibrary.wiley.com/doi/abs/10.1111/joms.12634
14. Jafari Navimipour, N., Charband, Y.: Knowledge sharing mechanisms and techniques in project teams: literature review, classification, and current trends. Comput. Human Behav. **62**, 730–742 (2016). https://doi.org/10.1016/j.chb.2016.05.003, https://www.sciencedirect.com/science/article/pii/S0747563216303211
15. Jamal, M., Anwar, I., Khan, N., Saleem, I.: Work during Covid 19: assessing the influence of job demands and resources on practical and psychological outcomes for employees. Asia-Pacific J. Business Adm. **13**, 293–319 (02 2021). https://doi.org/10.1108/APJBA-05-2020-0149
16. Keeler, J.B., Scuderi, N.F., Brock Baskin, M.E., Jordan, P.C., Meade, L.M.: How job resources can shape perspectives that lead to better performance: a remote worker field study. J. Org. Effective.: People Perform. (2023). https://doi.org/10.1108/joepp-04-2023-0154
17. Kermanshachi, S., Dao, B., Shane, J., Anderson, S.: An empirical study into identifying project complexity management strategies. Proc. Eng. **145**, 603–610 (2016). https://doi.org/10.1016/j.proeng.2016.04.050
18. Kermanshachi, S., Dao, B., Shane, J., Anderson, S.: Project complexity indicators and management strategies - a delphi study. Proc. Eng. **145**, 587–594 (2016). https://doi.org/10.1016/j.proeng.2016.04.048
19. Li, R., Yang, N., Zhang, Y., Liu, H.: Risk propagation and mitigation of design change for complex product development (cpd) projects based on multilayer network theory. Comput. Indust. Eng. **142**, 106370 (2020). https://doi.org/10.1016/j.cie.2020.106370, https://www.sciencedirect.com/science/article/pii/S0360835220301042

20. Luo, L., He, Q., Jaselskis, E.J., Xie, J.: Construction project complexity: Research trends and implications. J. Construct. Eng. Manage. **143**(7) (2017). https://doi.org/10.1061/(asce)co.1943-7862.0001306
21. Marle, F.: An assistance to project risk management based on complex systems theory and agile project management. Complexity **2020**, 1–0 (Oct 2020). https://doi.org/10.1155/2020/3739129
22. Nyman, H.J., Öörni, A.: Successful projects or success in project management - are projects dependent on a methodology? Int. J. Inform. Syst. Project Manage. **11**(4), 5-25 (Dec 2023). https://doi.org/10.12821/ijispm110401
23. Okpara, L., Werner, C., Murray, A., Damian, D.: The role of informal communication in building shared understanding of non-functional requirements in remote continuous software engineering. Require. Eng. **28**(4), 595–617 (2023). https://doi.org/10.1007/s00766-023-00404-z
24. Padalkar, M., Gopinath, S.: Are complexity and uncertainty distinct concepts in project management? a taxonomical examination from literature. Int. J. Project Manage. **34**(4), 688–700 (2016). https://doi.org/10.1016/j.ijproman.2016.02.009
25. Ramesh, B., Cao, L., Baskerville, R.: Agile requirements engineering practices and challenges: an empirical study. Inf. Syst. J. **20**(5), 449–480 (2010). https://doi.org/10.1111/j.1365-2575.2007.00259.x
26. Reed, A.H., Knight, L.V.: Effect of a virtual project team environment on communication-related project risk. Int. J. Project Manage. **28**(5), 422–427 (2010). https://doi.org/10.1016/j.ijproman.2009.08.002, https://www.sciencedirect.com/science/article/pii/S026378630900088X
27. Sathe, C.A., Panse, C.: Analyzing the impact of agile mindset adoption on software development teams productivity during Covid-19. J. Adv. Manage. Res. **20**(1), 96–115 (2022). https://doi.org/10.1108/jamr-05-2022-0088
28. Taleb, H., Ismail, S., Wahab, M.H., Rani, W.N.: Communication management between architects and clients. AIP Conf. Proc. (2017). https://doi.org/10.1063/1.5005469
29. Turk, D., Robert, F., Rumpe, B.: Assumptions underlying agile software-development processes. J. Database Manage. **16**(4), 62–87 (2005). https://doi.org/10.4018/jdm.2005100104
30. Umaji, K., Paireekreng, W.: A study of the remote working efficiency in it project implementation during the Covid-19 pandemic. WSEAS Trans. Bus. Econom. **20**, 400-409 (Feb 2023). https://doi.org/10.37394/23207.2023.20.37
31. Wang, C.M., Xu, B.B., Zhang, S.J., Chen, Y.Q.: Influence of personality and risk propensity on risk perception of Chinese construction project managers. Int. J. Project Manage. **34**(7), 1294–1304 (2016). https://doi.org/10.1016/j.ijproman.2016.07.004
32. Yan, K., Xia, E., Li, J., Gao, F.: Do teams need their employees to work remotely? a simulation analysis based on a multi-layer interactive system. Expert Syst. Appl. **236**, 121372 (2024). https://doi.org/10.1016/j.eswa.2023.121372

Virtual Reality for Home-Based Citizen Participation in Urban Planning – An Exploratory User Study

Martin Guler, Valmir Bekiri, Matthias Baldauf(✉), and Hans-Dieter Zimmermann

OST – Eastern Switzerland University of Applied Sciences, Rosenbergstrasse 59, 9001 St. Gallen, Switzerland
{martin.guler,valmir.bekiri,matthias.baldauf,hansdieter.zimmermann}@ost.ch

Abstract. Participatory urban planning fosters collaboration among residents, stakeholders, and decision-makers, enhancing urban development's inclusivity and residents' quality of life. Virtual Reality (VR) integration offers transformative potential, surpassing traditional engagement methods by immersing individuals in immersive virtual environments. This paper investigates VR's role in participatory urban planning, focusing on its increasingly accessible applications at home. We conducted an exploratory user study using an immersive VR protoype visualizing a real urban development project in Switzerland. 14 participants engaged with the prototype from home. Findings indicate a great potential of home-based VR in enhancing citizen engagement in urban planning. The study contributes insights into VR's efficacy for home-based participatory planning, derives practical recommendations, and outlines promising avenues for future research.

Keywords: Virtual Reality · Urban planning · Participatory planning · Smart government

1 Introduction

Participatory urban planning, i.e. the active involvement and collaboration of residents, stakeholders, and decision-makers, embodies a democratic approach to shaping urban spaces. Engaging citizens actively in the planning and decision-making processes not only enhances the inclusivity of urban development initiatives but also enriches the overall quality of life for residents [14]. In this context, the integration of Virtual Reality (VR) technology offers promising opportunities. VR provides a transformative platform that transcends the limitations of traditional citizen engagement methods, allowing individuals to immerse themselves in dynamic, interactive, and lifelike virtual environments. By bridging the gap between conceptual urban designs and tangible experiences, VR empowers citizens to visualize proposed changes, explore architectural nuances, and express their opinions in a manner that is both intuitive and engaging (cf. [4,5,10,12,16]).

F. F.-H. Nah and K. L. Siau (Eds.): HCII 2024, LNCS 14721, pp. 38–49, 2024.
https://doi.org/10.1007/978-3-031-61318-0_4

In many cases, prior related research focused on integrating VR technology in guided participatory processes, for example, by providing respective equipment at on-site information events. However, driven by technical advances such as stand-alone VR goggles and affordable hardware prices for headsets, an increasing penetration of headsets in private households is to be expected which makes VR-based participation in urban planning projects possible from home as well. To study citizens' impressions of VR technology applied in an urban planning scenario and empirically assess the potential of VR technology for "home-based" participatory planning, we conducted an exploratory user study. Using state-of-the-art VR tools, we created a functional VR prototype for a concrete current urban development project in a municipality in Switzerland. Overall 14 participants used the prototype at their homes to explore immersive interactive visualizations of this extensive construction project in planning.

The remainder of this paper is structured as follows. Section 2 provides a compact overview of prior research on applying mixed reality technology within the context of urban planning and citizen participation. In Sect. 3, we introduce methodological details of our study. Section 4 reports on the study results while Sect. 5 discusses the results and derives practical recommendations. We conclude the paper with an outlook on promising future work.

2 Related Work

In recent years, several studies have been exploring the application of mixed reality technology for urban planning and related participatory processes, often focusing on its potential effectiveness and challenges. Researchers have highlighted the benefits of using VR in urban planning, such as displaying comprehensive digital twins driven by smart city data, enabling efficient scenario testing, and providing a more shareable and engaging urban design experience for stakeholders (cf. [2,12]). Prior research also emphasizes the importance of citizen-centered design and the involvement of citizens in the decision-making processes through immersive VR experiences (cf. [2,9]).

An early example of using an interactive 3D model for citizen participation in urban planning is the Virtual Urban Simulation (VUS) method [18]. Applied for the concrete case of Vilnius city, the work showed that VR (even though without glasses and therefore without full immersion) integrated on gaming platforms can be used an innovative, involving and efficient tool for facilitating wider and more effective public involvement in the planning and design of urban environments. Van Leeuwen et al. [12] were among the first to study 3D visualizations on VR glasses integrated into a municipal decision-making process. Their case study on redesigning a public park with residents showed several benefits of immersive VR over 2D presentation technologies such as higher engagement and more vivid memory of the viewed content. More recently, Dembski et al. [8] presented an urban 3D model visualized through a portable VR setup (including a back-projection wall, tracking system, a 3D-projector, and active shutter glasses) for participatory processes. This digital twin of the German city Herrenberg was enriched with real data and simulation models and then put into VR to gain insights with and from citizens.

Besides VR technology, also Augmented Reality (AR) technology, i.e. overlaying virtual objects on the user's field of view in real-time, has been investigated for participatory urban planning. In early work, Allen et al. [1] presented a smartphone-based AR system to show virtual representations of architectural designs. They reported on an increased willingness to participate in urban planning events for younger study participants if they had access to such a system. More recent work by Saßmannshausen et al. [17] and Hunter et al. [11] report on similar findings and describe the entertaining and motivating function of AR in urban planning.

While the application of AR technology allows experiencing construction variants on-site (using a regular smartphone), VR is able to provide immersive experiences from remote and visualize complete redesigns of city squares, new road layouts, or even city districts, for example. Overall, prior research indicates a growing interest in leveraging VR technology to enhance participatory urban planning processes, while also recognizing the need to address various challenges associated with its implementation. With the increasing penetration of VR goggles in private households, citizens might also consume immersive urban representations through sophisticated headsets at home and thus participate from remote and independent from any on-site events. However, scientific knowledge on the application of VR technology for such home-based citizen participation is scarce.

3 Method

In the following, we describe the method of our user study. We report on the participation scenario, the VR prototype, the participants involved as well as the study procedure.

3.1 Participation Scenario

A genuine public urban development project was selected, and careful consideration was given to the choice of technology for creating the prototype. Among the various potential urban development projects in our university's region, it was crucial to select one that would directly or indirectly impact the potential participants of the study. Thus, the public development project "Bahnhofplatz/Allee" in Wil, Switzerland, was chosen, focusing on the redevelopment of the city's railway and bus station. This area is slated for a comprehensive overhaul, aimed at modernization and improved accessibility. As a vital nexus for mobility within the region, the railway and bus station plays a central role in the city's infrastructure. Therefore, this urban development initiative presents an ideal opportunity for studying the potential impact of VR technology on citizen participation. Details of the project including various visualizations and renderings are presented to the public on a website [6].

Fig. 1. The 3D model of selected parts of the urban development project as built in Arkio according to publicly available information material.

3.2 Prototype

For developing an interactive 3D model for the scenario described above, we chose *Arkio*. Arkio is a Web-based app, which enables users to build and experience 3D models directly in VR with many predefined building blocks and thus realize a prototype in a cost-saving way. Models built in Arkio can not only be viewed with VR headsets but also on computers and smartphones. While there are related VR prototyping and authoring tools such as *Forum8*, *Sketchbox*, and *SketchUp*, we chose Arkio due to its intuitive and simple modelling approach. To implement the model according to the current plans, we made use of publicly available information material on the development project such as renderings from the official project website [6]. As suggested by Stauskis [18], our virtual model was intentionally crafted to avoid distracting participants with an overload of details. Instead, it emphasizes essential features of the construction project, ensuring participants can focus on key elements without unnecessary diversion. Figure 1 shows the final model as rendered in Arkio.

The model was installed on *Meta Quest 2* VR goggles [15]. This headset is a popular and affordable standalone head-mounted VR device, typically used for entertainment purposes at home. Using a hand controller and head movements, users had the ability to navigate freely within the 3D model, enabling them to gain comprehensive insights from every angle.

3.3 Participants

Our explorative user study involved 14 participants, carefully selected to ensure heterogeneity across various demographic factors such as age, gender, experience with virtual reality (VR), and IT affinity. The participants exhibited an average level of IT affinity. The age range of the participants spanned from 19 to 83 years (M=41.4; SD=20.6) reflecting a diverse cross-section. The participants' professions and industries were strategically considered to present a representative sample of the general population, encompassing individuals from various walks of life, including students, professionals (from various disciplines) and retirees. (see Fig. 3).

Although the term "Virtual Reality" was familiar to all participants, nine participants had not gained any practical experience with VR headsets yet. For them, the technology was seen as a futuristic gadget that currently has little impact on their everyday lives. VR was associated with entertainment and, above all, gaming. Other areas of application were not known by this group. The remaining five participants already had experienced VR technology in different scenarios. These included business applications such as the planning of a supermarket or support in machinery maintenance. Applications in private life included a VR rollercoaster in an amusement park, a VR app within an escape room, VR gaming, and watching TV/Youtube in VR. All five participants described their experiences as positive, with the immersion (experience in VR as close to reality as possible) being particularly surprising.

7 out of the 14 participants had taken part in a public participation process at their local municipality/city in the past. The topics included flood protection, the construction of school buildings/multi-purpose halls or a municipal merger. The majority of these participation processes took place at local information events. Only two people were able to submit their opinions digitally in the form of a survey.

3.4 Procedure

The user tests were conducted exclusively at the test subjects' locations to ensure the experience closely mirrored reality (see Fig. 3). For each test, one research assistant equipped with standalone VR googles with the prototype installed visited a participant at home and guided her or him through a five-step-procedure.

First, the research assistant introduced himself, elaborated on the overall study topic and its goals, briefed the participant about the method in detail, and obtained the participant's written consent for recording the interviews, taking photos, and using the data for study purposes. This was followed by an introductory questionnaire on demographics, the participant's IT affinity and knowledge on VR technology, and his or her attitudes towards citizen participation and prior use of digital government services.

Having completed this introduction, the research assistant introduced the participation scenario (cf. Sect. 3.1). He presented the publicly available information material for this urban planning project such as renderings on a tablet computer. Then, the participants were briefed that the city administration in the future could also present the project to citizens in VR and invite them to experience it and participate from home. The research assistant introduced the VR goggles and the prototype and explained the usage (viewing the environment by turning and tilting the head, moving through the environment by using the controllers, etc.). In addition, the participants were informed about possible effects on well-being such as motion sickness and asked to report any complaints or to stop wearing the VR googles. Furthermore, it was emphasized that participants should focus on the participation scenario and the application of VR technology instead of technical implementation details or usability issues, for example. The participants then were allowed to try out the prototype. They were not accompanied but could freely explore the virtual representation to form their opinion about the project and the use of VR for the given purpose.

When the participants had finished using the prototype, the research assistant conducted a short interview on their experience. Questions included "What is your first impression of using the VR goggles in this scenario?", "How can VR technology be used to facilitate citizen participation?", and "How does VR technology affect the participation of citizens in the decision-making process?". Each test was completed with a final questionnaire to rate the overall experience of using VR for participatory urban planning and the participants' assessment of whether they would use this novel form of participation, both on 10-point scales.

The study was conducted in April 2023. Each test took between 45 and 60 min. To document both typical and unique behaviors, as well as usage patterns when trying out the prototype, photos were taken occasionally (Fig. 2).

Fig. 2. Example of the user's view in the VR prototype.

Fig. 3. User tests were conducted exclusively at the participants' premises to ensure the experience closely mirrored the envisioned usage scenario.

4 Results

In the following, we report on the results of our study. We group the results according to participants' overall impression and perceived support of the VR experience, its potential for increasing citizen participation, and their feedback on useful additional features for home-based participation.

4.1 VR for Visualizing Planning Projects

In our study, participants overwhelmingly expressed the superiority of VR over conventional photos when it came to comprehending the spatial intricacies of a construction project, including its scale and impact on the surrounding environment. The ability to navigate and explore the project model in VR provided participants with a profound sense of how the project would manifest in reality, enhancing their conceptualization of its look and feel.

Only one person did not see any major benefits of the VR approach over photos and renderings of the construction project. This participant rated his own spatial imagination as above average. Overall, it can be stated that the first impression of the scenario in VR and its support of the spatial imagination were experienced as very positive overall. This is supported by participants' statements such as *"this is overwhelming"* (P1), *"very impressive"* (P11) or *"I am totally thrilled"*. (P3)

One crucial element bolstering the capacity for imagination turned out to be the ability to explore diverse perspectives. Unlike conventional photos where viewers are confined to a single viewpoint, the VR prototype empowered users to assume their preferred vantage point, fostering the development of individual opinions. This flexibility extends to observing scenes from slightly elevated angles or adopting a bird's eye view.

Furthermore, feedback from participants highlighted that navigating within the VR prototype is notably more intuitive compared to static photos. The pivotal factor contributing to this ease of orientation lies in the freedom of independent movement within the model. As one participant expressed, *"You are not merely observing the model; you are inside it."* (P4).

The advantage of VR visualizations for participatory urban projects was also demonstrated through the rating task, where participants rated their overall impression on average with 8 out of 10.

4.2 VR for Increasing Citizen Participation

The participants' initial encounter with VR technology in participatory urban planning was notably positive. They held the belief that VR could increase citizen participation while potentially diminishing the prerequisites for involvement. It is crucial to note, however, that the participants did not perceive VR as a replacement for construction plans or traditional tools; instead, they regarded it as a supplementary resource. Construction plans and renderings offer distinct advantages, presenting a level of detail and precision not easily achievable in VR. Consequently, participants in a planning scenario find value in leveraging both formats to ensure a comprehensive and accurate understanding of the project.

One major advantage identified by the participants is that citizens who are not able to understand or read construction plans can also participate. Another aspect is the support of the imagination through VR. People who have a limited spatial imagination can form a much more meaningful opinion of the construction project through VR than through photos alone.

Furthermore, participants emphasized the advantage of "anytime-anywhere" participation. They appreciated that the VR solution can be used independently from home and does not require any on-site information events, while still conveying a realistic impression. Furthermore, participants appreciated the easy navigation within the immersive 3D environment.

Even though our prototype did not include any features for providing personal feedback or comments, participants mentioned the anonymity of sharing one's opinion in VR as another advantage. Participants found that not everyone has the self-confidence to give their full and honest opinion at a town hall meeting or information event, for example. VR allows opinions to be expressed anonymously and without prejudice by others.

Few respondents believed that citizen participation will not change significantly despite the use of VR. These people see the discrepancy between the different age groups and openness to new technologies as the reason for this. It is assumed that more younger people will participate, but that participation among older people will decrease as they will be overwhelmed by VR technology. This means that the number of people involved will even out in the end and citizen participation cannot be increased.

The participants' opinion towards increased citizen participation through VR applications was also reflected in the second rating task. Their inclination towards future participation in VR-based initiatives was remarkably high, receiving an average rating of 9 out of 10.

4.3 Additional Features for VR-Based Participatory Planning

During the study, participants identified a variety of additional features that they found useful for a respective VR application. 12 out of the 14 participants wished for additional information about the project in form of an audio guide to simplify the independent use of the app. Participants suggested to include suitable audio messages at relevant locations within the model. Care should be taken to ensure that the guide does not force the user to take a predefined route, but can be switched on or off as required. With this extension, information that would otherwise be documented in a project brochure can be processed while experiencing the prototype.

Regarding the visualization of the model, some participants suggested including different detail levels of the construction model that can be switched. Furthermore, animated sequences were mentioned as a helpful feature to better understand the impact of seasonal changes or varying traffic conditions on the construction project. However, two tech-affine people expressed their disapproval and believed that the programming effort (movement of traffic and pedestrians) would be too high to make the experience realistic. They also assumed that the necessary data for these animations are not available in the appropriate quality (e.g., traffic or pedestrian flow).

5 Discussion and Recommendations

In summary, participants were impressed with the VR prototype and engaged with it intuitively. Nearly all highlighted the advantages of its immersive presentation over conventional 2D information materials. These results confirm related findings from guided participation processes (cf. [12]). From the participants' feedback, we deduce that the ability to easily alter perspectives is crucial for gaining a comprehensive understanding of the construction project. This aspect holds particular significance for remote participation scenarios, where individuals consume content independently without any guidance through municipal representatives.

Based on feedback from our participants, it became clear that having various representations and levels of detail in the model enhances understanding of the construction project. This aligns with findings from related studies on location-based services (cf. [3]), which suggest that offering different forms of visualization aids in exploring a spatial model and its content. For instance, lower-detail views can offer a broad overview of the construction project, while higher-detail sections of specific model components can provide precise visualizations of their appearance.

It was important to our study participants that the VR prototype is not a complete replacement for traditional information materials, for example, but rather a useful extension. Similarly, Alpan et al. [2] proposed a human-centered VR experience as one component of the participatory design process, not the only one. For independent usage at home, a seamless integration of conventional materials with VR presentations is paramount. This could involve incorporating photos, detailed renderings, or, as suggested by participants, spoken information into the VR app. Alternatively, VR presentations could be developed using web technologies like *WebXR* or *A-Frame*, and then linked to in supplementary information materials through QR codes.

Several participants highlighted anonymous feedback as a benefit of a VR application. However, others pointed out the absence of interaction with fellow residents when using such a VR application independently at home. This lack of discussion undermines an essential aspect of the participation process. Although our prototype did not incorporate a discussion option, exploring the inclusion of such a feature and its design seems important. The potential implementations are diverse, ranging from simple note-taking features on model components to real-time conversations facilitated by 3D avatars akin to *Metaverse* environments.

6 Conclusion and Outlook

In this paper, we presented an exploratory user study investigating the use of VR for citizen participation in urban planning from the comfort of residents' homes. 14 participants evaluated a prototype VR application designed to provide an immersive experience of a municipal construction project using VR goggles. The study revealed the potential and impact of home-based VR in enhancing citizen engagement in urban planning. Results indicate that VR can serve as an additional incentive for citizen participation and facilitate opinion-forming through intuitive immersive visualizations. However, participants viewed the VR app as a supplementary tool rather than a replacement for conventional materials like information brochures and maps.

Based on the findings of this study, several avenues for future research in the realm of home-based VR citizen participation at home emerge. While we deliberately assessed the VR prototype in isolation with participants at home, it remains to be explored how such an application should be integrated into a comprehensive participation process and which factors positively impact its autonomous usage. Furthermore, the facilitation of discussions between citizens within the VR application in an intuitive manner poses an unresolved complex challenge. Moreover, the exploration of gamification elements holds promise. Game elements have repeatedly proposed and investigated to enhance citizen participation (cf. [7,13,19]). Similarly, integrating game elements into a VR application could stimulate further engagement with the project.

References

1. Allen, M., Regenbrecht, H., Abbott, M.: Smart-phone augmented reality for public participation in urban planning. In: Proceedings of the 23rd Australian Computer-Human Interaction Conference, OzCHI 2011, pp. 11–20 (2011). https://doi.org/10.1145/2071536.2071538
2. Alpan, Z.B.K., Çatak, G.: Use of Virtual Reality in Participatory Urban Design, p. 193–203. Springer International Publishing (2021). https://doi.org/10.1007/978-3-030-65060-5_16

3. Baldauf, M., Fröhlich, P., Masuch, K., Grechenig, T.: Comparing viewing and filtering techniques for mobile urban exploration. J. Location Based Serv. **5**(1), 38–57 (2011). https://doi.org/10.1080/17489725.2010.541161
4. Baldauf, M., Zimmermann, H.D., Baer-Baldauf, P., Bekiri, V.: Virtual reality for smart government – requirements, opportunities, and challenges. In: HCI in Business, Government and Organizations, pp. 3–13. Springer Nature Switzerland (2023). https://doi.org/10.1007/978-3-031-35969-9_1
5. Ceconello, M., Spallazzo, D.: Virtual reality for enhanced urban design. In: Proceedings of the 5th Intuition International Conference: Virtual Reality in Industry and Society: From Research to Application (Oct 2008)
6. City of Wil: Gesamtkonzept Wil Vivendo - Bahnhofplatz/Allee. https://wilvivendo.ch/projekte/bahnhofplatz-allee-2/. Accessed 21 Jan 2024
7. Contreras-Espinosa, R.S., Eguia-Gomez, J.L.: How to gamify e-government services?: a taxonomy of game elements, pp. 86–104. IGI Global (2022). https://doi.org/10.4018/978-1-7998-9223-6.ch004
8. Dembski, F., Wössner, U., Letzgus, M., Ruddat, M., Yamu, C.: Urban digital twins for smart cities and citizens: the case study of herrenberg, germany. Sustainability **12**(6) (2020). https://doi.org/10.3390/su12062307
9. Evers, S., Dane, G.Z., van den Berg, P.E.W., Klippel, A.K.A.J., Verduijn, T., Arentze, T.A.: Designing healthy public spaces: a participatory approach through immersive virtual reality. AGILE: GIScience Series **4**, 1–8 (Jun 2023). https://doi.org/10.5194/agile-giss-4-24-2023
10. Fegert, J., et al.: Ich sehe was, was du auch siehst. Über die Möglichkeiten von Augmented und Virtual Reality für die digitale Beteiligung von Bürger:innen in der Bau- und Stadtplanung. HMD Praxis der Wirtschaftsinformatik **58**(5), 1180–1195 (Aug 2021). https://doi.org/10.1365/s40702-021-00772-6
11. Hunter, M.G., Soro, A., Brown, R.A., Harman, J., Yigitcanlar, T.: Augmenting community engagement in city 4.0: Considerations for digital agency in urban public space. Sustainability **14**(16), 9803 (Aug 2022). https://doi.org/10.3390/su14169803
12. van Leeuwen, J.P., Hermans, K., Jylhä, A., Quanjer, A.J., Nijman, H.: Effectiveness of virtual reality in participatory urban planning: a case study. In: Proceedings of the 4th Media Architecture Biennale Conference, pp. 128–136. MAB18, Association for Computing Machinery, New York, NY, USA (2018). https://doi.org/10.1145/3284389.3284491
13. Lehner, U., Baldauf, M., Eranti, V., Reitberger, W., Fröhlich, P.: Civic engagement meets pervasive gaming: towards long-term mobile participation. In: CHI Conference on Human Factors in Computing Systems, CHI'14, Toronto, ON, Canada - April 26 - May 01, 2014, Extended Abstracts, pp. 1483–1488. ACM (2014). https://doi.org/10.1145/2559206.2581270
14. Leitner, M.: Digitale Bürgerbeteiligung. Springer Fachmedien Wiesbaden (2018). https://doi.org/10.1007/978-3-658-21621-4
15. Meta Platforms, Inc.: Meta Quest 2. https://www.meta.com/quest/products/quest-2/. Accessed 21 Jan 2024
16. Portman, M.E., Natapov, A., Fisher-Gewirtzman, D.: To go where no man has gone before: Virtual reality in architecture, landscape architecture and environmental planning. Comput., Environ. Urban Syst. **54**, 376–384 (Nov 2015). https://doi.org/10.1016/j.compenvurbsys.2015.05.001
17. Saßmannshausen, S.M., Radtke, J., Bohn, N., Hussein, H., Randall, D., Pipek, V.: Citizen-centered design in urban planning: How augmented reality can be used in

citizen participation processes. In: Designing Interactive Systems Conference 2021. DIS '21, ACM (2021). https://doi.org/10.1145/3461778.3462130
18. Stauskis, G.: Development of methods and practices of virtual reality as a tool for participatory urban planning: a case study of vilnius city as an example for improving environmental, social and energy sustainability. Energy, Sustain. Society **4**(7) (Apr 2014). https://doi.org/10.1186/2192-0567-4-7
19. Thiel, S.K., Ertiö, T.P., Baldauf, M.: Why so serious? the role of gamification on motivation and engagement in e-participation. Interact. Design Architect.(s) (35), 158-181 (Dec 2017). https://doi.org/10.55612/s-5002-035-008

From Use to Value: Monitoring the User Adoption Journey

Marvin Heuer(✉) and Philip Ostermann

Universität Hamburg, Vogt-Kölln-Straße 30, 22527 Hamburg, Germany
marvin.heuer@uni-hamburg.de,
philip.ostermann@studium.uni-hamburg.de

Abstract. In the era of hybrid work, collaborative, user-dependent IS like Microsoft 365 have gained popularity, profoundly shaping communication, value co-creation and cooperation within organizations. Recognizing their impact on organizations' and employee performance, strategic use of these tools is crucial for overall organizational success. This research aims to assess the user adoption journey of employees automatically and objectively, providing a foundation for targeted improvement measures. The central question guiding this investigation is: How can the user adoption of collaboration tools be evaluated both objectively and individually, facilitating tailored recommendations to enhance the user adoption journey? The research employs the Action Design Research approach, integrating current research and practical insights to derive universally applicable conclusions. Initial literature reviews cover system use, technical adoption, innovation adoption, and change management. Expert interviews contribute additional perspectives. Design principles for a 'User Adoption Monitoring Tool' are formulated, guiding the development of an initial version. Iterative refinement, interventions, and evaluations involving practical experts enhance the tool, evolving the principles for broader applicability and contributes to current research.

Keywords: User adoption · introduction · monitoring · action design research (ADR)

1 Introduction

Organizations have increasingly started introducing cloud-based, user-dependent IS, such as Microsoft 365, to facilitate collaboration and increase productivity [1]. As a central challenge the organization needs to ensure the intended type and scope of use of such collaborative IS. We call this process that a user undergoes until reaching this desired state the User Adoption Journey. The shorter the user adoption journey, the faster the benefits of the IS become effective for organizations and the less is invested in user adoption improvement measures until a productivity gain occurs. Simultaneously, current process tracking is very individual, visualized badly and depends (mostly) on subjective influences and attitudes [2]. Surveys are a well-established method to get a subjective impression of motivation and progress [3, 4]. However, this is unsatisfactory as it causes

F. F.-H. Nah and K. L. Siau (Eds.): HCII 2024, LNCS 14721, pp. 50–62, 2024.
https://doi.org/10.1007/978-3-031-61318-0_5

high efforts, and the intention to act and the actual action have a certain discrepancy. Here, an objective, automated monitoring of the actual user adoption enables improving the user adoption journey and facilitating the best possible software-supported, data-driven usage and collaboration. Due to the heterogeneous collective usage of software and the intended improvement of user adoption, we investigate how a 'User Adoption Monitoring Tool' (UAMT) must be designed to achieve a tool-supported, automated analysis of user adoption, implement this prototype, and evaluate it.

The central research question is: "How can a user's adoption journey be visualized and measured?"; the sub-questions were: "What recommendations can be derived to improve the user adoption journey?", "How does the tool has to be designed?" and "To what extent can the analysis and presentation be automated?" We focused on the user and their perceived usage.

To answer these questions, we followed an ADR research approach [5] in a multinational case organization in the eCommerce sector. The developed artifact can be used for different organization forms and contains a prototypical dashboard.

This paper is structured as follows: In the second section, the background on theoretical concepts regarding the developed artifacts is explained. The third section explains the followed research methodology, including the paradigm Action Design Research (ADR) and expert interviews. In the fourth section, the results are presented, explained and the evaluation is shown. The fifth section discusses the theoretical and practical implications of our work. While the sixth and final section contains the limitations of our work, a conclusion, and offers a brief outlook on current promising avenues of future research.

2 Research Background

In this section, several theoretical foundations are laid out for the following developed UAMT. In a first subsection the 'System Usage' is explained and further different forms of it. The second subsection comprises the Unified Theory of Acceptance and Use of Technology model regarding the adoption of technology.

2.1 System Usage

System usage' can be derived by describing the elements involved and how they interact, even though there is no definition [6]. A target-oriented system use is therefore determined by the combination of the following components: Firstly, by the users, who generally represent the subject that uses the information system [6, 7]. Secondly, the system used and thirdly, the tasks performed in this context [6, 7]. Our goal is through the system use to increase the performance of an employee by using the system in question to complete the defined tasks. Thus, system usage can be described as "the use of a system by a user to perform a task" with the overarching goal of improving performance [6].

At the collective system usage level, these individual usage goals are supplemented by 'interdependencies-in-use'. These refer to the interdependencies between users when processing their goals, for example in larger projects that influence performance and results [6]. Conflicts regarding management performance goals that are not compatible with the personal goals of the users, for example, also cause 'interdependencies-in-use', as they affect the overall performance of a unit, such as a team or a department [6]. Consequently, collaboration and coordination are needed to unite the different views into a goal-oriented result [6].

There are also different types of system usage. While individual usage is limited to the isolated performance of a specific task using a system by a single user, collective usage can be divided into different subtypes [6]. Collective use occurs when a group, such as a team or an entire organization, collaborates on the use of a system to work on a specific task or project [8]. Collective use occurs when a group, such as a team or an entire organization, collaborates on the use of a system to work on a specific task [6, 9]. Collective usage can occur in different forms, depending on various factors. The three forms of 'Collective Usage' are called 'Global', 'Shared', and 'Configural' + 'Collective Usage' [10].

At 'Global Collective Usage', we focus on different traits of users and their behavior on the 'level of origin' or the 'level of theory' towards each other [10, 11]. 'Shared Collective Usage' always occurs when there are very similar working or usage habits within a group [6, 11, 12]. This can be, for example, the homogeneous use of application or collaboration systems in a team. For example, it is often the case that the e-mail system is used by all department or group members with roughly the same intensity, frequency, and intention. This pattern of use is referred to as the group's 'shared level of use' [6].

'Configural Collective Usage' represents an opposite type of usage compared to 'Shared Collective Usage'. This can be, for example, the heterogeneous use of application or collaboration systems in a team with different intensity, frequency, and intention [6].

2.2 Unified Theory of Acceptance and Use of Technology (UTAUT)

So far, the importance of system use for service provision in organizations has been explained primarily regarding the partially aggregated, collective use. The question now is whether and to what extent there is a dependency between system use and the individual attitudes and characteristics of the users. In addition, it should be determined to what extent the system is used in a target-oriented manner to incorporate the findings into the design principles of the monitoring tool. There are various scientific models on this topic that show the factors on which the user acceptance of a newly introduced software tool in the corporate context depends. The Unified Theory of Acceptance and Use of Technology (UTAUT) is explained here.

As a well-known framework the Unified Theory of Acceptance and Use of Technology (UTAUT) plays an important role in the area of user adoption and behavior [13, 14]. Developed by Venkatesh et al. (2003) [15], UTAUT integrates various theories to provide a comprehensive understanding of individuals' acceptance and usage of technology [13, 15]. The model incorporates Performance Expectancy, Effort Expectancy, Social Influence, and Facilitating Conditions as key determinants influencing users' behavioral intentions [13, 15]. Performance Expectancy refers to the perceived benefits of using technology, while Effort Expectancy relates to the ease of use [15]. Social Influence considers the impact of societal factors, and Facilitating Conditions assess the support infrastructure [15]. UTAUT acknowledges moderating factors such as gender, age, and experience. This holistic approach offers researchers and practitioners a nuanced perspective, enhancing the predictive power and applicability of technology adoption models across diverse contexts [15]. As technology continues to evolve, UTAUT remains a valuable tool for analyzing and predicting user acceptance, guiding the development and implementation of innovative technological solutions [6].

3 Methodology

The research approach is shown in Fig. 1. We defined and analyzed user adoption problems, user's intentions, and motivations. Therefore, we performed five semi-structured expert interviews in a multinational case organization. Afterward, we developed what measures could be used for tracking user adoption. This took place at various levels, in the design of tools and by the mentioned expert and user interviews in the case organization. The derived requirements shall show whether the usage aligns with the organization's expectations. Further, it should facilitate targeted actions to improve each employee's user adoption journey or make cultural adjustments. Based on these insights, also shown in Fig. 1, we developed a prototype dashboard.

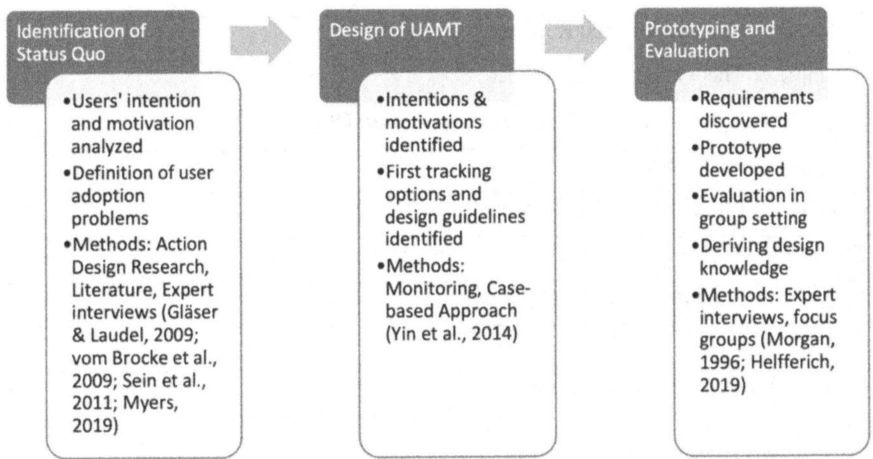

Fig. 1. Research Approach.

In general, we structured our work according to the proposed research method 'Action Design Research' by Sein et al. (2011) [5]. As a another main method, we followed expert interviews and focus groups in a case-based approach [16], which are explained in the following subsections.

3.1 Action Design Research

Sein et al. (2011) [5] stated that design science research is suitable for the development of innovations in the field of IS. However, there is a lack of recognition that the created artifact is decisively influenced by the interests, values, and assumptions of developers, investors, and users from the field. In design science process models, for example, it is not sufficiently appreciated that the artefact is decisively influenced by the organization in which it is created and that this is also intentional. Specifically, the strict separation between the creation of the artefact and the evaluation is criticized [5].

Therefore, we follow Sein's method to combine the creation of innovative IT artifacts in a practical context with the simultaneous learning from practice and thus contribute to the solution of a practical problem. This is done through artifact development with a mixture of design, use and continuous improvement [5], which is also appropriate here due to the peculiarities of the research questions. Combining action research with design research is called ADR [5]. The aim of the method is to create an artifact that not only reflects the theoretical antecedents and theoretical intentions of the researchers, but also shows the influence of the practical users and developers in the ongoing collaboration [5]. ADR is organized into four stages, which follow distinct principles and tasks, as shown in Table 1.

Table 1. Action Design Research Stages and Principles from [5].

Stage 1: Problem Formulation
Principle 1: Practice-inspired Research
Principle 2: Theory-ingrained Artifact
Stage 2: Building, Intervention, and Evaluation (BIE)
Principle 3: Reciprocal Shaping
Principle 4: Mutually Influential Roles
Principle 5: Authentic and Concurrent Evaluation
Stage 3: Reflection and Learning
Principle 6: Guided Emergence
Stage 4: Formalization of Learning
Principle 7: Generalized Outcome

3.2 Stage 1: Problem Formulation

The first stage focuses on defining the to-be-researched problem, which was here the user adoption problems. Therefore, we followed a literature search to achieve a theoretical foundation and understanding of the problem. Additionally, we conducted expert interviews with five experts to get a practical understanding. The expert interviews were conducted with experts from the case organization. We developed a semi-structured interview guide [17]. The interview varied between approximately 30 and 60 min. We transcribed and coded the conducted experts interviews. Based on seven identified theoretical problems and challenges, the gathered information of the expert interviews as well as the research on currently offered dashboards, we secured organizational commitment and established the research team, as proposed by [5]. The extracted, resulting insights are shown in Fig. 2.

Insight	Source
The UAMT must provide the ability to measure performance goals and demonstrate the outcome.	Literature
The KPIs of the UAMT must be based on the performance targets which are shown in the case organization by 'guardrails' for collaboration.	Literature
System usage should be presented in relation to employees on an individual basis. It should be possible to aggregate the individual values.	Literature
It must be possible to derive whether it is "Shared" or "Configured Collective Usage" at the department level to be able to appropriately implement the measures to improve UA.	Literature
The dashboards should be as universal as possible, but at the same time tailored to the application purpose and environment.	E1-3,5
To implement a system-oriented approach, it must be possible to assign the KPIs to the various collaboration systems.	Literature
It should be possible to show KPIs over time.	E1,3,5 Literature
The UAMT must not let employees feel that they are being monitored.	E4-5

Fig. 2. Extracted Insights From Literature and Experts.

3.3 Stage 2: Building, Intervention, and Evaluation

The second stage is about the (further) development of the artifact, the intervention, and its evaluation. The focus here is on the practical (further) development of the artifact, which was created based on theoretical findings. This is followed by a practical intervention and evaluation by stakeholders from the field. These steps (BIE for short) can also take place in parallel and represent a cycle that is repeated iteratively [5].

Once the defined BIE cycle has been completed, it is determined to what extent a further cycle is necessary or whether the further development of the artifact has ended with the completion of the current cycle.

A distinction is made between 'IT-dominant BIE' and 'organization-dominant BIE' forms of BIE [5]. The former means that the version of the artefact under consideration is still strongly IT- or technology-driven and therefore the first BIE cycle is run through in a limited, technology-savvy group of experts. 'Organization-dominant BIE' typically is conducted with a larger part of the organization. The design of a more technically mature version of the artefact is influence by a larger part of the organization.

We followed an IT-dominant BIE cycle and developed the first version of the artifact by using Microsoft PowerBI for the aggregation and dashboard interface. In the first BIE cycle, we developed a first dashboard version and evaluated this in focus group [18] setting with experts from the IT department and the business units. This allowed the gathering of technical and professional feedback. The Framework for Evaluation in Design Science [19] was used with the Human Risk & Effectiveness strategy for the evaluation. The first BIE cycle resulted in a total of three different versions and subsequent interventions on the part of the organization. The various intermediate stages are presented in the Results section.

In the second BIE cycle, we presented and evaluated the prototypical dashboard with end-users and introduced it for a small group in a naturalistic setting, which helped us refining the artifact.

3.4 Stage 3: Reflection and Learning and Stage 4: Formalization of Learning

The third stage transfers the specifically developed artifact solution to the general solution of problems and for this purpose the design and principles are evaluated [5].

In parallel to the previous phases, design principles were derived that make it easier for other researchers to develop the UAMT.

The fourth stage formalizes what has been learned from the process and describes the findings of the IT artifact. The generally valid, scientific research problem for which a solution was developed with the help of ADR comes from the area of research into system utilization. In the results section, we present the findings and classify them in the Discussion section.

4 Results

In the first stage, we searched the literature and conducted interviews as explained in Sect. 3.2. Our research revealed the insights shown in Fig. 2, which came from literature and expert interviews. Based on these first results, we started going into the second stage. We performed two Build, Intervene and Evaluate (BIE) cycles and used Microsoft PowerBI for the artifact creation. Based on the insights from Fig. 2, we developed a first version of the UAMT in form of a dashboard. The UAMT contains several sections to show and integrate the organization's individual context and performance targets. The dashboard version for the first BIE cycle is shown in the Fig. 3.

Especially in comparison to the later shown version after the first BIE evaluation steps, as shown in Fig. 4 several changes have been implemented. As proposed by Sein et al. (2011) [5], we first evaluated with experts and afterward with users. Within the first cycle, experts intervened in a focus group setting [18] by proposing dashboards containing further tools and more refined options for showing monthly results. Further, they proposed to include average values to be included in the dashboards, including more detailed graphs. Based on intervention and evaluation, we adapted the benchmarks to the 'Combined Model' [6, 20]. The implemented scoring model followed the 'Induced Categories' approach [21]. Moreover, experts stated to block any possibility to filter the results based on individual employees, as to comply with EU regulations. Therefore, only aggregated values with at least seven employees were now displayed to protect users' data and comply with EU regulations.

Figure 4 shows one possible aggregation of the developed version of the UAMT based on the results of the first BIE cycle.

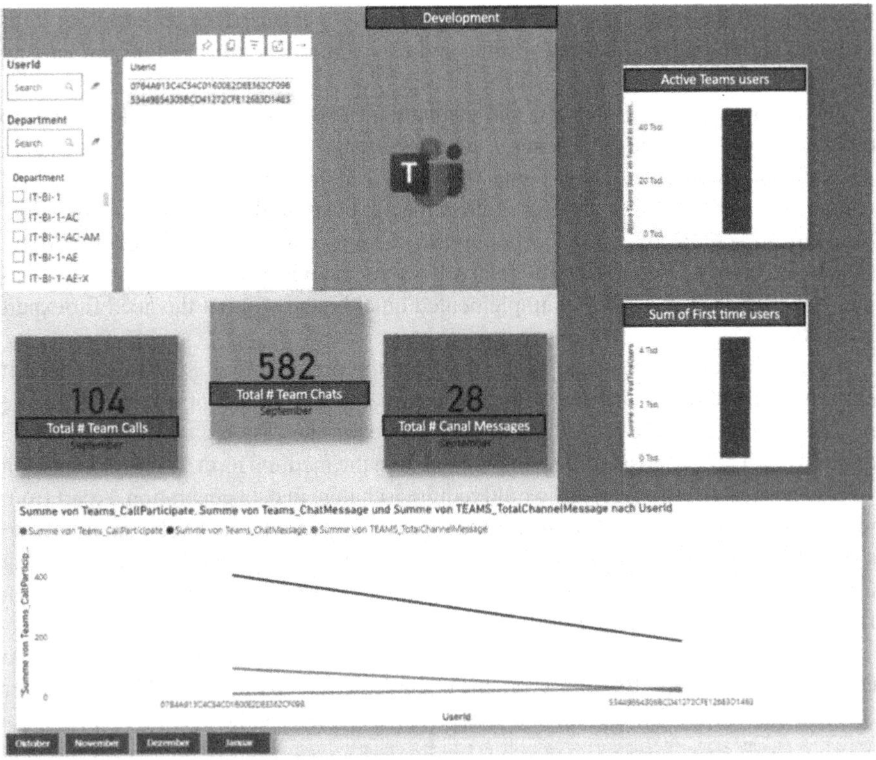

Fig. 3. Dashboard version for the first BIE cycle (translated to English).

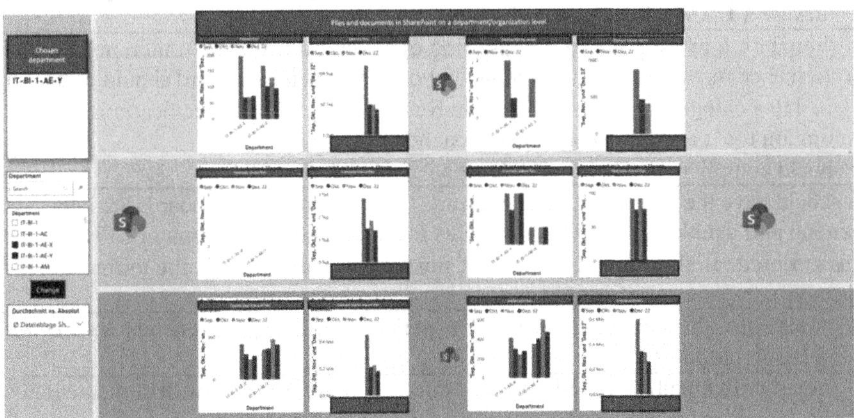

Fig. 4. Dashboard excerpt (translated to English).

The developed and evaluated dashboard design and logic is customizable and adaptable to an organization's values. The dashboard shows KPIs over time, as tool usage can vary. In addition, the UAMT shows benchmarks to rank KPI values and goal achievement.

In summary, it enables organizations to individually measure the user adoption journey while collaborating, interpret results, and derive future actions and accompanying concepts.

The results of the intervention and evaluation in the second BIE cycle are not to be classified as feedback for a general, long-term improvement of the UAMT. Rather, they serve the individual customizing of the UAMT based on the specific needs of the case organization. They therefore reflect the application of the framework function. These customizations already incorporate the 'design principle' of easy changeability of the tool. By using Power BI, which largely offers the possibility of no-code development, superficial changes can be implemented quickly and without the need for expert knowledge.

A general proposal was also made to enable the presentation of a KPI that cannot currently be displayed, which was not implemented due to a lack of interfaces. As the user adoption monitoring project is being continued in the case organization, there is a possibility that the proposal will be implemented in the medium term. A daily evaluation was also proposed. However, this would require a change in data generation. Apart from the suggestions for change, the feedback was positive and the UAMT was judged to be useful for evaluating the status of user adoption at departmental level and deriving possible improvement measures. This assessment as a 'basis' is reflected in the 'Design Principles' and thus in the intention of a framework-like implementation.

Based on the gathered insights in the second stage, design knowledge could be derived to gather steps for future practical implementations of the UAMT for other organizations. The following Table 2 contains an overview on the derived knowledge.

Table 2. Derived Design Knowledge in Stage 3.

To design a UAMT for an organization…
it should be a monitoring tool consisting of several dashboards that represent the performance measurement in the collaboration area. Each dashboard should be assigned to a category of KPIs that are derived from the usage specifications for collaboration tools and are recognizably assigned to a specific tool.
it should comply with country regulations.
it should have the character of a framework. To this end, the dashboards are kept general and usable so that details such as the adaptation of designs or KPIs can be implemented at short notice to make it possible to individualize the tool to the respective company specifications .
it should show the KPIs over several time periods and should enable benchmarks for comparison purposes .
it should individually derived and introduced a scoring model with the categories 'light', 'medium' and 'heavy'.
it should be embedded in a process to improve the user adoption. The use of the UAMT must be embedded in a process so that employees do not feel monitored.

5 Discussion

Burton-Jones and Gallivan [6] explain what the motive of system usage is and what types of usage there are. They elaborate that no scientific method has been identified to measure this usage once it is not 'Shared Collective Usage,' and thus, the usage in a group is not homogeneous [6]. We address this gap using collaboration tools in the case organization as an example by developing an artifact that measures performance and the user adoption journey based on the individual performance goals set by the organization for individuals, even at heterogeneous use.

Measurement with the developed UAMT is theoretically possible and can be implemented. Thus, we contribute directly to existing research. Furthermore, collaboration tools are used to perform collaboration processes in hybrid work environments. The use of these work scenarios is growing. Collaboration, and thus collaboration tools, are essential in all work scenarios where 'Collective Usage' is present. Here, the 'Interdependencies In Use' directly influence the performance. This represents the overall goal of any system usage. Therefore, the measurements that are possible with the UAMT provide an important contribution to improving user adoption of any software that is used collectively.

Further, we contribute to the emerging field of research on collaborative, user-dependent IS [22–25]. The UAMT offers practical implications for the data-driven design and implementation of tools that facilitate the introduction and operation of collaborative, user-dependent IS. In addition, the UAMT gives the opportunity to get a better data-driven understanding of the user adoption, their journey, and how to track it, which unravels new possibilities to get user faster to creating an actual value for organizations.

6 Limitations and Conclusion

Due to the increasing importance of collaborative, user-dependent IS, the aim of the paper was to find out how the use of this software can be analyzed in a target-oriented, automated and objective manner in order to derive improvement measures for the user adoption journey of employees from the results in a second step.

We answer this by designing a prototypically implemented artifact known as the UAMT. It was developed on a scientific basis and based on the practice of a major international corporation in the field of e-commerce. Although scientific research exists in the field of user adoption, the monitoring and implementation support of collaborative, user-dependent IS is still largely unexplored. For this reason, experts from the field were involved in the development process as part of the ADR research method.

In interviews with experts, we determined that the UAMT should be an objective software tool that displays various KPIs on dashboards that could show the use of the IS used and monitored on a user-specific basis. In practice, we restricted this function due to legal and organizational regulations, so that only an aggregated evaluation is possible. This limits the usefulness of the UAMT as a basis for individual measures to improve user adoption. The specific design of the dashboards is presented in the results section; the implementation is based on the 'Design Principles'.

Finally, it should be emphasized that the UAMT can be easily adapted to the individual requirements of the organization. Determining usage is only expedient if the individual specifications or guiding principles for the use of collaborative, user-dependent IS are implemented. Therefore, the UAMT has the character of a customizable framework that provides a standardized framework whose dashboards can be adapted to new organizations within a few working hours. It was also determined that the dashboards must show the KPIs over time, as the use of IS can vary over time. In addition, the UAMT must be able to show benchmarks and scoring that can be used to classify the values of the KPIs. This allows a suitable concept to be derived for the development of measures for the individual improvement of employee user adoption. This development would be very individually adapted to different employee needs.

In summary, the paper provides a data-driven basis that enables organizations to individually measure the status of the user adoption journey for collaborative, user-dependent IS and to interpret the results, as well as to derive perspective measures with the help of a concept to be developed.

By using the UAMT, even with heterogeneous system usage ('Configural Collective Use'), this usage can be analyzed individually, and suitable measures can be derived. This closes a research gap in the area of system use.

The development of the UAMT offers various approaches for further research. On the one hand, the UAMT can be further developed from its prototype status. Further technical evaluations with potential users are relevant here in order to adapt the layout, KPIs and dashboards. On the other hand, it is essential to develop a process as part of a service that shows exactly how the UAMT is used. It is important to define how the UAMT can be used at a team, department or even individual level and how it is linked to measures to improve user adoption.

Another problem is how to individually assign measures that serve to improve user adoption and result from the KPI values determined by the UAMT. Due to legal requirements, the results of the usage analysis are not presented individually, but in aggregated form. Theoretically, an individual score could be determined for each employee based on this analysis. This score provides the basis for an algorithm that weighs up which methods can be individually suggested to employees to improve user adoption. This would have the advantage that employees can use the most effective form of learning for their individual needs. If employees do not want to use the model due to various external factors, this is potentially recognizable by the fact that none of the suggested methods for improving personal user adoption are used.

This could be a cultural challenge within organizations if employees do not use a newly introduced system and do not take the opportunity to learn how to use a system due to e.g. lack of time or value.

References

1. Veerapandian, S., et al.: Reimagine remote working with microsoft teams: a practical guide to increasing your productivity and enhancing collaboration in the remote world, ed. S. Rogers. Packt Publishing Ltd., Birmingham, UK (2021)
2. Powell, A., Yager, S.E., Vandever, J.: Why Isn't everyone an early adopter? In: Americas Conference on Information Systems (AMCIS), Omaha, NE, USA, pp. 1180–1183 (2005)

3. McDonald, J.D.: Measuring personality constructs: the advantages and disadvantages of self-reports, informant reports and behavioural assessments. Enquire **1**(1), 75–94 (2008)
4. Rehouma, M.B., Hofmann, S.: Government employees' adoption of information technology: a literature review. In: Annual International Conference on Digital Government Research: Governance in the Data Age, Delft, The Netherlands (2018)
5. Sein, M.K., et al.: Action design research. MIS Q. **35**(1), 37–56 (2011)
6. Burton-Jones, A., Gallivan, M.J.: Toward a deeper understanding of system usage in organizations: a multilevel perspective. MIS Q. **31**(4), 657–679 (2007)
7. Burton-Jones, A., Straub Jr., D.W.: Reconceptualizing system usage: an approach and empirical test. Inf. Syst. Res. **17**(3), 228–246 (2006)
8. Devaraj, S., Kohli, R.: Performance impacts of information technology: is actual usage the missing link? Manag. Sci. **49**(3), 273–289 (2003)
9. Orlikowski, W.J.: Using technology and constituting structures: a practice lens for studying technology in organizations. Organ. Sci. **11**(4), 404–428 (2000)
10. Klein, K.J., Kozlowski, S.W.: From micro to meso: critical steps in conceptualizing and conducting multilevel research. Organ. Res. Methods **3**(3), 211–236 (2000)
11. Kozlowski, S.W., Klein, K.J.: A multilevel approach to theory and research in organizations. In: Klein, K.J., Kozlowski, S.W. (eds.) Multilevel Theory, Research, and Methods in Organizations, pp. 3–90. Jossey-Bass, San Francisco (2000)
12. Hofmann, D.A., Jones, L.M.: Some foundational and guiding questions for multi-level construct validation. In: Dansereau, F., Yammarino, F.J. (eds.) Multi-Level Issues in Organizational Behavior and Processes, pp. 305–315. Emerald Group Publishing Limited, Amsterdam (2004)
13. Venkatesh, V., Thong, J.Y., Xu, X.: Unified theory of acceptance and use of technology: a synthesis and the road ahead. J. Assoc. Inf. Syst. **17**(5), 328–376 (2016)
14. Williams, M., et al.: Is UTAUT really used or just cited for the sake of it? A systematic review of citations of UTAUT's originating article. In: European Conference on Information Systems (ECIS), Finland (2011)
15. Venkatesh, V., et al.: User acceptance of information technology: toward a unified view. MIS Q. **27**(3), 425–478 (2003)
16. Yin, R.K.: Case Study Research: Design and Methods (Applied Social Research Methods), 5th edn. SAGE Publications, Thousand Oaks (2014)
17. Helfferich, C.: Leitfaden- und experteninterviews. In: Baur, N., Blasius, J. (eds.) Handbuch Methoden der empirischen Sozialforschung, pp. 669–686. Springer, Wiesbaden (2019). https://doi.org/10.1007/978-3-658-21308-4_44
18. Morgan, D.L.: Focus groups. Ann. Rev. Sociol. **22**(1), 129–152 (1996)
19. Venable, J., Pries-Heje, J., Baskerville, R.: FEDS: a framework for evaluation in design science research. Eur. J. Inf. Syst. **25**(1), 77–89 (2016)
20. Straub, D., Rai, A., Klein, R.: Measuring firm performance at the network level: a nomology of the business impact of digital supply networks. J. Manag. Inf. Syst. **21**(1), 83–114 (2004)
21. Lee, J.-N., Miranda, S.M., Kim, Y.-M.: IT outsourcing strategies: universalistic, contingency, and configurational explanations of success. Inf. Syst. Res. **15**(2), 110–131 (2004)
22. Heuer, M., Kurtz, C., Böhmann, T.: Towards a governance of low-code development platforms using the example of microsoft powerplatform in a multinational company. In: Hawaii International Conference on System Sciences (HICSS), Hawaii, HI, USA (2022)
23. Lewandowski, T., et al.: Design knowledge for the lifecycle management of conversational agents. In: International Conference on Wirtschaftsinformatik (WI): A Virtual Conference (2022)

24. Heuer, M., et al.: Rethinking interaction with conversational agents: how to create a positive user experience utilizing dialog patterns. In: Marcus, A., Rosenzweig, E., Soares, M.M. (eds.) HCII 2023. LNCS, vol. 14033, pp. 283–301. Springer, Cham (2023). https://doi.org/10.1007/978-3-031-35708-4_22
25. Heuer, M., et al.: Towards effective conversational agents: a prototype-based approach for facilitating their evaluation and improvement. In: Marcus, A., Rosenzweig, E., Soares, M.M. (eds.) HCII 2023. LNCS, vol. 14033, pp. 302–320. Springer, Cham (2023). https://doi.org/10.1007/978-3-031-35708-4_23

Virtual Reality: Curse or Blessing for Cultural Organizations and Its Consequences on Individuals' Intentions to Attend

Kai Israel and Christopher Zerres(✉)

Offenburg University, Badstraße 24, 77652 Offenburg, Germany
christopher.zerres@hs-offenburg.de

Abstract. Virtual reality (VR) provides many marketing opportunities for cultural organizations to promote their offerings. Besides many advantages, VR poses the threat that VR experiences could inhibit individuals' intention to attend their events in reality. Using the performing arts industry as an example, this study examines the influence of two independent variables—satisfaction with a VR experience, and an individual's intrinsic motivation—on the degree of substitutability of a real theater attendance by a VR experience. In addition, the influence of these two independent variables on individuals' attendance intention is investigated. The results of the study indicate that the potential threat, that VR will inhibit an individual's intention to attend a theater performance, does not materialize in reality. The study's findings contribute to the existing literature and practice by demonstrating that VR does not induce substitution effects, but rather has a positive influence on the intention to attend a theater and helps to increase audience numbers in reality.

Keywords: Virtual Reality · Cultural Organizations · Visit Intention · Satisfaction · Substitutability

1 Introduction

Declining audience numbers are a widespread, serious problem for cultural organizations [1, 2]. Especially, interest in performing arts among younger audiences is steadily declining [3]. This negative trend has been further intensified by the global COVID-19 pandemic, which forced cultural institutions to cease operations [4]. As a result, most performing arts organizations are seeking solutions to increase audience numbers and secure their existence [1]. The internet and the use of novel technologies have been identified in research as potential solutions to this problem that can help organizations target young audiences, create awareness, arouse interest in their offerings, and create innovative cultural experiences [5–7]. In particular, virtual reality (VR) technology is changing the way that individuals can gain experiences [8]. VR experiences are becoming more realistic and offer users a strong sense of presence [9]. VR is defined as *"a completely immersive virtual and aural world that a user experiences, usually through a head-mounted display"* [10, p. 2], *"that simulates a user's physical presence and environment in a way that allows the user to interact with it"* [11].

F. F.-H. Nah and K. L. Siau (Eds.): HCII 2024, LNCS 14721, pp. 63–78, 2024.
https://doi.org/10.1007/978-3-031-61318-0_6

VR offers the great advantage of conveying information, feelings and emotions to consumers in an extraordinary way that was previously only possible through a real-life experience [12–14]. This advantage of VR is also exploited by cultural organizations to promote their offerings [9]. With the help of VR, cultural organizations provide virtual tours of their institutions [15], realistically retell the stories of deceased historical persons such as Anne Frank [16], and make performing arts accessible to a global audience [17]. VR use cases already exist that feel realistic and evoke similar reactions to a real visit [14]. Whether attending a theater performance by Cirque du Soleil [18], William Shakespeare's world-famous tragedy Hamlet [19], or Puccini's opera Tosca [20] all of these real-world experiences can be digitally replicated and virtually consumed.

The consequences of VR marketing are controversially discussed in research. While researchers have observed the positive effects of VR on individuals' (re)visit intentions [21–23], other researchers suggest there is a threat that the use of VR could make real experiences obsolete for an individual [12, 24–26]. The ubiquitous and low-cost availability of digital experiences (e.g., Cirque du Soleil €6.99), poses such a threat to performing arts organizations, as theater, opera, or ballet performances could be replaced by digital replicas in the future. Especially for these organizations, which very often operate in the nonprofit sector, it is important to be aware of this threat, as millions of people are employed in the performing arts worldwide [27–30]. As Yung and Khoo-Lattimore [31] noted, despite the controversial discussion, determining the extent of this threat empirically has received little attention. The aim of this study is to explore the extent of this threat, by examining the impact of virtual consumption of theater performances on the substitutability of the real experience and the individuals' intention to attend. In the present study, we therefore investigate the effects of intrinsic motivation and satisfaction with a virtual experience on the degree of substitutability of a real theater experience by a VR experience, and on individuals' intentions to attend a theater performance in reality. Thus, the study makes a valuable contribution by examining the potential threat of VR to the theater sector and by providing empirical evidence of the consequences of using this technology on individuals' intention to attend. Based on these findings, managers can better determine whether it is a good idea to use this novel technology to promote their performances.

2 Theoretical Framework and Hypotheses

Performing arts provide an experience to participants. The acquisition of such an experience has different aspects in which an individual passes through different stages [32]. Based on Gnoth's [32] model, which describes the process of acquiring an experience, an individual's motivation and satisfaction with an experience are important elements of this process that influence the perceived quality of an experience. Motivation explains the underlying motives of an individual to perform an action. Entertainment, curiosity, or the need for fun are common motivations for individuals to attend a performance [33]. These reasons are characteristic of intrinsic motivation [34, 35]. Intrinsic motivation is defined as doing an activity *"for its own sake"* [34, p. 114] which is an important predictor that influences behavioral intentions such as an individual's intention to visit [12, 36–38].

Satisfaction is the result of an evaluation process in which an individual assesses the usage experience of a product, experience, or service [39]. After consumption, individuals evaluate whether the use of a product, service, or experience has satisfied their needs, desires, and expectations [39]. If a product, service or experience succeeds in fulfilling or even exceeding the expectations of the individual through its use, this leads to satisfaction [39]. By using VR technology, individuals can consume performing arts easily, conveniently, and independent of time and place. Without having to leave their home, individuals can discover performing arts worldwide. Sitting in the front row at Cirque du Soleil [18], having a private performance of the Dutch National Ballet [40], or becoming involved in Hamlet's harrowing story [19] are exemplary VR experiences that may even exceed an individual's needs, desires, and expectations of a theater performance, leading to substitution effects [41]. As a result, the need for a real visit may become obsolete for individuals, as the real visit can be replaced by a VR experience.

In the following, the different factors are examined in more detail and their influence on an individual's visit intention as well as on the degree of substitutability of a real experience by a VR experience is derived. Furthermore, whether the degree of substitutability of a real experience influences individuals' visit intention will be investigated. Figure 1 illustrates the theoretical research model.

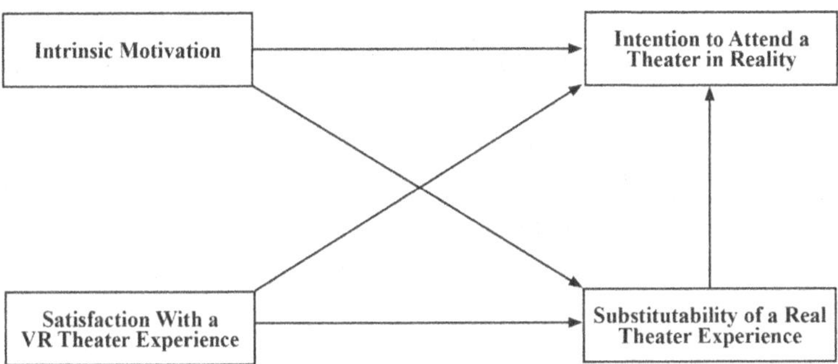

Fig. 1. Theoretical framework.

2.1 Intrinsic Motivation

In accordance with Self-Determination Theory, for intrinsically motivated individuals, the performance of the activity offers a kind of reward such as experiencing pleasure, feeling entertained, or learning something new and interesting [35]. This kind of motivated action is often observed in cultural activities. Whether it is visiting Roman heritage sites in Europe [37], UNESCO-protected gardens in China [38], or art museums in Spain [36], for all these cultural experiences, researchers have demonstrated that individuals participate because of intrinsic motivation. Experiencing new things, having fun, or being entertained are also some of the main reasons why individuals attend theater performances [33, 42]. Consistent with these research findings, we hypothesize that intrinsic motivation has a positive influence on individuals' intention to attend a theater performance.

Hypothesis 1 (H_1): Intrinsic motivation positively influences the intention to attend a theater performance.

Furthermore, intrinsically motivated individuals engage intensively with the entire activity process, in which every detail is important [35]. For example, attending a theater cannot be reduced only to the stage performance [42]. Rather, this experience also has many facets that complete the overall theater experience. Dressing smartly, social interaction with other theater-goers, or the feeling of being part of a special community of interest are elements that shape the special theater atmosphere [42]. Virtually recreating such an experience in its entirety is nearly impossible with VR technology [43, 44]. In other words, VR technology cannot (yet) be used to replicate a holistic experience process. For example, Mura et al. [44] concluded in their survey that VR technology could not replace a real experience. *"I like more non-virtual experiences because of physical contact"* [44, p. 154], *"I cannot get to feel the atmosphere"* [44, p. 154] or *"I want to go there and feel the place fully [...] Those little details are very important, that makes everything real"* [44, p. 154] are exemplary statements made by the travel-savvy study participants that contradict the substitutability of a real experience. Furthermore, in the context of a UNESCO world heritage site, Losada et al. [43] found that a VR experience is not capable of conveying the emotions that arise in users during a real visit. Due to the lack of physical mobility, rudimentary (social) interaction possibilities, and the absence of sensory stimuli (e.g., smelling, tasting, feeling), the adequate replication of a holistic experience process in VR is hardly possible [43–45]. Li and Chen [25] also found that VR experiences of destinations do not inhibit intention to travel for intrinsically motivated individuals. Consequently, intrinsic motivation is assumed to inhibit the degree to which a real experience can be substituted by a VR experience.

Hypothesis 2 (H_2): Intrinsic motivation negatively influences the substitutability of a real theater experience.

2.2 Satisfaction with the VR Experience

According to Expectation Confirmation Theory, satisfaction is the result of an evaluation process in which individuals assess the ratio between their expectations (before consumption) and the perceived performance (after consumption) of a product, experience, or service [46]. As research suggests in this regard, high levels of satisfaction increase the likelihood that an individual will develop a behavioral intention such as recommending or using a service [47]. Likewise, in the case of cultural activities such as visiting museums [48], heritage sites [49], or performing arts events [50], researchers have demonstrated that satisfaction with the cultural experience has a positive influence on individuals' intention to visit. In the context of virtual applications, Jung et al. [23] showed in their study that VR applications can enhance a real museum experience. VR applications arouse curiosity, are enjoyable, engage users, and can convey information in an entertaining way [23]. These positive attributes contribute to satisfaction with a museum visit and increase an individual's willingness to revisit a museum [23]. Lee et al. [21] come to a similar conclusion, stating that an appealing virtual museum tour is a suitable marketing tool to attract potential visitors. Accordingly, VR applications offer the opportunity to present cultural experiences in an attractive way [21]. These new possibilities increase satisfaction with a virtual museum experience and positively affect individuals' visit intentions [21]. The same applies to the visualization of UNESCO world heritage sites. As An et al. [22] demonstrated in their study using the Great Barrier Reef in Australia as an example, an individual's intention to visit increases when the VR experience meets their expectations. The ability to virtually explore foreign places enhances the awareness of cultural activities and increases individuals' desire to visit them in reality [22]. Based on the findings of the mentioned studies, the hypothesis to be tested is as follows:

Hypothesis 3 (H_3): Users' satisfaction with a VR theater experience positively influences the intention to attend a theater performance in reality.

On the other hand, there is also the possibility that a satisfying VR experience could be a substitute for a real experience. Substitutability is defined as *"the tendency of people to switch from one product to another that fulfils the same purpose"* [41, p. 796]. Consequently, substitution effects could occur when two products can satisfy the same needs, desires, and expectations of an individual in different ways [41]. As Wagler and Hanus [14] noted, this applies to VR applications and real experiences. In their study, the authors examined the experience of a guided tour of a state capitol building to determine if there were differences between an immersive VR experience and the experience in reality. The researchers found that whether participants took the tour in real life or used a VR application was irrelevant when evaluating the experience. Consequently, Wagler and Hanus [14] concluded that a VR application is a strong analog to a real experience. A similar conclusion was reached by Sarkady et al. [26], who asked study participants to view various tourist destinations (e.g., Amsterdam, Dubrovnik, Cadiz) using a VR application. As the researchers revealed, VR applications are suitable for getting a realistic impression of a destination, which could reduce an individual's travel intention [25, 26]. In addition, study results by Sarkady et al. [26] suggest that such VR experiences could serve as a travel substitute for study participants. Deng et al. [24] also showed in their study that realistic and highly interactive VR applications can reduce an individual's interest in visiting a museum in real life. Furthermore, VR applications eliminate many of the traditional barriers such as financial costs, visa restrictions, physical limitations, health concerns, family obligations, or transportation requirements [9, 26] that in reality inhibit individuals' willingness to travel [51] in general and to attend theaters, operas, or ballets in particular [52]. Given the study results presented, we hypothesize that satisfaction with a VR theater experience has a positive influence on the degree of substitutability of a real theater experience.

Hypothesis 4 (H_4): Users' satisfaction with a VR theater experience positively influences the substitutability of a real theater experience.

2.3 Substitutability and Intention to Attend

The substitutability of a real experience by a VR experience depends on whether a VR experience is accepted by individuals as an adequate substitute [9]. One main argument against the substitution of a real experience by a VR experience is that a VR experience is simply not a real experience [53, 54]. Rather, a VR experience is a form of fictional artificial entertainment [53]. As researchers have stated in qualitative studies, a VR experience primarily lacks the social component that distinguishes a real experience [53, 54]. Even the trip or attendance itself often constitutes a social experience [54]. In most cases, an individual does not undertake such an activity alone, and those individuals who do participate in cultural experience alone usually enjoy the company of others [54]. Social interaction with other theater-goers, experiencing the unique theater atmosphere, and feeling deeper emotions are some of the core characteristics of a theater attendance [33, 42] that VR applications would struggle to provide [53, 54]. Sensory stimuli such as smelling or tasting are also difficult to reproduce [9]. Likewise, the discomfort when wearing head-mounted displays limits the quality of the VR experience [55], and can lead to cybersickness [56] and health problems [57]. Considering these limitations, it will take a long time before users can fully experience a real-world experience in VR in a pleasurable way [53]. Taking into account the previous discussion, the research findings, and the current technological possibilities, we assume that the degree of substitutability of a real theater experience by a VR experience is not (yet) sufficient to influence individuals' intention to attend a theater performance in reality.

Hypothesis 5 (H_5): The degree of substitutability of a real theater experience by a VR experience does not influence the intention to attend a theater performance in reality.

Figure 2 summarizes the proposed research framework and hypotheses to be tested.

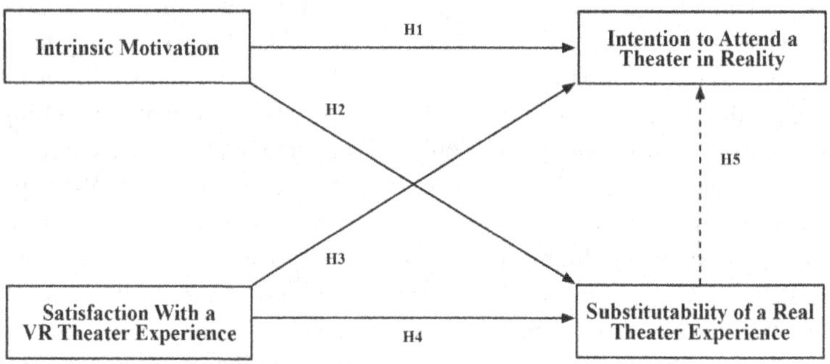

Fig. 2. Research framework and hypotheses.

3 Method

3.1 Operationalization and Stimulus

The operationalization of the constructs was based on multi-item scales adapted from previous research (Table 1). All items were measured using a seven-point Likert scale (1 = I strongly disagree; 7 = I strongly agree).

The official VR application *Inside the Box of Kurios* by Cirque du Soleil [58] served as the stimulus (Fig. 3). The Cirque du Soleil performance combines multiple aspects of theater performances, because plays, music, dance, and artistry are combined. During the ten-minute VR experience, users can change the viewing perspective in real time by moving their heads, giving them the feeling of becoming part of the performance themselves.

Fig. 3. Screenshots of the Cirque du Soleil VR application.

3.2 Design of the Study

Each participant went through three phases during the study. In the first phase, the participants were asked to provide information about their sociodemographic characteristics by using a computer-assisted questionnaire. In order to determine the actual affinity to attend theater performances, the intrinsic motivation of the participants was measured. Following the questionnaire, participants were asked to read through the scenario description.

In the scenario description, participants were asked to imagine that during a dinner, friends were telling them about their attendance at a theater performance. The friends really enjoyed the theater performance and recommended that the study participant also should attend the show. They said that the theater was nearby and the performance would be running for a further two years. Furthermore, the participants were asked to imagine that after the dinner, they searched for information on the internet about the theater performance. During their search, they came across a VR application that they loaded onto their own VR system to experience the theater performance virtually.

At the beginning of the second phase, the participants were familiarized with the handling of the VR system (Oculus Go). In addition to the VR system, the participants were also provided with noise-canceling headphones. Subsequently, the participants were asked to sit on a chair and start the VR application. After the VR theater performance ended, the participants were asked in phase three to evaluate the VR experience using a computer-assisted questionnaire.

3.3 Data Collection and Sample

The sample size was determined a priori based on the recommendations of Hair et al. [59]. Accordingly, at least 103 participants are required to detect small effects at a 5% significance level [59]. Various channels were used to recruit study participants. Companies, public institutions, as well as associations agreed to make their employees, members and students aware of the study. As an incentive, vouchers (20 euros) for a theme park were raffled among the participants. The study period ranged from November 2019 to March 2020. During this period, 142 data sets were collected under laboratory conditions.

The age of the participants ranged from 16 to 74 years (M = 35.0, SD = 15.0). The gender distribution (61 female, 73 male, one no information) was balanced. The majority of participants were employees (40.0%). The term *"virtual reality"* was known to almost all participants (97.0%). Of those participants who were familiar with the term *"virtual reality"* (n = 131), over two-thirds (69.5%) reported having used a VR system in the past.

3.4 Data Analysis

A total of 142 participants took part in the study. The initial data screening was performed with SPSS 26. Based on the recommendation of Pituch and Stevens [60], those data sets whose z-scores exceeded the critical value of ±3 were eliminated. Seven outliers were identified, which were excluded from further analysis. The data basis for the structural model thus comprised 135 valid data sets. The PLS-SEM method was used to analyze the structural model. This method is particularly useful for testing the predictive power of the theoretical research framework and is well established for the modeling of latent variables in the context of novel technologies [61].

4 Results

4.1 Assessment of the Measurement Model

To determine the suitability of the measurement model, the internal consistency, convergence validity, and discriminant validity of the measurement model were examined. For the verification of the internal consistency, the Cronbach's alpha coefficient (α), the factor reliability (ρc) as well as the Dijkstra-Henseler's coefficient (ρA) were considered. As shown in Table 1, all three criteria met the minimum requirements [59].

Factor loadings, indicator reliability, and average variance extracted (AVE) were used as criteria to determine convergence validity. As shown in Table 1, convergence validity was confirmed by all three criteria.

Table 1. Validity and reliability of the constructs.

Construct and items	α	ρA	ρc	AVE	Loading
Criteria	> 0.7	> 0.7	> 0.7	> 0.5	> 0.7
Satisfaction (SAT)	**0.783**	**0.802**	**0.873**	**0.696**	
adapted from van Ittersum et al. [62]					
I am satisfied with my experience in the virtual world.					0.790
I enjoyed the VR theater experience.					0.855
Overall, I am satisfied with the VR theater experience.					0.856
Intrinsic Motivation (MOT)	**0.934**	**0.935**	**0.950**	**0.791**	
adapted from Center for Self-Determination Theory [63]					
Going to a theater performance is fun.					0.896
I think it is boring to attend a theater performance.*					0.857
It is very interesting to attend a theater performance.					0.867
Going to a theater performance is quite enjoyable.					0.905
I enjoy going to a theater performance very much.					0.921
Substitutability (SUB)	**0.811**	**0.877**	**0.885**	**0.720**	
adapted from Evrard and Krebs [64]					
Watching a theater performance in VR is as good as seeing it in a theater.					0.735
Thanks to VR, you no longer need to go to a theater to see the performance.					0.891
Using VR for a theater visit is more fun than visiting the theater in reality.					0.910
Intention to Attend (ITA)	**0.864**	**0.867**	**0.917**	**0.787**	
adapted from Jung et al. [23]					
I intend to attend the theater performance after experiencing the VR application.					0.900
I will attend the theater performance after experiencing the VR application.					0.917
I want to recommend the theater performance to others after experiencing the VR application.					0.843

*reverse coded

Discriminant validity was determined based on the Fornell-Larcker criterion [65] and the heterotrait-monotrait correlation ratio (HTMT) of Henseler et al. [66] using the conservative threshold of 0.85. As the analysis revealed, discriminant validity was confirmed by both the Fornell-Larcker criterion and the $HTMT_{.85}$.

4.2 Assessment of the Structural Model

In the first step, the variance inflation factor (VIF) values of all predictor constructs were examined to elicit any collinearity problems of the structural model. In this regard, the analysis revealed that all VIF values were in the range of 1.048 to 1.340, well below the recommended threshold of five [59]. The coefficient of determination (R^2) was used to further assess the structural model. According to Cohen [67], R^2 values of 0.25 and above are considered as high. The R^2 values of the two endogenous variables of the structural model were above this value (Table 2), indicating a good model fit.

Table 2. Mean values and assessment criteria of the structural model.

Construct	Mean	SD	R^2	Q^2
ITA	5.25	1.27	0.264	0.163
SUB	3.26	1.46	0.251	0.192
SAT	6.40	0.67		
MOT	5.77	1.04		

To determine the predictive relevance of the indicators ($Q^2_{predict}$) and constructs (Q2), both the $PLS_{predict}$ procedure [68] and the blindfolding procedure [69, 70] were performed. The analysis showed that the $Q^2_{predict}$ values of all indicators were above the critical value of zero [68]. Furthermore, the prediction errors of the individual indicators were examined. All indicators had lower root mean squared error values as well as lower mean absolute error values in the PLS-SEM analysis compared to the linear regression model. The structural model thus indicated a high predictive power [68]. This result was also supported by the blindfolding procedure at the construct level (Table 2), as Q2 of all endogenous variables was above zero [69, 70]. The standardized root mean square residual of the structural model was 0.070, confirming a good model fit [59].

The significance of the path coefficients was determined using the bootstrapping procedure with 5,000 subgroups [59]. As the results indicate (Table 3, Fig. 4), four of the five research hypotheses were significant, providing statistical support for the hypothesized relationships.

Table 3. Results of hypothesis testing.

	Relationships	Path coefficient	CI (Bias Corrected)	t-Value	p-Value	Supported
H1	MOT→ ITA	0.159*	[0.015, 0.320]	2.042	0.041	Yes
H2	MOT→ SUB	-0.468**	[-0.580, -0.331]	7.424	0.000	Yes
H3	SAT→ ITA	0.442**	[0.280, 0.573]	5.894	0.000	Yes
H4	SAT→ SUB	0.307**	[0.171, 0.424]	4.820	0.000	Yes
H5	SUB→ ITA	-0.145	[-0.338, 0.054]	1.434	0.152	Yes

Note: ** $p < 0.001$, * $p < 0.05$.

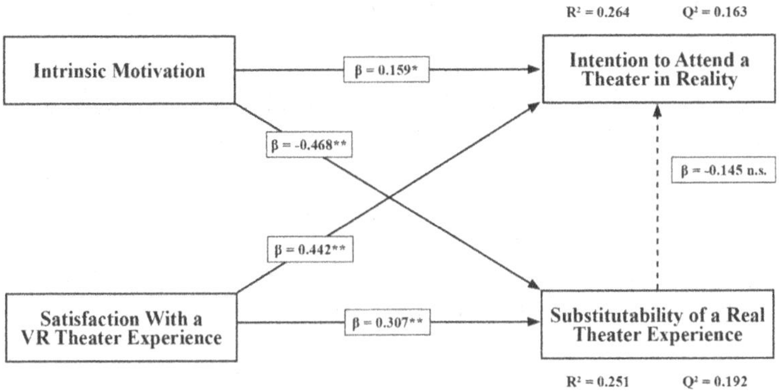

Fig. 4. PLS results of the structural model.

5 Discussion and Implications

The aim of the present study was to investigate the effects of two independent variables—satisfaction with a VR experience, and an individual's intrinsic motivation—on the degree of substitutability of a real theater performance by a VR theater experience. In addition, the influence of these two independent variables on an individual's intention to attend was investigated. Another objective of the study was to examine whether the degree of substitutability influences the intention to attend a theater in reality.

The first major finding of the study is that satisfaction with a VR theater experience positively influences an individual's intention to attend a theater in reality ($H_3, \beta = 0.442$, $p < 0.001$). This finding extends prior research conducted in the context of museums and heritage sites [21–23]. Moreover, in contrast to the study by Jung et al. [23], the participants in the present study were not surveyed during their real participation in a cultural experience, but instead both their level of satisfaction with the VR theater experience and their intention to attend prior to a potential attendance were assessed. The results from our ex-ante survey provide evidence that VR applications could not only enhance a cultural experience and increase the intention to revisit [23], but that the VR theater experience could also be an initial reason for an individual to attend a theater performance. This is another indication that marketing managers should embrace VR technology as an effective marketing tool that can lead to increased attendance [21]. The extraordinary visualization capabilities of VR technology can be used to attractively present and promote theater performances. This finding is further supported by the high agreement scores ($M = 6.40$, $SD = 0.67$), according to which a VR theater experience almost completely fulfills the users' expectations (Table 2).

On the other hand, the results also reveal that there is a significant positive relationship between users' satisfaction with a VR theater experience and the degree to which a real theater performance can be substituted by a VR experience (H_4, $\beta = 0.307$, $p < 0.001$). This result is consistent with prior research findings that individuals might consider VR applications good enough to provide them with a realistic impression of a service, which could inhibit real consumption intentions [24–26]. However, the ability to get a realistic impression of a theater performance virtually does not mean that the real theater experience is completely replaced by a VR experience. As mentioned above, the satisfaction of the study participants with the VR theater experience was particularly high ($M = 6.40$, $SD = 0.67$). In comparison, the degree of substitutability ($M = 3.26$, $SD = 1.46$) was rated lower by the participants. Although participants were highly satisfied with the VR theater experience, they tended to reject the VR experience as a substitute for a real theater experience. Furthermore, the direct effect between the degree of substitutability and individuals' intention to attend a theater in reality was also not statistically significant (H_5, $\beta = -0.145$, $p = 0.152$). This is further evidence that VR technology is highly suitable for marketing purposes. As the results of this study demonstrate, the general fear that a VR experience reduces the likelihood of real participation in cultural activities seems to be rather unfounded in the case of theaters.

Furthermore, the present study demonstrates that an individual's intrinsic motivation has a strong negative influence on the degree to which a real theater experience can be substituted by a VR experience (H_2, $\beta = -0.468$, $p < 0.001$). The study provides quantitative evidence and complements the qualitative research findings of Mura et al. [44] by demonstrating that for intrinsically motivated individuals, a VR theater experience is not an adequate substitute for attending a real theater performance. As the research results illustrate, no substitution effects occur through the use of VR technology among individuals who associate theater attendance with enjoyment, fun and pleasure.

Another finding that emerges from the results of the present study confirms the positive relationship between an individual's intrinsic motivation and intention to attend a theater in reality (H_1, $\beta = 0.159$, $p < 0.05$). Extending previous studies conducted in the context of museums [36], heritage sites [37, 38], concerts, and cinemas [33], the present study provides evidence that individuals are more likely to attend theater performances when attendance is linked to positive associations such as enjoyment, fun and pleasure.

6 Conclusion, Limitations, and Future Research

The central objective of this study was to provide insights for cultural organizations concerning the consequences of the use of VR technology on individuals' intention to attend their offerings in reality. On the one hand, the use of VR technology provides the possibility for the managers to create extraordinary experiences. On the other hand, the omnipresent availability and convenient access to VR cultural experiences could lead individuals to use such digital experiences as a substitute for a real experience, and consequently they might attend fewer cultural events in reality. VR technology could thus be both a curse and a blessing for the cultural sector. Based on the research findings within the context of theater performances, we can conclude that VR applications should rather be seen by marketing managers as a blessing to attractively present and promote their cultural offerings. The potential threat that the use of VR applications will inhibit an individual's interest in attending theaters in reality has not (yet) materialized. Rather, the study results demonstrate that VR technology is a suitable marketing tool that has a positive influence on individuals' intention to attend theaters and can help to increase theater attendance in reality.

Despite the findings of this study, there are further questions that could be addressed in future research. As VR technology continues to evolve, a variety of new technical possibilities will emerge. For example, future VR systems may be capable of simulating olfactory and gustatory stimuli in addition to visual and auditory stimuli. Due to these additional features, users could feel more involved in the VR experience, which could influence the degree of substitutability of a real experience by a VR experience. In the future, these aspects could be investigated by further studies. Furthermore, in our study the participants were surveyed under laboratory conditions to ensure the internal validity of the study. Since the number of VR systems in private households will increase, future research could verify the presented results in field experiments.

References

1. Pompe, J., Tamburri, L., Munn, J.: Marketing strategies for performing arts audiences: characteristics of ticket purchasers. J. Nonprofit Public Sect. Mark. **32**(5), 453–464 (2020)
2. White, C.J., Tong, E., Schwartz, M.: Psychological needs, passion, and consumer re-patronage intentions to the performing arts. J. Nonprofit Public Sect. Mark. **31**(5), 451–463 (2019)
3. National Endowment for the Arts: A decade of arts engagement: Findings from the survey of public participation in the arts, 2002–2012, Washington, D.C. (2015)
4. Betzler, D., Loots, E., Prokůpek, M., Marques, L., Grafenauer, P.: COVID-19 and the arts and cultural sectors: investigating countries' contextual factors and early policy measures. Int. J. Cult. Policy **27**(6), 796–814 (2021)
5. Hume, M.: To technovate or not to technovate? Examining the inter-relationship of consumer technology, museum service quality, museum value, and repurchase intent. J. Nonprofit Public Sect. Mark. **27**(2), 155–182 (2015)
6. Gomes, R., Knowles, P.A.: Strategic internet and e-commerce applications for local nonprofit organizations. J. Nonprofit Public Sect. Mark. **9**(1–2), 215–245 (2001)
7. Turrini, A., Soscia, I., Maulini, A.: Web communication can help theaters attract and keep younger audiences. Int. J. Cult. Policy **18**(4), 474–485 (2012)
8. Flavián, C., Ibáñez-Sánchez, S., Orús, C.: The impact of virtual, augmented and mixed reality technologies on the customer experience. J. Bus. Res. **100**(4), 547–560 (2019)
9. Guttentag, D.: Virtual reality and the end of tourism? A substitution acceptance model. In: Xiang, Z., Fuchs, M., Gretzel, U., Höpken, W. (eds.) Handbook of e-Tourism, pp. 1–19. Springer, Cham (2020). https://doi.org/10.1007/978-3-030-05324-6_113-1
10. PwC: Growing VR/AR companies in the UK. https://www.pwc.co.uk/intelligent-digital/vr/growing-vr-ar-companies-in-the-uk.pdf. Accessed 06 Dec 2023

11. Isaac, J.: Step into a new world: virtual reality. https://www.completegate.com/2016070154/blog/virtual-reality-explained. Accessed 27 July 2023
12. Israel, K., Zerres, C., Tscheulin, D.K.: Virtual reality — substitute for a real experience? The role of user motivation, expectations and experience type. Int. J. Innov. Technol. Manag. **20**(3) (2023)
13. Chen, C., Yao, M.Z.: Strategic use of immersive media and narrative message in virtual marketing: understanding the roles of telepresence and transportation. Psychol. Mark. **39**(3), 524–542 (2022)
14. Wagler, A., Hanus, M.D.: Comparing virtual reality tourism to real-life experience: effects of presence and engagement on attitude and enjoyment. Commun. Res. Rep. **35**(5), 456–464 (2018)
15. Evans, C.: IL DIVINO: Michelangelo's Sistine Ceiling in VR. https://store.steampowered.com/app/1165850/IL_DIVINO_Michelangelos_Sistine_Ceiling_in_VR/. Accessed 06 Dec 2023
16. Force Field: Anne Frank House VR. https://www.oculus.com/experiences/quest/1958100334295482. Accessed 06 Dec 2023
17. Felix & Paul Studios: Inside the Box of Kurios. https://www.felixandpaul.com/?kurios. Accessed 06 Dec 2023
18. Felix & Paul Studios: Cirque du Soleil VR. https://www.oculus.com/experiences/quest/2833307773392914/. Accessed 06 Dec 2023
19. Maler, S.: Hamlet 360: Thy Father's Spirit. https://commshakes.org/production/hamlet-360-thy-fathers-spirit/. Accessed 06 Dec 2023
20. Relative Motion: Tosca VR | World's First Made-For Virtual Reality Opera. https://www.relativemotion.co.uk/tosca-vr.html. Accessed 06 Dec 2023
21. Lee, H., Jung, T.H., tom Dieck, M., Chung, N.: Experiencing immersive virtual reality in museums. Inf. Manag. **57**(5), 103229 (2020)
22. An, S., Choi, Y., Lee, C.-K.: Virtual travel experience and destination marketing: effects of sense and information quality on flow and visit intention. J. Destin. Mark. Manag. **19**, 100492 (2021)
23. Jung, T., tom Dieck, M.C., Lee, H., Chung, N.: Effects of virtual reality and augmented reality on visitor experiences in museum. In: Inversini, A., Schegg, R. (eds.) Information and Communication Technologies in Tourism 2016, pp. 621–635. Springer, Cham (2016). https://doi.org/10.1007/978-3-319-28231-2_45
24. Deng, X., Unnava, H.R., Lee, H.: "Too true to be good?" when virtual reality decreases interest in actual reality. J. Bus. Res. **100**(1), 561–570 (2019)
25. Li, T., Chen, Y.: Will virtual reality be a double-edged sword? Exploring the moderation effects of the expected enjoyment of a destination on travel intention. J. Destin. Mark. Manag. **12**(4), 15–26 (2019)
26. Sarkady, D., Neuburger, L., Egger, R.: Virtual reality as a travel substitution tool during COVID-19. In: Wörndl, W., Koo, C., Stienmetz, J.L. (eds.) Information and Communication Technologies in Tourism 2021, pp. 452–463. Springer, Cham (2021). https://doi.org/10.1007/978-3-030-65785-7_44
27. Baker, W.F., Gibson, W.C., Leatherwood, E.: The World's Your Stage. How Performing Artists Can Make a Living While Still Doing What They Love. AMACOM American Management Association, New York (2016)
28. OECD: Economic and social impact of cultural and creative sectors. Note for Italy G20 Presidency Culture Working Group (2021)
29. UNESCO: Promoting the Diversity of Cultural Expressions and Creative Economy. https://www.unesco.org/en/articles/promoting-diversity-cultural-expressions-and-creative-economy. Accessed 06 Dec 2023

30. European Commission: Data on the cultural sector. https://culture.ec.europa.eu/policies/selected-themes/data-on-the-cultural-sector. Accessed 06 Dec 2023
31. Yung, R., Khoo-Lattimore, C.: New realities: a systematic literature review on virtual reality and augmented reality in tourism research. Curr. Issue Tour. **22**(17), 2056–2081 (2019)
32. Gnoth, J.: Tourism motivation and expectation formation. Ann. Tour. Res. **24**(2), 283–304 (1997)
33. Manolika, M., Baltzis, A.: Concert hall, museum, cinema, and theater attendance: what difference do audience motivations and demographics make? Empir. Stud. Arts **40**(1), 37–56 (2022)
34. Csikszentmihalyi, M., Graef, R., Gianinno, S.M.: Measuring intrinsic motivation in everyday life. In: Csikszentmihalyi, M. (ed.) Flow and the Foundations of Positive Psychology, pp. 113–125. Springer, Dordrecht (2014). https://doi.org/10.1007/978-94-017-9088-8_8
35. Ryan, R.M., Deci, E.L.: Intrinsic and extrinsic motivations: classic definitions and new directions. Contemp. Educ. Psychol. **25**(1), 54–67 (2000)
36. Forgas-Coll, S., Palau-Saumell, R., Matute, J., Tárrega, S.: How do service quality, experiences and enduring involvement influence tourists' behavior? An empirical study in the Picasso and Miró Museums in Barcelona. Int. J. Tour. Res. **19**(2), 246–256 (2017)
37. Kolar, T., Zabkar, V.: A consumer-based model of authenticity: an oxymoron or the foundation of cultural heritage marketing? Tour. Manag. **31**(5), 652–664 (2010)
38. Shen, S., Schüttemeyer, A., Braun, B.: Visitors' intention to visit world cultural heritage sites: an empirical study of Suzhou, China. J. Travel Tour. Mark. **26**(7), 722–734 (2009)
39. Oliver, R.L.: A Cognitive Model of the Antecedents and Consequences of Satisfaction Decisions. J. Mark. Res. **17**(4), 460–469 (1980)
40. Leung, P.: Night Fall. A Virtual Reality Dance Film for Dutch National Ballet. http://www.peter-leung.com/dance-film-for-dutch-national-ballet. Accessed 06 Dec 2023
41. Flavián, C., Gurrea, R.: Perceived substitutability between digital and physical channels: the case of newspapers. Online Inf. Rev. **31**(6), 793–813 (2007)
42. Walmsley, B.: Why people go to the theatre: a qualitative study of audience motivation. J. Cust. Behav. **10**(4), 335–351 (2011)
43. Losada, N., Jorge, F., Teixeira, M.S., Melo, M., Bessa, M.: Could virtual reality substitute the 'real' experience? Evidence from a UNESCO world heritage site in Northern Portugal. In: Abreu, A., Liberato, D., González, E.A., Garcia Ojeda, J.C. (eds.) ICOTTS 2020. SIST, vol. 209, pp. 153–161. Springer, Singapore (2021). https://doi.org/10.1007/978-981-33-4260-6_14
44. Mura, P., Tavakoli, R., Pahlevan Sharif, S.: 'Authentic but not too much': exploring perceptions of authenticity of virtual tourism. Inf. Technol. Tour. **17**, 145–159 (2017)
45. Cheong, R.: The virtual threat to travel and tourism. Tour. Manag. **16**(6), 417–422 (1995)
46. Oliver, R.L.: Effect of expectation and disconfirmation on postexposure product evaluations: an alternative interpretation. J. Appl. Psychol. **62**(4), 480–486 (1977)
47. Lin, C.-H., Kuo, B.Z.-L.: The behavioral consequences of tourist experience. Tour. Manag. Perspect. **18**, 84–91 (2016)
48. Duantrakoonsil, T., Lee, H.Y., Reid, E.L.: Museum service quality, satisfaction, and revisit intention: evidence from the foreign tourists at Bangkok National Museums in Thailand. Culin. Sci. Hosp. Res. **23**(6), 127–134 (2017)
49. Chen, C.-F., Chen, F.-S.: Experience quality, perceived value, satisfaction and behavioral intentions for heritage tourists. Tour. Manag. **31**(1), 29–35 (2010)
50. Hume, M., Sullivan Mort, G.: Understanding the role of involvement in customer repurchase of the performing arts. J. Nonprofit Public Sect. Mark. **20**(2), 299–328 (2008)
51. Hudson, S., Gilbert, D.: Tourism constraints: the neglected dimension in consumer behaviour research. J. Travel Tour. Mark. **8**(4), 69–78 (2000)

52. Kay, P.L., Wong, E., Polonsky, M.J.: Marketing cultural attractions: understanding non-attendance and visitation barriers. Mark. Intell. Plan. **27**(6), 833–854 (2009)
53. Sussmann, S., Vanhegan, H.: Virtual reality and the tourism product substitution or complement? In: ECIS 2000 Proceedings, pp. 117.1–117.7 (2000)
54. Rauscher, M., Humpe, A., Brehm, L.: Virtual reality in tourism: is it 'real' enough? Academica Turistica **13**(2), 127–138 (2020)
55. Herz, M., Rauschnabel, P.A.: Understanding the diffusion of virtual reality glasses: the role of media, fashion and technology. Technol. Forecast. Soc. Change **138**(6), 228–242 (2019)
56. Israel, K., Zerres, C., Tscheulin, D.K.: Reducing cybersickness: the role of wearing comfort and ease of use. In: Frontiers in Optics, Washington, D.C, JTu2A. 113. OSA, Washington, D.C. (2017)
57. Rebenitsch, L., Owen, C.: Estimating cybersickness from virtual reality applications. Virtual Real. **25**(1), 165–174 (2021)
58. Felix & Paul Studios: Cirque du Soleil VR. https://www.felixandpaul.com/?kurios. Accessed 06 Dec 2023
59. Hair, J.F., Hult, T.M., Ringle, C.M., Sarstedt, M.: A Primer on Partial Least Squares Structural Equation Modeling (PLS-SEM). SAGE Publications, Thousand Oaks (2017)
60. Pituch, K.A., Stevens, J.P.: Applied multivariate Statistics for the Social Sciences. Analyses with SAS and IBM's SPSS. Routledge Taylor and Francis Group, New York (2016)
61. Henseler, J., Hubona, G., Ray, P.A.: Using PLS path modeling in new technology research: updated guidelines. Ind. Manag. Data Syst. **116**(1), 2–20 (2016)
62. van Ittersum, K., Wansink, B., Pennings, J.M., Sheehan, D.: Smart shopping carts: how real-time feedback influences spending. J. Mark. **77**(6), 21–36 (2013)
63. Center for Self-Determination Theory: Intrinsic Motivation Inventory. http://selfdeterminationtheory.org/intrinsic-motivation-inventory/. Accessed 06 Dec 2023
64. Evrard, Y., Krebs, A.: The authenticity of the museum experience in the digital age: the case of the Louvre. J. Cult. Econ. **42**(3), 353–363 (2018)
65. Fornell, C., Larcker, D.F.: Evaluating structural equation models with unobservable variables and measurement error. J. Mark. Res. **18**(1), 39–50 (1981)
66. Henseler, J., Ringle, C.M., Sarstedt, M.: A new criterion for assessing discriminant validity in variance-based structural equation modeling. J. Acad. Mark. Sci. **43**(1), 115–135 (2015)
67. Cohen, J.: Statistical Power Analysis for the Behavioral Sciences. Lawrence Erlbaum Associates Publishers, Hillsdale (1988)
68. Shmueli, G., et al.: Predictive model assessment in PLS-SEM: guidelines for using PLSpredict. Eur. J. Mark. **53**(11), 2322–2347 (2019)
69. Stone, M.: Cross-validatory choice and assessment of statistical predictions. J. R. Stat. Soc. Ser. B (Methodol.) **36**, 111–147 (1974)
70. Geisser, S.: A predictive approach to the random effect model. Biometrika **61**(1), 101–107 (1974)

Maturity Measurement Framework for Evaluating BIM-Based AR/VR Systems Adapted from ISO/IEC 15939 Standard

Ziad Monla(✉), Ahlem Assila, Djaoued Beladjine, and Mourad Zghal

LINEACT, CESI Engineering School, Reims, France
{zmonla,aassila,dbeladjine,mzghal}@cesi.fr

Abstract. This article proposes a framework for evaluating the maturity of BIM-based augmented and virtual reality systems by defining indicators based on the ISO/IEC 15939 standard. These indicators aim to provide a more comprehensive assessment of system maturity by combining results from various measurements, while differentiating between sub-criteria, criteria, and maturity aspects. This framework not only defines the indicators but also presents a process for interpreting the results, while preserving the raw measurement data to identify maturity-related issues and recommend mechanisms for transitioning between different levels. To illustrate the application of the framework, the article presents an example of defining a maturity indicator related to the technology criterion. Furthermore, a feasibility study was conducted by academic experts to validate the proposed framework. The results demonstrate the effectiveness of the indicators in assessing the maturity of BIM-based AR/VR systems. This research contributes to the field by providing a comprehensive framework for evaluating system maturity and offering new perspectives. Future works are focused on implementing the proposed indicators in real-world scenarios and exploring additional dimensions of maturity assessment to further enhance the evaluation process.

Keywords: Augmented reality · Virtual reality · Building Information Modeling · Maturity · Indicator · Evaluation

1 Introduction

Building Information Modeling (BIM) has revolutionized the Architecture, Engineering, and Construction (AEC) industry, providing numerous benefits and transforming traditional construction practices [1, 2]. BIM has been praised for its ability to facilitate collaboration, improve coordination, reduce errors, enhance project quality, and lower costs throughout the project life cycle [3, 4]. Despite its promising potential, BIM tools have yet to fully deliver on their capabilities [5, 6]. To address this challenge and further enhance the construction process, immersive technologies such as Augmented Reality (AR) and Virtual Reality (VR) have been integrated with BIM, giving rise to BIM-based AR/VR systems [7, 8].

F. F.-H. Nah and K. L. Siau (Eds.): HCII 2024, LNCS 14721, pp. 79–95, 2024.
https://doi.org/10.1007/978-3-031-61318-0_7

Using AR and VR technologies in conjunction with BIM offers a range of benefits in the construction industry [9]. These systems provide an engaging and immersive environment for creating and visualizing multi-dimensional simulations of construction projects, allowing stakeholders to experience the project virtually [7]. BIM-based AR/VR systems facilitate effective communication, improve design visualization, enable automatic document management, enhance collaboration, support cost estimation, and aid in schedule modeling based on the concept of 4D [7]. Moreover, the integration of AR and VR technologies can help address conflicts between contractors and consultants by enhancing coordination between design and construction, thus improving efficiency and preventing data loss [4].

Despite the potential benefits of BIM-based AR/VR systems, it is crucial to evaluate their maturity. Evaluating the maturity of these systems allows companies to make informed decisions about technology investment and develop effective strategies for adopting augmented and virtual reality in the AEC industry [10].

To date, the literature on the maturity evaluation of BIM-based AR/VR systems is limited, with only two research studies identified [11, 12]. The first study [11] focuses only on the evaluation of BIM-based VR systems and does not consider the broader scope of AR and VR integration. On the other hand, the second study [12] proposes an approach for evaluating the maturity of BIM-based AR/VR systems but lacks essential evaluation concepts such as measures and evaluation criteria. However, current methods for assessing the maturity of BIM-based AR/VR systems lack uniformity and suffer from diversity and variation, making it challenging to compare and standardize evaluation results.

To address this issue, the aim of this paper is to propose a systematic approach for evaluating the maturity of BIM-based AR/VR systems using the ISO/IEC 15939 standard [13] as the foundation. The adoption of this international standard will provide defined guidelines for measurement processes, ensuring consistency, comparability, and reliability in the assessment of these innovative systems.

By developing a structured evaluation process, we seek to address the limitations of existing approaches and overcome the challenges posed by the diverse and varied methods currently used to assess the maturity of BIM-based AR/VR systems. Our proposed approach will enable companies in the AEC industry to make more informed decisions about technology investment and develop effective strategies for incorporating AR and VR technologies into their construction projects.

The remaining part of this article is structured as follows: Sect. 2 offers an in-depth overview and the latest research findings concerning our subject, specifically BIM-based AR/VR systems and the importance of maturity evaluation. It explores the established maturity evaluation methodologies within the construction sector and underscores the importance of ISO/IEC 15939 in enhancing the evaluation process through the adoption of the indicators concept. Section 3 presents a proposed framework derived and adapted from the ISO/IEC 15939 standard for the maturity assessment process, incorporating the use of indicators concept. Additionally, this section introduces various models for the interpretation of results and the representation of scores. Following this, we present a practical example of the specification of a maturity Sect. 4 of the paper conducts a feasibility study on the proposed framework, aiming to validate its applicability within an academic context. It then examines the main findings and outlines future perspectives for the framework. Finally, the paper is concluded in Sect. 5.

2 Background

2.1 BIM-Based AR/VR Systems and Maturity Evaluation

The integration of the BIM with AR and/or VR technologies has emerged as a crucial solution for the increasingly complex and challenging construction projects in the current industry [14]. BIM was initially developed to absorb information within a 3D visualization environment, supporting various life-cycle activities and aiding decision-making for project managers [15]. However, the construction phase often leads to incomplete or incorrect information in the BIM model, resulting in inefficiencies in construction management.

Integrating AR and VR technologies with BIM can address these challenges by providing a reliable coordination and communication tool. AR overlays digital information onto the real world, allowing for visualization and identification of progress discrepancies on construction sites [16]. In the AEC industry, AR has found applications in site inspection, construction simulation, safety management, design review, and facility maintenance, among others [17]. On the other hand, VR offers a complete virtual environment that enables professionals to visualize project designs even before construction begins [18]. The immersive experience provided by VR helps simulate scenarios throughout the project life cycle, including onsite construction and emergency evacuations. The integration of BIM with VR has been utilized in indoor lighting design, healthcare design, collaborative remote project review, and evacuation scenario simulations [17].

Several examples of using BIM-based AR/VR systems demonstrate their potential in enhancing perception, inspection, maintenance, and workforce training [19]. In this context, researchers have developed BIM-based AR systems for the inspection and maintenance of fire safety equipment, enabling real-time visualization and reducing potential risks. VR training programs have significantly improved on-site safety awareness by simulating dangerous scenarios for construction workers [20].

Evaluating the maturity of BIM-based AR/VR is of paramount importance for construction companies [5]. Maturity models can provide insights into a company's current status, helping prioritize tasks and identify areas for improvement [12]. Considering the complexity involved, assessing maturity is a sophisticated and challenging process, involving various considerations. Drawing insights from existing studies and proposed maturity evaluation frameworks in the literature, such as the one suggested by Kim et al. [11] and Hazar Dib et al. [21], and referring to Capability Maturity Model Integration (CMMI) and the Project Management Maturity Model (PMMM), these models underscore the importance of defining specific areas of interest for evaluation, focusing on particular aspects of the studied system. As the adoption of BIM, AR, and VR technologies continues to grow, these evaluation models become vital to optimize the benefits of these technologies according to the specific needs of construction companies [12].

2.2 Maturity Measurement Concepts in the Construction Field

Key elements are specified for evaluating the maturity of BIM-based AR/VR systems [22]. These crucial components include aspects, criteria, sub-criteria, and measures. Aspects represent areas of interest to be covered and evaluated [23, 24]. Criteria specify the particular elements within the studied aspect, providing a better understanding of the intended goal [25]. To improve accuracy and comprehension, criteria are typically subdivided into sub-criteria, providing additional details and enhancing clarity in the evaluation process. The evaluation of these criteria and sub-criteria can be carried out by defining measures, typically suggested or addressed by standards and research. Measures serve as a way of achieving the intended objective by providing defined metrics applicable to reaching a final decision on the maturity of AR/VR systems based on BIM [26].

A second consideration is the presence of well-defined data collection methods, defined as the way used to gather information needed for the intended purpose, which is, in our case, the maturity evaluation of BIM-based AR/VR systems [27]. Once data has been collected following the specified measures, the subsequent step of data interpretation becomes pivotal. This critical phase allows for precise and effective analyses, enabling a comprehensive understanding of the measurement results. However, in most existing studies [11, 27], the interpreting method is based on simple calculation methods, such as mean calculation. This simplicity is attributed to the consideration of only subjective measures relying on personal opinions, leading to the uniformity of the scale of obtained results, instead of using a theoretical approach. In our study, we dedicate attention to the heterogeneity of measures. The evaluation of considered aspects can be conducted in a subjective or objective manner.

Interpreting data from BIM-based AR/VR systems evaluation can be challenging due to the variety of areas of interest and different aspects that need evaluation. This complexity arises from variations in the nature of obtained results, such as different scales, units, and relevance to various criteria and sub-criteria. Additionally, the data may come from the application of different tools and methods, further complicating the interpretation process.

To ensure success in this task, it is necessary to implement a thorough and well-structured approach based on a solid foundation to guarantee the delivery of effective results. One of the most suitable solutions is the adoption of the ISO/IEC 15939 standard which enables the merging and integration of heterogeneous data and defines a robust method for interpreting the results that will be discussed in the upcoming section.

2.3 The ISO/IEC 15939 Standard Measurement Concepts

The ISO/IEC 15939 standard, also recognized as the International Standard for Software Measurement and Functional Size Measurement, establishes a framework of rules and procedures for the application of functional size measurement methods. It underscores the importance of employing specific criteria and measures for evaluation purposes, providing comprehensive guidelines for their development, and facilitating the interpretation of results [13, 28].

Moreover, this standard offers recommendations and guidelines for the elaboration of an evaluation process. It defines various concepts, with a critical emphasis on incorporating the indicator concept, which is presented as a measure providing an estimation or evaluation of specified attributes [13].

The measurement indicator is organized based on an information measurement model that delineates fundamental concepts essential for constructing an indicator. This model further establishes a structured connection between the requirements and the relevant entities, defining the attributes of interest [13]. The visual representation of this relationship is illustrated in the Fig. 1 below.

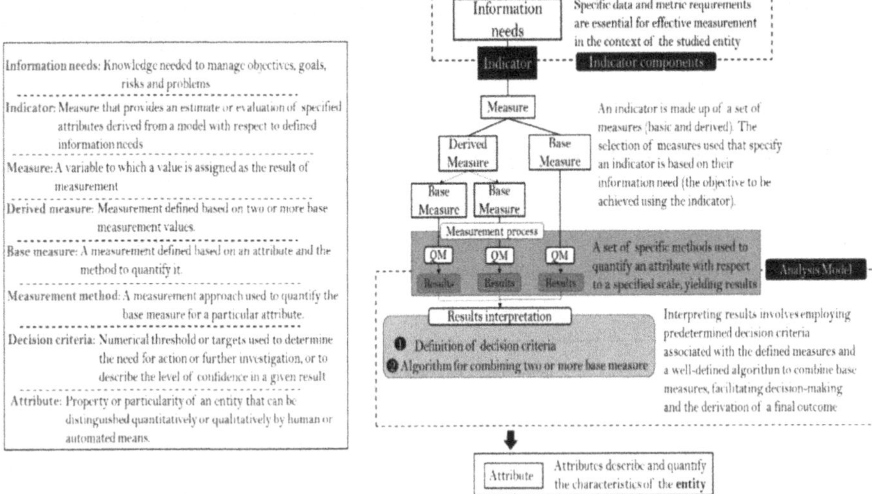

Fig. 1. Structure and Key Relationships of Information Measurement Model of ISO/IEC 15939 standard (ISO/IEC 15939,2007)

Referring to Fig. 1 above, an indicator encompasses a set of measures, both basic and derived. The choice of measures employed to define an indicator is driven by specific information needs aligned with the intended goal for using the indicator. In this phase, measurement methods are applied to quantify the attributes under examination, with respect to specific scales generating results. The results obtained are then subjected to analysis using a well-defined analytical model. Subsequently, this analysis model puts forth a suggested algorithm or model for combining one or more basic and/or derived measures, ensuring their correlation with decision criteria that must be clearly defined. Furthermore, this model presents a structure connecting the needs to the relevant entities that characterize the attributes of interest [28].

The primary objective of the ISO/IEC 15939 standard is to assist organizations in enhancing their software and system development processes through effective and consistent measurement practices [13]. Despite its potential benefits, this standard has not seen widespread application in the AEC industry. Presently, its predominant use is within the domain of information technology for evaluating human-machine interaction usability [28], demonstrating considerable success. Nevertheless, extending its application to the assessment of BIM-based AR/VR systems holds significant promise. Such an expansion can greatly enhance the overall evaluation process by ensuring consistency, comparability, and reliability of results.

3 Maturity Indicator Framework Adapted from ISO/IEC 15939 Standard for Evaluating BIM-Based AR/VR Systems

This section is dedicated to introduce an adapted concept of indicators, with the goal of establishing maturity indicators that align with the ISO/IEC 15939 standard for assessing BIM-based AR/VR systems. Within this section, we explore the description of the proposed framework and provide an illustrative general example of maturity indicator specification.

3.1 The Proposed Framework Description

As illustrated in Fig. 2, the proposed maturity indicator definition process comprises three primary stages, namely: specifying the information needs, defining the maturity indicators, and generating the maturity levels.

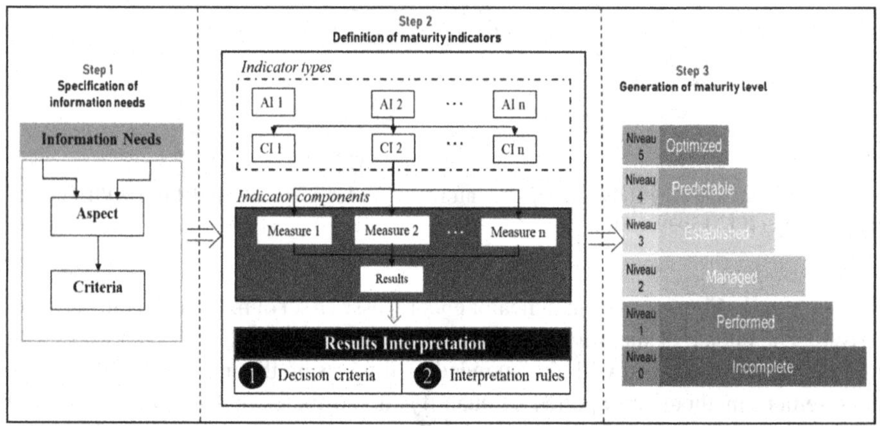

Fig. 2. Measurement model of maturity indicators adapted from ISO/IEC 15939 standard.

The first stage of the maturity assessment procedure requires identifying the information needs for evaluating BIM-based AR/VR systems. This involves consulting relevant literature on the evaluation of these systems. Within this literature, existing studies emphasize the importance of delineating essential elements in the evaluation of maturity, such as the aspects and criteria to be evaluated [22].

Consequently, as a natural progression in the second step of this process, we align with these established concepts by adapting the ISO/IEC 15939 standard. Through this adaptation, a maturity indicator is formulated, comprising two pivotal types: the aspect indicator and the criteria/sub-criteria indicator.

Within this context, the maturity indicator serves as a comprehensive measure that encompasses all the defined measures used to evaluate the maturity of attributes within the BIM-based AR/VR system. Furthermore, the aspect indicator focuses only on measures designated to assess the maturity of attributes related to a specific aspect within the BIM-based AR/VR system. While the criteria/sub-criteria indicator is more specific, including only the measures intended to evaluate the maturity of particular attributes related to specific criteria or sub-criteria within a specific aspect within the overall studied entity. It is important to highlight that the measures employed in this context may exhibit subjectivity or objectivity, depending on the nature of the evaluation criteria. This acknowledgment underscores the diversity in the assessment methodologies, and obtained results.

Based on the aforementioned points, it can be deduced that the maturity indicator, which evaluates the maturity of the BIM-based AR/VR system under study, is derived from the sum of indicators corresponding to the defined aspects. Therefore, we can express it as follows:

$$MI = (AI1, AI2, ..., AIn).$$

Similarly, following the same principle, the aspect indicator encompasses the indicators related to the specified criteria and their respective sub-criteria. This can be represented as $AI = (CI1/SCI1, CI2/SCI2, \ldots, CIn/SCIn)$.

Furthermore, since the criteria/sub-criteria indicator comprises the measures of CI/SCI = (M1, M2, ..., Mn), it indicates that the maturity indicator is derived from the set of measures defined for the overall evaluation of the system. Hence, we can express it as $MI = (M1, M2, \ldots, Mn)$.

In summary, In the context of key elements, indicators represent the defined measures for evaluating the maturity of BIM-based AR/VR systems. Figure 1 below visually depicts the hierarchical structure and interdependencies among these indicators (Fig. 3).

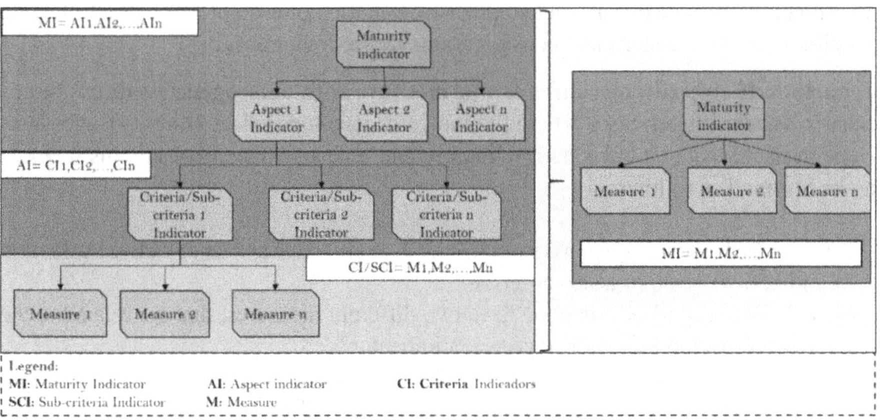

Fig. 3. Hierarchical structure and interdependencies of defined indicators

Moving on to the next phase within this step, this phase involves an interpretation method. Its primary aim is to address the variability in results obtained from measurements due to their reliance on different measurement techniques. This method involves defining decision criteria, which necessitates the establishment of thresholds, ultimately leading to more uniform results.

This step can be achieved by utilizing an algorithm that combines the relevant measures that characterize the indicator with their associated decision criteria. According to the ISO/IEC 15939 standard [13], the use of this concept allows the definition of rules to interpret the results of one or more measures at a time, and therefore identify, in our study, the level of maturity of the evaluated element.

In order to ensure consistency and a comprehensive assessment of the results of obtained measures, it is crucial to establish decision criteria. This objective can only be achieved by standardizing the results of these measures through the definition of uniform thresholds, making them comparable. To accomplish this, we draw inspiration from ISO 33061, which proposes six maturity levels (ranging from 0 to 5) to gauge the progress of processes [13]. These levels are categorized as Incomplete, Performed, Managed, Established, Predictable, and Optimized, representing a continuum from basic process organization to continuous improvement and refinement at the highest levels. The implementation of these maturity levels constitutes the core of the third step in the proposed maturity measurement process, which aims to attribute or assign a definitive maturity level to the studied system.

Based on these maturity levels, we identify five specific thresholds, labeled as X, Y, Z, K, and W, for each measure. These thresholds create six distinct zones, each corresponding to one of the maturity levels (see Fig. 2). By categorizing the results of measures within these zones, we can easily evaluate the level of maturity achieved in each process [22].

To effectively assess the maturity levels of BIM-based AR/VR system implementations, considering the defined decision criteria and uniform thresholds, we proposed a structured approach comprising two distinct evaluation scenarios. Each scenario encompasses a set of cases with their respective proposed methods of evaluation. These cases have been developed to address various potential situations that may arise during the assessment, ensuring a consistent and precise evaluation of each process. The following guidelines outline the steps and considerations for both scenarios.

Scenario 1: If only two measures or one measure with heterogeneous data. In this scenario, we encounter cases where the indicator is composed of either two measures or one measure that combines both subjective and objective data. Cases in this scenario can be resumed as follows:

- *Case 1:* The first case is when conducting both a subjective and an objective evaluation for either a criterion or a sub-criterion.
- *Case 2:* The second case is if we have two different measures, that are related to the same sub-criterion indicator or criterion indicator
- *Case 3:* The third case is if we have two different single measures related to different criteria or sub-criteria or to combine results obtained from the two criteria or two sub-criteria indicators.

- *Case 4:* If we deal only with two aspects:

3. When we have only one measure in each aspect
4. More than one measure in each aspect (overall score for each aspect should be calculated in advance based on the corresponding case).

Proposed Evaluation Method. The evaluation process based on ISO/IEC 15939 considers the diverse nature of measures, encompassing both subjective and objective ones, whether they are derived or base measures. To handle this diversity, the standard presents two distinct methods for combining them. The first method involves using mathematical calculations to derive an overall score that represents the combined evaluation results. The second method consists of defining an algorithm or model that enables the combination of these measures while preserving the raw data associated with each measure. This approach eliminates the need for data aggregation, facilitating a more comprehensive analysis and interpretation. Importantly, this innovative method also allows for the combination of two distinct measures, enhancing the model's analytical capabilities. In line with this second method, and with the aim of applying the indicator specification concept in the AEC industry based on ISO/IEC 15939, we propose an integrated model. This model effectively combines the results of two different measures (Fig. 4).

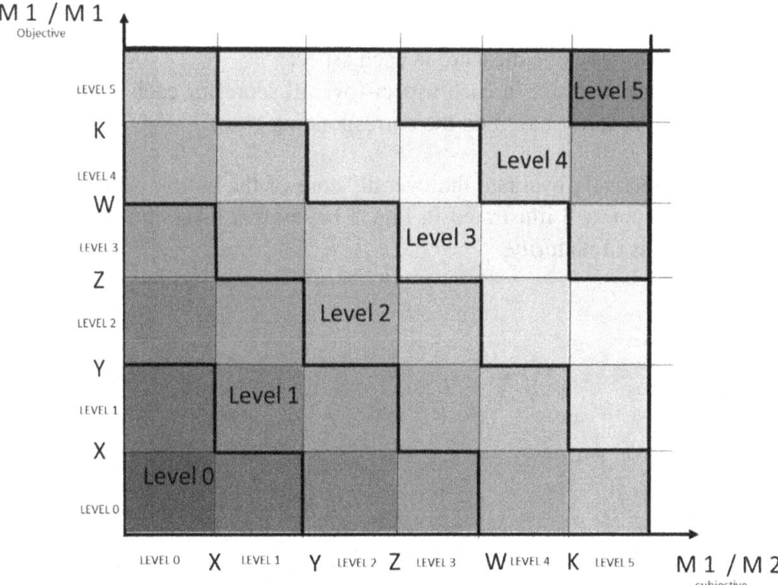

Fig. 4. Qualitative and quantitative combination of two different or objective and subjective measures based on decision criteria and maturity levels.

To represent the proposed approach, we have created a graph. The graph uses axes to represent the decision criteria defined for each of the two studied measures. Each criterion is linked to evaluation thresholds that indicate different levels of maturity, consisting of six distinct levels. The graph showcases an intersecting grid, with delineating lines marking the boundaries of these maturity levels based on the rounded-down calculation of the mean of the crossed levels.

For ease of understanding, we have employed a color-coded system to represent the results. Each level is assigned a color based on a gradient scheme, ranging from red (indicating lower levels of maturity) to green (signifying higher levels of maturity). This qualitative approach visually illustrates the obtained results and provides a quick understanding of the maturity levels achieved. Each level, except level 5, includes two colors representing the non-rounded level and the rounded one which is qualitatively slightly higher. This dual representation, combining both visual and numerical aspects, enhances the model's ability to present and interpret the evaluation outcomes effectively.

Scenario 2: If indicator is composed of three or more measures. In this scenario, we encounter situations where the evaluation involves three or more measures. These cases can be summarized as follows:

- *Case 1:* If we have, three measures or more that are related to the same sub-criterion or criterion.
- *Case 2:* If we have three or more single measures that are related to different criteria or sub-criteria but pertain to the same aspect, or to combine results obtained from three and more criteria or sub-criteria indicators
- *Case 3:* If we deal only with three or more aspects:

  1. When we have only one measure in each aspect
  2. More than one measure in each aspect (overall score for each aspect should be calculated in advance based on the corresponding case).

In order to effectively evaluate the overall score of the entity, we have developed a comprehensive approach illustrated in Fig. 5 below that takes into account various indicators and levels of maturity.

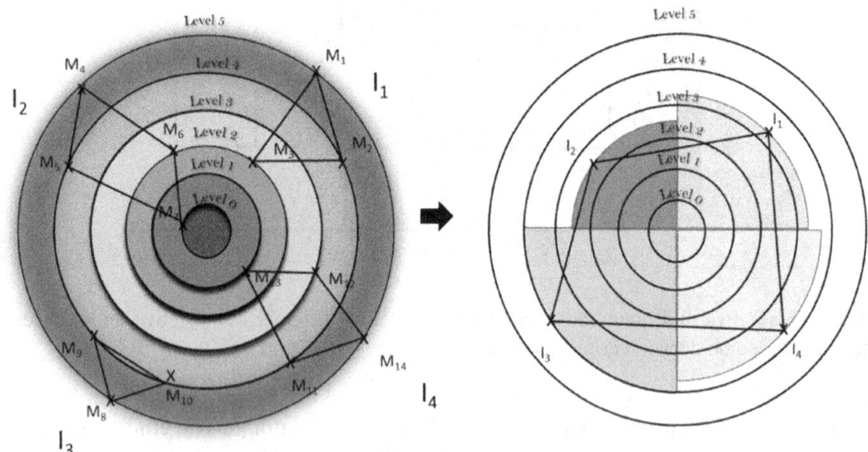

Fig. 5. Qualitative and quantitative combination of three or more different measures based on decision criteria and maturity levels.

The proposed diagram of our approach consists of six concentric circles representing defined levels of maturity. Each circle represents a specific level, with the innermost circle indicating the lowest one.

Within the diagram, we present results derived from specific measures, organized into distinct parts, aligned with a particular type of indicator. Each measure is linked to its corresponding level, and subsequently, the results are interlinked to form a radar representation illustrating the overall outcome.

To compute the overall score for the main indicator under study, we calculate the average score of the associated defined elements. These average scores are denoted by the black "x" symbol within the diagram. These (x) symbols are connected through lines, providing a comprehensive overview of the results and signaling the final score of the main indicator being studied. This final score encompasses the maturity indicator, which will be determined at a later stage.

3.2 Specification of a Maturity Indicator for Evaluating the Technological Aspect of BIM-Based AR/VR Systems

In this section we present the specifications of the principal concepts of the information measurement models proposed to define a maturity indicator. Specifically, we have chosen the indicator of technology to dissect, given that technology is a highly essential aspect in the experience of AR/VR systems coupled with BIM. Alacron et al. [30] identified two main criteria within the technology aspects, namely hardware and software. This distinction leads to the formulation of two separate criteria indicators: the hardware indicator and the software indicator. Further, based on the work of Kim et al. [11], visual immersion has been identified as a significant sub-criterion within the software domain. Consequently, we introduce a sub-criteria indicator named the visual immersion indicator. The evaluation of visual immersion has been addressed in studies such as that by Meherfard et al. [31], which proposed various measures related to image quality, including resolution, interpupillary distance adjustability, brightness and color accuracy (B&CA), text readability, field of view, and perceived contrast. In this section, our focus will be narrowed to only two specific measures which are the resolution and the B&CA.

As mentioned in the study by Meherfard et al. (2019), display resolution pertains to the quantity of pixels found in a screen and is closely tied to the overall quality of the visual experience, immersion level, and level of detail. A higher resolution is also linked to improved legibility of text in virtual environments. Conversely, brightness is defined as the amount of light emitted by the display, essentially representing the luminosity, and it is measured in nits, which is equivalent to the number of candles in a square meter ($cd/m2$). On the other hand, color accuracy signifies the difference between the displayed color and the actual color, and it is assessed by calculating the color difference (Δe) in the CIELAB color space.

In Table 1 below we present a detailed overview of the measurement information model related to our suggested indicator. As shown in this table (Decision criteria), we have put forward rules for understanding the results of the indicator based on defined thresholds.

Table 1. Specifications of the information measurement model for the technology aspect indicator (resolution measure)

Information need	Maturity evaluation of BIM-based AR/VR systems	
Indicator	Technology (immersion) within the environment of a BIM-based AR/VR System	
Analysis model	Model encompasses one measure related to display resolution of used headset	
Decision Criteria	**Measure Interpretation:** – For a Display resolution, the higher the value the better the performance – Resolution < X: The display system at this level demonstrates very basic and limited resolution capabilities. Potentially leading to a blurry or pixelated BIM-based AR/VR experience – X < Resolution < Y: At this stage, the display system is still in the early phases of development – Y < Resolution < Z: User can expect reasonably clear and detailed BIM models in the AR/VR environment – Z < Resolution < W: The display system offers good resolution, enhancing the overall quality of BIM model on AR/VR – W < Resolution < K: The display system provides excellent clarity and detail of BIM models in the AR/VR, resulting in a highly immersive experience – Resolution > K: At this exceptional level, the display system delivers unparalleled clarity and detail, making the BIM-based AR/VR experience exceptionally immersive and realistic	
	Level 0 Resolution < X	**Beginner maturity level** **Recommendations:** – Invest in higher-resolution display hardware – Explore more advanced rendering techniques to improve visual quality
	Level 1 X < Resolution < Y	**Under development level** **Recommendations:** – Continue upgrading hardware components to provide better resolution – Collaborate with developers: Work closely with software and content developers to create AR/VR content optimized for improved resolution

(*continued*)

Table 1. (*continued*)

	Level 2 Y < Resolution < Z	**Intermediate level** **Recommendations:** – Quality control in content creation: Maintain strict quality control measures during content creation to ensure high-resolution visuals – Monitor performance: Continuously monitor the performance of the display system and address any bottlenecks or issues affecting resolution
	Level 3 Z < Resolution < W	**Advanced level** **Recommendations:** – Explore the development of custom hardware and software solutions that are specifically tailored for BIM-based AR/VR, with a strong emphasis on achieving exceptional resolution – Continue to invest in the latest display technologies that offer even higher resolution and visual quality compared to Level 3
	Level 4 W < Resolution < K:	**High level** **Recommendations:** – Allocate resources for ongoing research and development to create custom display solutions that can provide even higher resolutions and unprecedented visual quality – Partner with experts in display technology, augmented reality, virtual reality, and content creation to ensure your system remains at the forefront of the field
	Level 5 Resolution > K	**Exceptional level**
Derived measure	Image quality	
Measurement	Resolution	
Base measure		
Measurement method	Finding the display resolution of the headset used in the project can be accomplished through several methods: by checking the settings of the headset in use, verifying it on the supplier's website, or consulting the manual that accompanies it	

4 Feasibility Study of the Proposed Framework and Discussion

To assess the validity and feasibility of our proposed framework, we carried out a survey comprising a comprehensive report on the framework and a set of Likert scale statements for feedback collection. We invited academic and professional experts with experience in BIM and AR/VR systems to participate in the survey and comment on their choices. These statements were rated on a seven-point Likert scale, from 'strongly disagree' to 'strongly agree'. The survey addressed various aspects of the approach, including its conformity with ISO/IEC 15939, particularly its relevance to the addressed problem, and the alignment with this standard. One key focus was the evaluation of indicators, emphasizing the suitability of selected indicators for measuring specific attributes of BIM-based AR/VR systems. The survey also covered the framework's ease of use, clarity, relevance of the combined measurement approach, and the Maturity Assessment Diagram, along with general aspects related to ease of implementation and understanding. The statements we proposed for the survey are as follows:

1. The approach aligns well with the ISO/IEC 15939 standards.
2. The integration of ISO/IEC 15939 standards in the evaluation process addresses well the challenges related to assessing the maturity of BIM-based AR/VR systems.
3. The measure combination approach offers a clear method for combining objective and subjective measures, providing a comprehensive qualitative and quantitative perspective of the resulting score.
4. The Maturity Assessment Diagram provides a clear visual representation of the maturity assessment process.
5. The indicators used in the approach appropriately measure the specified attributes of BIM-based AR/VR systems.
6. The proposed framework is easy to understand.
7. The proposed framework is practical and feasible to implement.

The collected scores from the defined statements are shown in the Fig. 6 below.

Among the 7 experts we sent the survey to, feedback was received from only 5. One expert, an AR and VR specialist from "Bertrandt Group," chose to provide comments only, without completing the 7-point Likert scale questions. Additionally, another expert did not answer the first question, citing a lack of familiarity with ISO/IEC 15939. Overall, the feedback was positive, as depicted in Fig. 6. The majority of responses ranged between 6 and 7 on a 7-point Likert scale, indicating a favorable view of the framework. The feedback was organized into several key observations:

A point of critique was the unclear definition of 'maturity'. It was noted that although the term was frequently used, it lacked a precise definition. The expert suggested including a specific example of maturity within BIM to enhance understanding of the study.

Another comment highlighted the absence of quantitative data when discussing resolution measurement. Our study intentionally omitted specific numbers to focus on broadly validating the evaluation framework, rather than verifying each discussed aspect. This approach aimed to assess the evaluation method itself rather than the details it encompassed.

The third feedback focused on the recommendation to invest in high-quality headsets as a single expense rather than upgrading incrementally. The success of Apple Vision as an example of effective improvement was mentioned. While the suggestion was valuable, we believe such a significant quality shift requires extensive financial analysis, especially for companies that utilize a significant number of headsets.

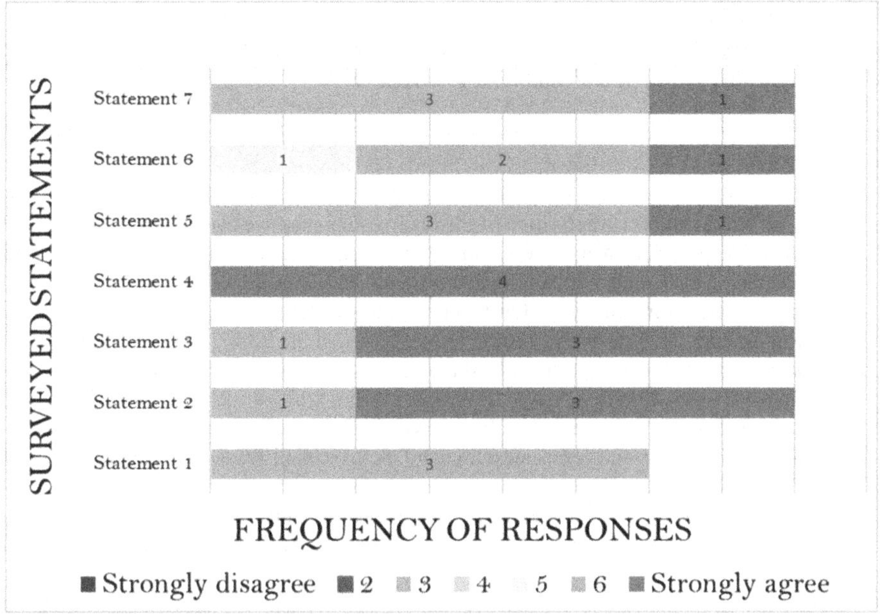

Fig. 6. Expert findings based on a seven-point Liker scale for the seven defined statements.

Additionally, further criteria have been proposed for maturity evaluation, including comfort, realism, 3D model detail, and immersion which is highlighted by the use of technology like gloves that simulate touching real objects. It is worth mentioning that this study is not just a part of an ongoing larger study, and of course, many other aspects have been addressed and are in the process of validation. As a perspective, a decision-support tool to automate our framework is a good improvement that will be included in our study.

5 Conclusion and Perspectives

In conclusion, this article presents an innovative approach for assessing the maturity of BIM-based AR/VR systems, addressing the limitations of existing maturity assessment methods. This approach is based on the guidelines of ISO/IEC 15939 standard, which place a critical emphasis on the concept of indicators that we have adopted in our study. Based on this concept, a measurement model has been established to determine the process of specifying it. Inspired from key concepts found in existing maturity evaluation methods and adapted based on the latter standard, a maturity evaluation framework has been proposed. Real examples of indicator specifications have been discussed, particularly in relation to the technology aspect in a BIM-based AR/VR environment. Within the framework of this approach, a method combining qualitative and quantitative measures, both objective and subjective, has been adapted and applied. A maturity evaluation diagram has been developed to synthesize all the scores obtained. Following this proposal, we prepared a comprehensive survey containing evaluation statements, which were subsequently reviewed by experts. The feedback received from experts reflects a positive opinion regarding our approach and its effectiveness.

References

1. Meng, Y., Li, X, Ma, C.: Application of the fuzzy comprehensive evaluation based on AHP in the BIM application maturity evaluation. In: ICCREM 2014: Smart Construction and Management in the Context of New Technology, pp. 280–286 (2014)
2. Lu, W., et al.: Measuring building information modeling maturity: a Hong Kong case study. Int. J. Constr. Manag. **21**(3), 299–311 (2021)
3. Bouška, R.: Evaluation of maturity of BIM tools across different software platforms. Procedia Eng. **164**, 481–486 (2016)
4. Abbasianjahromi, H., Ahangar, M., Ghahremani, F.: A maturity assessment framework for applying BIM in consultant companies. Iranian J. Sci. Technol. Trans. Civ. Eng. **43**, 637–649 (2019)
5. Morlhon, R., Pellerin, R., Bourgault, M.: Building information modeling implementation through maturity evaluation and critical success factors management. Procedia Technol. **16**, 1126–1134 (2014)
6. El Mounla, K., Beladjine, D., Beddiar, K., Mazari, B.: Lean-BIM approach for improving the performance of a construction project in the design phase. Buildings **13**(3), 654 (2023)
7. Ahmed, S., Hossain, M.M., Hoque, M.I.: A brief discussion on augmented reality and virtual reality in construction industry. J. Syst. Manag. Sci. **7**(3), 1–33 (2017)
8. Wang, P., Wu, P., Wang, J., Chi, H.L., Wang, X.: A critical review of the use of virtual reality in construction engineering education and training. Int. J. Environ. Res. Public Health **15**(6), 1204 (2018)
9. Ververidis, D., Nikolopoulos, S., Kompatsiaris, I.: A review of collaborative virtual reality systems for the architecture and engineering (virtual tour). Int. J. Eng. Res. Technol. (IJERT), **11**(11) (2022)
10. Monla, Z., Assila, A., Beladjine, D., Zghal, M.: Maturity evaluation methods for BIM-based AR/VR in construction industry: a literature review. IEEE Access (2023)
11. Kim, J.I., Li, S., Chen, X., Keung, C., Suh, M., Kim, T.W.: Evaluation framework for BIM-based VR applications in design phase. J. Comput. Des. Eng. **8**(3), 910–922 (2021)
12. Assila, A., Beladjine, D., Messaadia, M.: Towards AR/VR maturity model adapted to the building information modeling. In: Product Lifecycle Management Enabling Smart X: 17th IFIP WG 5.1 International Conference, PLM 2020, Rapperswil, Switzerland, 5–8 July (2020)
13. ISO/IEC 15939: International Organization for Standardization/International Electrotechnical Commission. Systems and software engineering — Measurement process (ISO / IEC 15939:2007 (E)), Geneva, Switzerland (2007)
14. Khan, S., Panuwatwanich, K., Usanavasin, S.: Integrating building information modeling with augmented reality: application and empirical assessment in building facility management. Eng. Constr. Archit. Manag. (2023)
15. Machado, R.L., Vilela, C.: Conceptual framework for integrating BIM and augmented reality in construction management. J. Civ. Eng. Manag. **26**(1), 83–94 (2020)
16. Adebowale, O.J., Agumba, J.N.: Applications of augmented reality for construction productivity improvement: a systematic review. Smart Sustain. Built Environ. (ahead-of-print) (2022)
17. Chen, K., Chen, W., Cheng, J.C., Wang, Q.: Developing efficient mechanisms for BIM-to-AR/VR data transfer. J. Comput. Civ. Eng. **34**(5), 04020037 (2020)

18. Zaker, R., Coloma, E.: Virtual reality-integrated workflow in BIM-enabled projects collaboration and design review: a case study. Vis. Eng. **6**(1), 1–15 (2018)
19. Yigitbas, E., Nowosad, A., Engels, G.: Supporting construction and architectural visualization through BIM and AR/VR: a systematic literature review. arXiv preprint arXiv:2306.12274 (2023)
20. Chen, H., Hou, L., Zhang, G.K., Moon, S.: Development of BIM, IoT and AR/VR technologies for fire safety and upskilling. Autom. Constr. **125**, 103631 (2021)
21. Dib, H., Chen, Y., Cox, R.: A framework for measuring building information modeling maturity based on perception of practitioners and academics outside the USA. In: Proceedings of the CIB W, vol. 78, p. 2012, October 2012
22. Monla, Z., Assila, A., Beladjine, D., Zghal, M.: A conceptual framework for maturity evaluation of BIM-based AR/VR systems based on ISO standards. In: De Paolis, L.T., Arpaia, P., Sacco, M. (eds.) XR Salento 2023. LNCS, vol. 14218, pp. 139–156. Springer, Cham (2023). https://doi.org/10.1007/978-3-031-43401-3_9
23. Crawford, J.K.: Project Management Maturity Model. Auerbach Publications (2006)
24. Constantinescu, R., Iacob, I.M.: Capability maturity model integration. J. Appl. Quant. Methods **2**(1), 31–37 (2007)
25. Wu, C., Xu, B., Mao, C., Li, X.: Overview of BIM maturity measurement tools. J. Inf. Technol. Constr. (ITcon) **22**(3), 34–62 (2017)
26. Yilmaz, G., Akcamete, A., Demirors, O.: A reference model for BIM capability assessments. Autom. Constr. **101**, 245–263 (2019)
27. Liang, C., Lu, W., Rowlinson, S., Zhang, X.: Development of a multifunctional BIM maturity model. IEEE Access **4**, 5363–5373 (2016)
28. Assila, A., Marçal de Oliveira, K., Ezzedine, H.: Integration of subjective and objective usability evaluation based on IEC/IEC 15939: a case study for traffic supervision systems. Int. J. Hum. Comput. Interact. **32**(12), 931–955 (2016)
29. ISO/IEC TS 33061: 2021 Information technology — Process assessment — Process assessment model for software life cycle processes
30. Alarcon, R., et al.: Augmented Reality for the enhancement of space product assurance and safety. Acta Astronautica **168**, 191–199 (2020)
31. Mehrfard, A., Fotouhi, J., Taylor, G., Forster, T., Navab, N., Fuerst, B.: A comparative analysis of virtual reality head-mounted display systems. arXiv preprint arXiv:1912.02913 (2019)

Evaluating the Usability of Online Tools During Participatory Enterprise Modelling, Using the Business Model Canvas

Anthea Venter(✉) and Marné de Vries

Department of Industrial and Systems Engineering, University of Pretoria, Pretoria, South Africa
{anthea.venter,marne.devries}@up.ac.za

Abstract. Participatory enterprise modelling (PEM) is focused on actively involving enterprise stakeholders and modelling experts in the modelling process. Many online/digital tools exist that could facilitate this process, but limited re-search has been done to evaluate these tools for the purpose of PEM, and specifically the ease-of-use or usability of these tools. We examine the usability of an online tool, MURAL, during PEM when participants use a business model canvas (BMC) to describe and visualize a business. By using a standardized usability questionnaire, the System Usability Scale (SUS), we determine that MURAL scores above average and is considered an acceptable tool during PEM using the BMC. Additional insights, such as the functional features from MURAL that contribute to the use-in-context, are also obtained through user feedback, interviews and observing recordings of the modelling sessions.

Keywords: Collaboration · Ease-of-use · Enterprise modelling · Mural · Participatory modelling · Standardized usability questionnaire · SUS · Usability

1 Introduction

The need for collaborative software is growing considering that team members working together from different locations, has become an ordinary phenomenon. The role of collaborative software in virtual teamwork is essential as it contributes to team efficiency and productivity [1]. Limited research has however been done to evaluate virtual tools for the purpose of Enterprise Modelling (EM), and specifically the ease-of-use or usability of these tools.

According to Stirna and Persson [2], EM is "an integrated and multi-perspective way of capturing and analyzing enterprise solutions". Since different enterprise design domains are (re-)designed concurrently [3], multiple enterprise models may be required as representations. The term "model" is used in various ways as highlighted by Dietz and Mulder [4]. In their chapter on The MU Theory: Understanding Models and Modelling, it is stated that the only common denominator in the plethora of model notions is that a model is "a simplified representation of a thing, made for the purpose of studying those aspects of the thing that one is interested in" [4]. While models can be used for simulation and deployment, their use often lies in sense-making, communication, and

F. F.-H. Nah and K. L. Siau (Eds.): HCII 2024, LNCS 14721, pp. 96–114, 2024.
https://doi.org/10.1007/978-3-031-61318-0_8

improvements within the organization [5]. EM is used to document the current or a desired future state of a company in a graphical and comprehensible way [6] and the process of EM often calls for participation of everyone involved [7].

To address this, innovative approaches such as participative enterprise modelling (PEM) have been introduced. Gutschmidt [6] describes PEM as a "fruitful method of eliciting information and creating enterprise models at the same time" in which stakeholders actively take part in the enterprise modelling sessions. EM can also be classified as collaborative [2]. In collaborative enterprise modeling, several experts work together in a coordinated effort, whereas PEM also involves users or enterprise stakeholders that contribute their knowledge and experience of the company [8].

In Sect. 1.1, we introduce the Participatory Modelling (PM) tool used in our study: MURAL. Section 1.2 introduces the Business Model Canvas (BMC), used during our study's EM in MURAL. A brief overview of the methods that can be used to evaluate the usability of MURAL is provided in Sect. 1.3. Lastly, previous research relevant to our study is summarized in Sect. 1.4 and our research objectives are outlined in Sect. 1.5.

1.1 MURAL

MURAL is a PM tool that is a "secure, flexible, visual work platform purpose-built for collaboration" [9]. In MURAL, users can collaboratively work on a digital whiteboard, using features such as sticky notes and text, icons and images, and readily available templates. Jackson, van der Hoek, Prikladnicki and Ebert [10] have classified MURAL into the "Artifact management: virtual whiteboard software" technology type, along with tools such as Miro [11] and Google Jamboard [12]. According to Jackson, van der Hoek, Prikladnicki and Ebert [10], the primary purpose of these types of tools is "an online space for multiple participants to collaboratively create and edit content".

There are many digital tools available that can be used during collaborative sessions. De Vries and Opperman [13], for example, experimented with two digital PM tools, MURAL and Miro, later discovering a third no-cost platform called FigJam [14]. Gutschmidt [6] investigated modelling tools that are suitable for distributed collaboration, selecting freely available tools that can facilitate the drawing of both goal models and process models. Draw.io [15] and Google Drawings [16] fulfilled their criteria and were therefore the tools evaluated, with Draw.io being rated better in most cases. In this study, MURAL was used as this was the tool in which the limitations regarding ease-of-modelling and usability during PEM were initially identified [13].

1.2 The Business Model Canvas (BMC)

Schwarz and Legner [17] define a Business Model (BM) as "the blueprint of an organization's logic for value creation, delivery, and capture". According to Saebi, Lien and Foss [18], there is no generally agreed upon definition, but most define it in terms of "the firm's value proposition and market segments, the structure of the value chain required for realizing the value proposition, the mechanisms of value capture that the firm deploys, and how these elements are linked together in an architecture" [18]. When referring to BM tools, Schwarz and Legner [17] suggest distinguishing three interrelated perspectives, namely (1) BM tools as conceptual models, (2) BM tools as methods, and

(3) BM tools as IT support. The purpose of BM tools as conceptual models, e.g. the Business Model Canvas (BMC), is to share conceptualizations, describe, visualize, and discuss BMs [17].

The BMC consists of nine basic building blocks that summarize the main areas of a business [19]. These blocks include key activities, key partners, key resources, cost structure, customer relationships, customer segments, value propositions, channels, and revenue streams [19]. By using the BMC, stakeholders have a "a shared language for describing, visualizing, assessing, and changing business models" [19]. According to Ladd [20], however, even though the BMC is appropriate for describing or visualizing business models, it lacks guidance on the related innovation processes.

Business model innovation (BMI) takes place when BM elements and/or architecture is changed in a deliberate and observable way [21] and certain methods have been developed to address specific tasks during the BM design and implementation [17]. For example, an integrative conceptual framework for systematic BMI is proposed by Hoch and Brad [22] and a method that creates a visual roadmap of a business model's evolution that assists in navigating the versions created, is proposed by Fritscher and Pigneur [23].

1.3 Usability Evaluation Methods

Techniques or methods are used to evaluate usability, defined by the International Organization for Standardization [24] as "the extent to which a product can be used by specified users to achieve specified goals with effectiveness, efficiency, and satisfaction in a specified context of use". In a Usability Evaluation Method (UEM) taxonomy created by Ivory and Hearst [25], which remains highly relevant today [26], method classes and method types per class are defined. As an example, in the inquiry method class, users provide feedback on an interface which could include methods such as interviews, surveys, or user feedback.

According to Fernandez, Insfran and Abrahão [27], although several UEM taxonomies exist, the methods can be classified into two different types: empirical methods and inspection methods. Empirical methods are based on a tester (human and/or software) capturing and analyzing usage data from real end-users completing a predefined set of tasks, whereas inspection methods consist of experts reviewing usability aspects and how they conform to a set of guidelines [27]. In a study by Sagar and Saha [28], 37 different UEMs were identified, and the extensive use of usability testing, heuristic evaluation, and questionnaires were noted.

Standardized usability questionnaires, a type of questionnaire designed from metrics based on the responses obtained from participants, assess user satisfaction with perceived usability of a system [29]. The most widely used standardized usability questionnaires for the assessment of usability at the end of a study, according to Sauro [30], include the Questionnaire for User Interface Satisfaction (QUIS), the Software Usability Measurement Inventory (SUMI), the Post-Study System Usability Questionnaire (PSSUQ), and the System Usability Scale (SUS).

1.4 Previous Work

Previous work evaluated the usability of MURAL (introduced in Sect. 1.1) when participants used PEM in a controlled experiment [31]. In this previous experiment, participant pairs used an extended story-card method (eSCM) [13] to perform co-modelling remotely. A cooperation structure diagram (CSD) and a transaction product table (TPT) were constructed, in accordance with guidance from Dietz and Mulder [4]. The modelling of a CSD and TPT falls in the Organization Domain of Enterprise Engineering (EE), which incorporates interactions and communications between actor roles when production acts are performed at an enterprise [3].

In Venter and de Vries [31], we recommend the evaluation of MURAL in an alternative EM context. Therefore, in this study, instead of evaluating the usability when the eSCM is used to create a CSD and TPT in MURAL i.e. the Organization Domain, we evaluate the usability when participants model a BMC in MURAL i.e. constructing an enterprise model that represents multiple design domains, classified as a Cross-Domain enterprise model [3].

1.5 Research Objectives

The objective of our research is to evaluate MURAL's usability when participants work together in a different PEM context. In our current study, the participants were tasked with using the BMC to create a blueprint of their business idea, an example of a Cross-Domain enterprise model [3].

Our research questions (RQ) are therefore as follows:

RQ1: How user friendly is the online tool MURAL during PEM, and specifically when creating a BMC; and
RQ2: What functional features from MURAL contribute to this use-in-context?

2 Research Method

According to vom Brocke, Hevner and Maedche [32], if knowledge is already available to solve a problem identified, this knowledge can be applied following routine design, which does not constitute Design Science Research (DSR). The surveys used in the previous studies [13, 31], as well as UEMs such as standardized usability questionnaires were already available. This knowledge was therefore used in our study to create a new survey that extracts both quantitative and qualitative data, i.e. mixed methods research [33].

In our previous MURAL usability study [31], additional forms of inquiry, like interviews and observing the recorded modelling sessions, provided additional insights into the usability of MURAL. It was therefore decided that survey participants would be contacted for interviews if additional content/context was required based on their survey feedback. The recorded modelling sessions of the interviewees would also be studied to extract any additional insights.

The study participants are described in Sect. 2.1, the modelling process followed is included in Sect. 2.2, and the usability evaluation is detailed in Sect. 2.3.

2.1 Participants

The primary researcher introduced MURAL and the requirements to the initial group of 104 members, enrolled for a tertiary module named Business Engineering. The primary researcher then created 22 groups on the web-based learning management system 'Blackboard' [34]. Members could then either join one of the 22 groups themselves, or be allocated to a random group if they had not joined a group by a specified date. Each group consisted of between 4 and 5 members, with the exception of one group that consisted of three members after a member could no longer participate. As included in Sect. 2.3, 35 of the 104 members (about 34%) agreed to participate in our survey, and 4 of the 104 members (about 4%) agreed to be interviewed.

In the survey, when participants were asked to multiselect the modelling tools they had used in the past to represent process/activity/logic/business modelling, or any other collaborative work, most participants (about 83%) answered MURAL. Other tools provided as text answers, that weren't available in the survey's options, included Canva, Google Docs, Microsoft PowerPoint and "Traditional modelling tools; that is booking a boardroom to meet in that has a whiteboard with markers and making the use of paper charts in the size of A1–A3". Only one participant had not used any modelling tool before.

2.2 Modelling Process

For the first group deliverable, each group was asked to select a company and apply a strategic analysis tool, e.g. Porter's Five Forces [35]. Participants were asked to use MURAL during this deliverable, with the purpose of exposing the members to the tool prior to the usability evaluation. Another reason was to familiarize the participants with their selected company (i.e. improving their knowledge and experience in their selected business context).

For the second deliverable, groups were asked to use business model pattern cards [36] during business model ideation, and then the BMC to describe the company's business model, ensuring a shared understanding. At the time that the study was conducted, MURAL had four BMC templates available, enabling stakeholders to complete the BMC remotely and digitally. Students were asked to use a specific template (i.e. method) in MURAL, named the Business model canvas workshop by Product School [37]. This template had the most guidance in terms of how to complete the BMC at the time of the study. The sequence in which the BMC blocks are completed in, is prescribed, and each block contains a few guiding questions to support brainstorming. There is also an outline per block that includes facilitator tips. The tips are basic, and the same for each block (i.e. encourage participants to add sticky notes, vote on the best option, move the top-voted item to the top).

Note that each group modelled their specific case/company and the context per team therefore differed. A facilitator was also not assigned per group, i.e. all team members received the same brief. The reason for this is that most participants had similar levels of experience/exposure to MURAL and the BMC, and the assumption was that a team member would naturally take the lead in the modelling session (i.e. follow the facilitator role in the outline instructions of the prescribed template).

During this second deliverable, team members were familiar to each other as they had worked together before in the first deliverable. Teams were told that they could create their BMC over more than one session, if required, and were asked to work remotely. This meant that each group member was requested to contribute to the modelling on their own device and they could not, for example, sit next to each other with one person performing all the modelling while the other member advises. In addition, they were asked to record all of their modelling sessions in a specific meeting platform, i.e. an application embedded within the Blackboard learning system, called Blackboard Collaborate.

2.3 Usability Evaluation

Once the BMC deliverable (i.e. deliverable 2) was submitted, the 104 members were provided access to a usability survey on Blackboard. Members were informed that their participation in the survey is voluntary, that they could withdraw at any time without penalty, that their privacy would be protected throughout the survey, and that their participation would remain confidential.

The survey presented was similar to our previous study's survey [31], but was tailored to the BMC context and had slight text modifications to avoid misalignment regarding the purpose of the usability test, based on learnings from the previous round of evaluation. There were 29 questions in total, of which 15 were free-text questions encouraging participants to motivate their answers.

In the survey, the first question was to obtain consent, linked to the conditions for our approved application for ethical clearance. The next few questions were to obtain feedback on participant modelling tool experience and the duration spent on the modelling exercise. After these initial questions, the survey included the 10 System Usability Scale (SUS) questions from [38], as well as free-text fields in which motivations could be provided per question. We use the SUS in our study as it is quick to complete (i.e. consists of less than 20 questions) and is focused on usability, as discussed in Venter and de Vries [31]. Lastly, a few additional questions with supporting motivations were included, such as whether the participant would recommend the tool, whether they preferred face-to-face modelling, and whether they experienced any tool frustrations.

Of the 104 members, 35 agreed to participate in our survey (about 34%). The primary researcher evaluated the survey-results, obtaining a usability score from the standardized usability questionnaire used, as well as supporting text feedback provided by the participants. The survey results are included in Sect. 3.2 and Sect. 3.3.

Of the 35 participants, 18 (about 51%) were contacted via email, requesting an interview to provide additional content and context on the usability issues included in their feedback. Only 4 participants agreed to an interview (about 4% of all members). The interview results are included in Sect. 3.4.

In this article, we only include and report on the interviewees' modelling sessions. Two interviewees were from the same group, and one interviewee's modelling recording was not available. Therefore, two modelling sessions were observed to gain additional insight into the usability of MURAL during the modelling. The results from observing the recordings are included in Sect. 3.5.

3 Results

In Sect. 3.1, we provide findings from the method used in the study. The survey feedback is included in two sections: In Sect. 3.2 we present the SUS feedback and in Sect. 3.3 we present the results from the other questions, such as the types of frustrations encountered. Interview results are included in Sect. 3.4, and observations made by watching the recorded modelling sessions are included in Sect. 3.5.

3.1 Method

Students were asked to use a specific template (i.e. method) in MURAL, named the Business model canvas workshop by Product School [37]. When evaluating the groups' BMCs, however, it was found that only 7 groups (about 32%) used this template, and most groups (about 59%) used the template by Strategyzer [39]. The main difference between these templates, according to the researcher, is that the Strategyzer template has more questions per building block, potentially providing more assistance in formulating the components. As an example, in the Strategyzer template, Revenue streams has the following guidelines: "What are our sources of revenue? For what value are our customers willing to pay? How much does each revenue stream contribute to the overall revenues? How can we innovate to diversify our revenue streams?" whereas the Product School template only has "Where will revenue come from? What sources?". All the templates have the primary purpose of completing the nine building blocks.

3.2 The System Usability Scale (SUS)

The survey data, obtained from the 35 participants, were analyzed first. In this section, we discuss the SUS section of the survey, as well as the additional text feedback provided with the SUS answers. It should be noted that one participant left one SUS question unanswered, i.e. not responding to "I thought MURAL was easy to use during this deliverable". We updated this answer to "neither agree nor disagree", in order to calculate a SUS score per participant.

The average SUS score, when combining all participants' scores, is 74,29. When comparing all 35 participants' SUS scores to the acceptability descriptor [40], most participants (about 74%) found MURAL acceptable when modelling a BMC (see Fig. 1).

In the 10 SUS questions, participants provide feedback on how they feel regarding the statement, by providing a ranking of strongly disagree, disagree, neither agree nor disagree, agree or strongly agree [38]. The odd-numbered SUS questions are positive-oriented. In these questions, we would expect participants to agree or strongly agree to the statements, if they are satisfied with MURAL in the evaluation context. The positive-oriented SUS question feedback is summarized in Fig. 2.

Only one participant strongly disagreed with a positive-oriented question. The supporting text feedback provided was "The integration and functionality of MURAL is great, however, as a person who gets easily distracted, there is too much movement on the screen when using MURAL at the same time, it affects my productivity".

Of the 35 participants, 4 (about 11%) disagreed when asked whether they would use MURAL frequently to create a BMC. Reasons provided by the four participants

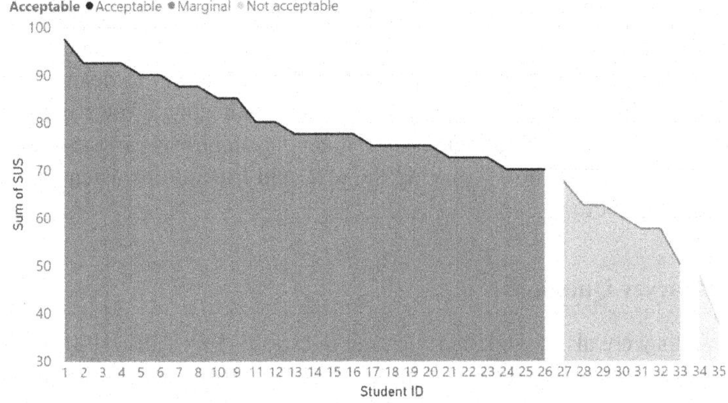

Fig. 1. SUS score per participant

Question	1. Strongly Disagree	2. Disagree	3. Neither Agree nor Disagree	4. Agree	5. Strongly Agree
Q01: I think that I would like to use MURAL frequently to create a business model canvas.		4	4	20	7
Q03: I thought MURAL was easy to use during this deliverable.		1	5	19	10
Q05: I found the various functions in MURAL were well integrated for the purpose of this deliverable.	1	1	5	22	6
Q07: I would imagine that most people would learn to use MURAL very quickly.		1	5	19	10
Q09: I felt very confident using MURAL during this deliverable.		3	6	15	11
Total	1	7	13	28	18

Fig. 2. SUS positive-oriented question summary

included difficulties with simple acts like resizing a textbox, lagging, shifting templates, navigations issues ("I found Mural slightly hard to navigate, as everyone can work anywhere and the canvas is very big") and that other better options are available ("Canva is easier and more user friendly").

The even-numbered SUS questions are negative-oriented. In Fig. 3 we summarize the answers per negative-oriented SUS question.

Question	1. Strongly Disagree	2. Disagree	3. Neither Agree nor Disagree	4. Agree	5. Strongly Agree
Q02: I found MURAL unnecessarily complex during this deliverable.	6	19	9	1	
Q04: I think that I would need the support of a technical person to be able to use MURAL.	18	15	1	1	
Q06: I thought there was too much inconsistency in MURAL during this deliverable.	6	23	5	1	
Q08: I found MURAL very frustrating to use when creating the business model canvas.	9	12	8	6	
Q10: I needed to learn a lot of things before I could get going with MURAL.	9	18	5	1	2
Total	20	34	17	7	2

Fig. 3. SUS negative-oriented question summary

Of the 35 participants, 2 strongly agreed that they needed to learn a lot of things before they could get going with MURAL. Supporting text feedback included "Some tools of Mural are 'hidden'…" and "There are so many different tools to get to know before being able to use Mural efficiently". The participant that agreed with this statement, added "The interface was new to me, and not being a tech-savvy person made it slightly more difficult to use".

About 17% of participants agreed that MURAL was frustrating to use when creating a BMC. Supporting feedback included the creation of unwanted sticky notes (e.g. "Clicking on something would create a sticky note, and having to delete it each time was frustrating"). Feedback also included difficulties connecting arrows and lines due to the auto-snap function, text alignment issues, that you sometimes can't read the text you are typing due to the "tool box" covering the text, and frustrations when adding object shapes and editing pictures.

3.3 Other Survey Questions

The participants were also asked the following question: "From an ease-of-use perspective, if I had to participate with other team members in future, doing online participative modelling, I would recommend that MURAL is used". Most participants (about 34%) neither agreed nor disagreed with the statement (see Fig. 4).

One participant that disagreed added "I think there are better tools to use, that I am more comfortable with", and the other "Having a few members working on Mural simultaneously causes the program to glitch, and not always save the alterations made".

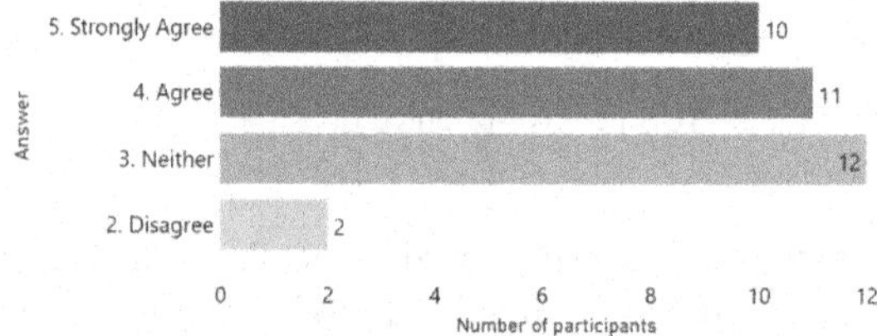

Fig. 4. Feedback for question: "I would recommend that MURAL is used in the future"

Participants were then asked "If I had to create a business model canvas in future, I would prefer to use face-to-face collaboration and drawing on a physical whiteboard, rather than using online modelling like MURAL". A large percentage (about 29%) neither agreed nor disagreed with the statement. The feedback obtained is summarized in Fig. 5.

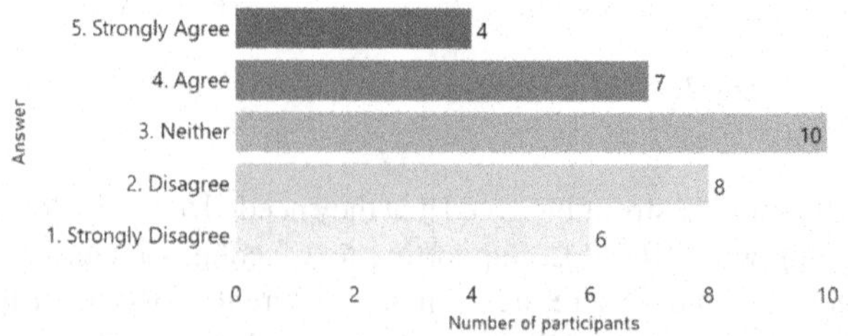

Fig. 5. Feedback for question: "I would prefer face-to-face collaboration rather than using MURAL"

Some feedback received from participants per answer is provided in Table 1. There was no accompanying text feedback for participants that strongly disagreed with the statement.

Table 1. Feedback when participants were asked whether they would prefer face-to-face collaboration rather than using an online modelling like MURAL

Answer	Participant feedback examples
Strongly agree	"I believe that face-to-face interaction is likely to yield more involvement and interaction. It also ensures that everyone present is focused on the task at hand. Whereas on an online platform, people could be reading emails or doing other things because there is less supervision"
Agree	"It is much easier to see the entire business model canvas at once without a screen constantly moving around and then you need to zoom in or out etc.", "Because then people won't talk over each other"
Neutral	"depends on the type of model I want, sometimes MURAL is helpful but other times it's easier to just physically do it"
Disagree	"I would like to be face-to-face and use MURAL at the same time. This allows for infinite board space, the recording of the board, send an image in the form of a PDF or a PNG. The tool works and can be used in a non-face-to-face to face setting, but will be best when used collaboratively in person"

The last question reads as follows: "You and/or your group members may have experienced some frustrations related to the tool functionality in MURAL. Please elaborate on these frustrations". Of the 35 participants, 10 (about 29%) did not provide an answer to this question, and two reported no MURAL frustrations, e.g. "No frustrations with using MURAL, only with following the dedicated time limit".

Frustrations identified from this specific question were grouped according to the categories or themes identified from the previous studies [13, 31]. Most frustrations could be linked to editing frustrations ("The only issue we had was neatness and getting the sticky notes and the font sizes to be the same as there is no number to measure what the font is and relate to the other") and issues when moving, duplicating, and selecting objects ("Moving things around is difficult and frustrating"). Automatic sticky note addition with double-click (e.g. "One frustration was when you would try click on something and a sticky note would appear instead. This disrupted the flow of my work") and auto-refresh problems (e.g. "It sometimes froze when I was zooming") were each mentioned three times.

Four categories of frustrations were only mentioned once in this question. Of these, two were frustrations initially identified in de Vries and Opperman [13], namely (1) Using an "area" as a container for constructs created problems ("All that was frustrating is that you have to be very precise when dragging shapes, and sometimes you would drag the whole business model canvas by mistake"), and (2) MURAL should have the ability to verbally communicate while modeling ("Not being able to be on a call with team members in MURAL was very frustrat[ing]. There is a way to make a call on MURAL but it seems as though that function is not for everyone").

A frustration identified in the eSCM usability study [31], i.e. user interface learnability, was also mentioned once ("At the beginning, it was difficult to figure out where everything is on Mural, but as soon as that is figured out, it is easy to use"). This frustration was however also mentioned by two different participants to support a SUS selection ("There are so many different tools to get to know before being able to use MURAL efficiently" and "certain features require explanation").

A new frustration was identified relating to the type of device, i.e. "Mainly we just found it difficult to use on tablets. No other complications were experienced". This device frustration was also mentioned by a different participant in SUS supporting text feedback. The participant neither agreed nor disagreed that they found MURAL very frustrating to use when creating the BMC, adding that "If you use a laptop or PC, it is easy to use, but when using an iPad or something similar, it is very difficult and frustrating".

Another frustration that couldn't be linked to a previously identified theme was control issues, e.g. "The mouse and keyboard controls could be better. Possibly aligning the controls with other popular software would be useful" and "In some of the other apps I've used, including AnyLogic, when you right click and drag the screen you can pan around. Having built this habit made MURAL tough since right-clicking continuously opens a menu". This was also mentioned by a different participant in the SUS supporting feedback, i.e. "To initially familiarize myself with this tool, I think having a technical person would be immensely helpful. The interface is not very universal (not similar to other digital tools) and certain features require explanation".

The third and final new frustration type identified was frustrations linked to having many modelers present on the mural, i.e. a crowded and inefficient workspace. This was mentioned by four participants: "It is very easy to move pictures/tables/text boxes on the canvas, especially when there are multiple users at once. This could make it very messy and hard to work with", "Our group struggled with everyone using the canvas at once. We eventually allocated the actual editing of the canvas to a single person in the group, and the rest contributed verbally", "Multiple people does not work on MURAL, the screen moves each time a member wants to work in a different section on their screen", and "Since everyone was supposed to work on MURAL together - at once - it was difficult because so many changes were being made at the same time. When it is one person typing or writing or making the changes, and the others can see clearly what is happening, it is easier to spot out problems".

3.4 Interviews

The interviews conducted ranged between about 7 and 12 min in duration. Interviewee 1 reiterated their previously listed frustrations (e.g. Auto-snap and sticky note concerns) and mentioned that they prefer Discord [41] above Blackboard Collaborate for team communication. Interviewee 2 mentioned that they liked that one could 'follow' participants to see what other people are working on without having to navigate themselves. The interviewer inquired about one of their concerns: When they right click and drag the screen to pan, a menu would constantly open, making panning difficult. The interviewee was however unable to replicate this problem. An issue that could however be replicated was that a toolbar popped up over the textbox, making it difficult to see the typed text. This was not mentioned in the interviewees' survey feedback, but was mentioned in another participants' feedback, e.g. "There are some frustrating aspects of Mural, for example, the 'tool box'; often covers some of the text you are adding and then you can't see what you are typing".

Interviewee 3 elaborated on their statement "The biggest issue was making things fit in a specified area with a specified font size" adding that the font size would auto adjust when adjusting the object size. The interviewee mentioned that they used icons to vote for ideas, functionality that the researcher had seen on other groups' MURALS as well. This interviewee also mentioned auto-refresh problems that hadn't been mentioned in his survey feedback, stating that work had been lost as a result of this.

Interviewee 4 stated that working on MURAL on an iPad causes frustrations. As an example, when wanting to update the text in a textbox, the interviewee would double click, but then sticky notes were created. Moving and resizing things on an iPad is also difficult, and she said that the application did not include any guidelines or tutorials to teach her how to do this. An issue mentioned in the interview that wasn't included in this participants' survey feedback was that the BMC would sometimes shift/move out under the sticky notes. She however said that they were able to learn how to no longer do this, also mentioning that they eventually discovered the 'lock' functionality. Another issue identified that wasn't listed in their survey, was that some items/objects were removed by other team members. By using the activity log, however, they were able to identify who made the changes.

3.5 Modelling Sessions

The first group that was observed recorded about five hours of content. This group, that consisted of five members, included two interviewees. One team member was however absent for most of the recording due to unforeseen circumstances. Frustrations identified from the recording included difficulties finding a specific MURAL in the team's room, connect problems (only one member), and an instance where an arrow was unintentionally added. This group did not follow the prescribed method/template. Instead, each team member was assigned one or more building blocks prior to the session, and each member performed their allocated analysis alone. In the recorded session, the individual work was copied and pasted onto sticky notes, and then briefly discussed and moved around. There was very little collaboration between members.

The second group that was observed recorded about four hours of content. This group, that consisted of five members, included one interviewee. The first recorded session (about 50% of their modelling) wasn't recorded in Blackboard Collaborate as requested, and was sent to the researcher via email on request. The second session (the remaining 50%) was however recorded on Blackboard Collaborate. This group's approach was as follows: all of the team members added their ideas on sticky notes for an allocated amount of time, after which each member could explain/present their ideas. Team members then voted for their favorite ideas, initially using the MURAL voting system, but then rather adding icons per sticky note to vote. The prescribed template was then used to create the BMC. It should however be noted that the specific instructions weren't followed (i.e. all components were completed without voting on/discussing the contributions). An arrow was accidently added once, and the components and their descriptions were moved around accidently on more than three occasions. A feature that one participant wanted, but wasn't available, was seeing who had added an icon on each sticky note (e.g. by hovering over the 3 hearts on a sticky note, see which three members had added the hearts/voted for the idea).

In both groups, one person was primarily responsible for facilitating the session. In the second group, in which the prescribed method was used, this person also read the component descriptions aloud, called on other team members to answer, and read the MURAL template outline descriptions.

4 Discussion

In this section we answer the research questions initially presented in Sect. 1.5. RQ 1, i.e. *How user friendly is the online tool MURAL during PEM, and specifically when creating a BMC?*, is presented in Sect. 4.1. RQ 2, i.e. *What functional features from MURAL contribute to this use-in-context?* is presented in Sect. 4.2. Additional findings on the UEMs used are included in Sect. 4.3.

4.1 MURAL Usability

According to Vlachogianni and Tselios [42], questionnaire data is difficult to interpret unless there is a basis for comparison and reference, and usability can also only be defined by reference to specific contexts. By using subjective metrics, such as the SUS, similar and dissimilar systems can be compared, "without regard to specific tasks, as would be required for comparisons of most quantitative metrics" [43].

Other tools' usability benchmarks using SUS have been published, e.g. Kortum and Bangor [43] presented the usability ratings from the SUS for everyday products like Google Search and Microsoft Excel. However, to the best of our knowledge, not many studies are available that focus on the usability of online tools in the context of PEM. In a usability study by Venter and de Vries [31], MURAL scored 74.23 when participant pairs used an extended story-card method (eSCM) [13] to perform co-modelling remotely. In the BMC study, MURAL still scored above average with 74.28, representing a marginal difference of 0.05. MURAL is therefore an acceptable, "good" and B grade system [40] during PEM when using the BMC as well as when using the eSCM.

From our other survey questions, we found that about 34% participants neither agreed nor disagreed that from an ease-of-use perspective, if they had to use PEM in future, they would recommend that MURAL is used. However, 60% either agreed or strongly agreed, that they would recommend MURAL from an ease-of-use perspective, showing interest/support towards the tool. A few participants also mentioned that they would like to combine the use of MURAL with in-person collaboration, e.g. "I would like to be face-to-face and use MURAL at the same time" and "We would however possibly use MURAL while sitting together in person".

4.2 Important Functional Features

Frustrations previously identified and grouped by de Vries and Opperman [13] that were not mentioned in this study, included the 'undo' function not working properly, lack of auto-alignment, and activity tracking. Work loss due to connection problems was mentioned in a SUS supporting answer ("Having a few members working on MURAL simultaneously causes the program to glitch, and not always save the alterations made") as well as in one of the interviews by a different participant. Connection problems were also observed in both observed groups. This could, however, be due to other factors, e.g. the internet stability of the participants.

Activity tracking was mentioned in an interview as a nifty feature when trying to determine who made changes (e.g. deleted something), especially considering the number of group members. Other nifty functionality mentioned/observed included the "follow" feature (where you automatically follow a collaborator) and the voting system (where you can add icons to sticky notes, e.g. a thumbs up to support a suggestion).

A frustration category identified in Venter and de Vries [31] that was not mentioned in any survey feedback was issues regarding team member awareness, i.e. participants had difficulty knowing where their co-modeler was, working on the whiteboard space. One interviewee, however, mentioned the usefulness of being able to follow a co-modeler around, which could be linked to team awareness. In this study, the number of people modelling at the same time became a frustration, something that hadn't previously been mentioned. This could be due to larger teams (the previous studies consisted of pairs), or it could be method-related, e.g. the facilitator did not create a structured and sequenced process and participants therefore provided inputs whenever it suited them.

The frustrations linked to each usability study conducted thus far, including [13, 31], is depicted in Fig. 6.

4.3 UEM Findings

We found in this study that, similar to Venter and de Vries [31], the supporting text feedback in the SUS questions was sometimes helpful, but often still lacked depth and context about the frustration itself. As an example, a survey participant said that "on several occasions, actions we could previously do on MURAL, did not allow us to do the same in the next session. The tools and acts we were able to utilize, changed, making it very frustrating". This participant did not agree to an interview, and therefore the researcher could not confirm which actions or tools were the problem. Similarly, even though additional insights could be obtained from the interviews, without the problem being displayed first hand/replicated, it is difficult to fully grasp the problem.

The SUS provides a subjective measurement of usability and not diagnostic information [44]. Without the text feedback, however, it is difficult to gauge the participants' understanding/interpretation of the statement, which could impact the measurement. As an example, one participant strongly disagreed that the various functions in MURAL were well integrated for the purpose of this deliverable. In the supporting text feedback, he added that "The integration and functionality of MURAL is great, however, as a person who gets easily distracted, there is too much movement on the screen when using MURAL at the same time, it affects my productivity". The concern is therefore not the integration itself.

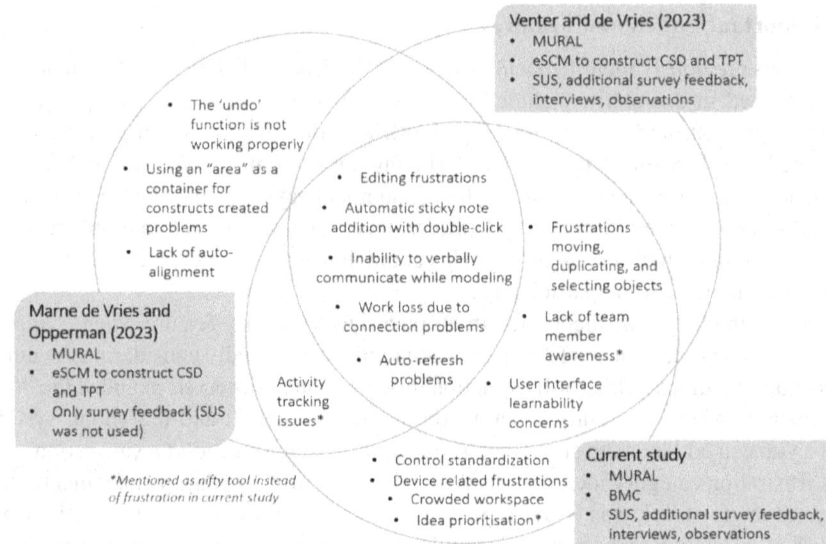

Fig. 6. Venn diagram depicting frustrations identified in different studies

5 Conclusion

It can be concluded from the SUS score that MURAL scores above average and is considered an acceptable tool during PEM using the BMC. Previously identified frustrations were also relevant to this study's context (e.g. the unwanted creation of sticky notes), and new frustrations were identified (e.g. the lack of standardized controls). In Sect. 5.1 we discuss the study's limitations and in Sect. 5.2 we propose future research opportunities.

5.1 Limitations

Even though most participants were introduced to BMs and the BMC before the evaluation via an in-person class and content/notes posted on Blackboard, there is no guarantee that participants were exposed to the information. Attendance was not compulsory and content accessed was not tracked. Therefore, some participants could have used the BMC in our study without prior knowledge. Similarly, participants were asked to use MURAL during the first deliverable, with the purpose of exposing the members to the tool prior to the usability evaluation, but there is no way of confirming that members familiarized themselves with the tool. To address this, training should be prioritized. This should include BM/BMC, PEM, MURAL, and method-related training, with additional training for members that will act as facilitators. Suggestions include a small-scale collaborative exercise to involve all members in the creation of a BMC on MURAL, a demonstration of PM between 4 or 5 members linked to a real-world case, and participants attending the 1-h free training session offered by MURAL.

Many groups did not use the required method when creating the BMC (refer to Sect. 3.1). Therefore, even though a BMC was used and the nine building blocks are the same when creating a BM, slight differences in the process took place. In future studies, participants should be incentivized to use a specific method, e.g. via marks, with the goal of reducing variability.

There was also group size variability, which could have impacted the participants' experience. As an example, a frustration was identified when many modelers were working together at the same time, but could be irrelevant to the group that consisted of only three members. In addition to this, each group could select and model a business of their choice, introducing case context variability. This could have impacted the duration of the modelling (e.g. that more intricate businesses took longer to model), which could have impacted the perceived usability of the tool.

Groups were asked to use business model pattern cards, introduced by [36], on an existing company's business model, to develop a new idea or solution. This new idea had to be visualized on the BMC. Since participants were designing a new idea, they could represent the "company" (i.e. enterprise stakeholders) as well as the experts (seeing as they understood the BMC). The level of company knowledge and experience of the participants could however differ per group and hence create an additional knowledge barrier for some groups.

Lastly, it should be noted that the possibility exists that MURAL developers address and resolve some of the issues (i.e. frustrations) identified in our study. The findings presented in this study therefore reflect the status of the system at the time of our analysis and may not reflect the updates/improvements already implemented by MURAL.

5.2 Future Research

In the prescribed template/method used, there are "instructions" for the facilitator. The guidance provided in this outline is minimal, e.g. members are only told to add ideas and upvote the ideas. The current method used in the prescribed BMC template did not enable participants to systematically and collaboratively transform a traditional business model into a new one. For future research, methods that address specific tasks during BM design and implementation (e.g. the systematic approach/steps provided by Euchner and Ganguly [45]) should be investigated and incorporated into the method prescribed to participants. Participants should be trained on the new method, and additional training should be provided to members that will act as facilitators. Training for facilitators could include awareness of the "summons" feature in MURAL and how the use of two screens can simplify the PM process.

Considering that three iterations of the usability evaluation of tools during PEM have been conducted, the features and functionality that were identified per study could be categorized and compared. Thus, a more in-depth analysis of frustrations to identify key features, as depicted in Fig. 6, can be conducted, building towards the development of a usability measurement instrument. This instrument, focused on and relevant to PEM specifically, was initially suggested in Venter and de Vries [31]. The categories/themes could also be clarified. As an example, in this study, frustrations were identified when using an iPad. Instead of creating a device related frustration, all frustrations could potentially be grouped per device type, per context (as some frustrations may, for example, only be relevant to iPads).

The usability of alternative modelling tools, such as FigJam or Miro should be assessed during PEM when using the BMC. The SUS scores could then be compared, building towards the usability benchmarking of online tools in the context of PEM. Since the frustrations identified in our study could be updated and improved by the system developers, we could analyze the system releases, determining which frustrations have been resolved. Seeing as new frustrations could also be introduced through system updates, each context should regularly be reevaluated to ensure relevance and accuracy. It is also advised that in these future iterations, the limitations of this study are addressed.

Disclosure of Interests. The authors have no competing interests to declare that are relevant to the content of this article.

References

1. Geszten, D., Hamornik, B.P., Hercegfi, K.: Exploring awareness related usability problems of collaborative software with a team usability testing approach. In: 9th IEEE International Conference on Cognitive Infocommunications (CogInfoCom), Hungary, pp. 45–50. IEEE (2018). https://doi.org/10.1109/CogInfoCom.2018.8639865
2. Stirna, J., Persson, A.: Enterprise Modeling: Facilitating the Process and the People. Springer, Cham (2018). https://doi.org/10.1007/978-3-319-94857-7
3. de Vries, M.: Towards consistent demarcation of enterprise design domains. In: de Cesare, S., Frank, U. (eds.) ER 2017. LNCS, vol. 10651, pp. 91–100 (2017). Springer, Cham. https://doi.org/10.1007/978-3-319-70625-2_9
4. Dietz, J., Mulder, H.: Enterprise Ontology: A Human-Centric Approach to Understanding the Essence of Organisation. Springer, Cham (2020). https://doi.org/10.1007/978-3-030-38854-6
5. Krogstie, J.: Model-Based Development and Evolution of Information Systems: A Quality Approach. Springer, London (2012). https://doi.org/10.1007/978-1-4471-2936-3
6. Gutschmidt, A.: An exploratory comparison of tools for remote collaborative and participatory enterprise modeling. In: 29th European Conference on Information Systems - Human Values Crisis in a Digitizing World, ECIS, Morocco (2021)
7. Gutschmidt, A., Sauer, V., Sandkuhl, K., Kashevnik, A.: Identifying HCI patterns for the support of participatory enterprise modeling on multi-touch tables. In: Gordijn, J., Guédria, W., Proper, H. (eds.) T PoEM 2019. LNBIP, vol. 369, pp. 118–133. Springer, Cham (2019). https://doi.org/10.1007/978-3-030-35151-9_8
8. Fellmann, M., Sandkuhl, K., Gutschmidt, A., Poppe, M.: Structuring participatory enterprise modelling sessions. In: Grabis, J., Bork, D. (eds.) PoEM 2020. LNBIP, vol. 400, pp. 58–72. Springer, Cham (2020). https://doi.org/10.1007/978-3-030-63479-7_5
9. Visual collaboration made easy. https://www.mural.co/features. Accessed 11 Nov 2023
10. Jackson, V., van der Hoek, A., Prikladnicki, R., Ebert, C.: Collaboration tools for developers. IEEE Softw. **39**(2), 7–15 (2022). https://doi.org/10.1109/MS.2021.3132137
11. Enter with a dream. Exit with the next big thing. https://miro.com/. Accessed 29 Jan 2024
12. Bring learning to life with Jamboard. https://edu.google.com/jamboard/. Accessed 10 Jan 2024
13. de Vries, M., Opperman, P.: Improving active participation during enterprise operations modeling with an extended story-card-method and participative modeling software. Softw. Syst. Model. **22**(4), 1341–1368 (2023). https://doi.org/10.1007/s10270-023-01083-8
14. Turn possibilities into plans. https://www.figma.com/figjam/. Accessed 11 Jan 2024

15. Draw.io. https://app.diagrams.net/. Accessed 29 Jan 2024
16. Lightweight drawing program for PC. https://google-drawings.en.softonic.com/. Accessed 10 Jan 2024
17. Schwarz, J.S., Legner, C.: Business model tools at the boundary: exploring communities of practice and knowledge boundaries in business model innovation. Electron. Mark. **30**(3), 421–445 (2020). https://doi.org/10.1007/s12525-019-00379-2
18. Saebi, T., Lien, L., Foss, N.J.: What drives business model adaptation? The impact of opportunities, threats and strategic orientation. Long Range Plan. **50**(5), 567–581 (2017). https://doi.org/10.1016/j.lrp.2016.06.006
19. Osterwalder, A., Pigneur, Y., Clark, T., Smith, A.: Business Model Generation: A Handbook for Visionaries, Game Changers, and Challengers. Wiley, Hoboken (2010)
20. Ladd, T.: Does the business model canvas drive venture success? J. Res. Mark. Entrep. **20**(1), 57–69 (2018). https://doi.org/10.1108/JRME-11-2016-0046
21. Ancillai, C., Sabatini, A., Gatti, M., Perna, A.: Digital technology and business model innovation: A systematic literature review and future research agenda. Technol. Forecast. Soc. Change **188** (2023). https://doi.org/10.1016/j.techfore.2022.122307
22. Hoch, N.B., Brad, S.: Managing business model innovation: an innovative approach towards designing a digital ecosystem and multi-sided platform. Bus. Process. Manag. J. **27**(2), 415–438 (2021). https://doi.org/10.1108/BPMJ-01-2020-0017
23. Fritscher, B., Pigneur, Y.: Visualizing business model evolution with the business model canvas: concept and tool. In: 16th Conference on Business Informatics, Switzerland. IEEE (2014)
24. International Organization for Standardization: Ergonomics of human-system interaction (ISO 9241-11:2018) (2018). https://www.iso.org/standard/63500.html
25. Ivory, M.Y., Hearst, M.A.: The state of the art in automating usability evaluation of user interfaces. ACM Comput. Surv. (CSUR) **33**(4), 470–516 (2001). https://doi.org/10.1145/503112.503114
26. Lu, J., Schmidt, M., Lee, M., Huang, R.: Usability research in educational technology: a state-of-the-art systematic review. Educ. Technol. Res. Dev. **70**(6), 1951–1992 (2022). https://doi.org/10.1007/s11423-022-10152-6
27. Fernandez, A., Insfran, E., Abrahão, S.: Usability evaluation methods for the web: a systematic mapping study. Inf. Softw. Technol. **53**(8), 789–817 (2011). https://doi.org/10.1016/j.infsof.2011.02.007
28. Sagar, K., Saha, A.: A systematic review of software usability studies. Int. J. Inf. Technol. (2017). https://doi.org/10.1007/s41870-017-0048-1
29. Gonçalves da Silva e Souza, P., Canedo, E.D.: Improving usability evaluation by automating a standardized usability questionnaire. In: Marcus, A., Wang, W. (eds.) DUXU 2018. LNCS, vol. 10918, pp. 379–395. Springer, Cham (2018). https://doi.org/10.1007/978-3-319-91797-9_27
30. Sauro, J.: Quantifying the User Experience: Practical Statistics for User Research, 2nd edn. Elsevier, Amsterdam (2016)
31. Venter, A., de Vries, M.: Usability of virtual tools for participatory enterprise modelling. S. Afr. J. Ind. Eng. **34**(3), 1–12 (2023). https://doi.org/10.7166/34-3-2942
32. vom Brocke, J., Hevner, A., Maedche, A.: Introduction to design science research. In: vom Brocke, J., Hevner, A., Maedche, A. (eds.) Design Science Research. Cases. Progress in IS, pp. 1–13. Springer, Cham (2020). https://doi.org/10.1007/978-3-030-46781-4_1
33. Creswell, J.W., Creswell, J.D.: Research Design: Qualitative, Quantitative, and Mixed Methods Approaches, 5th edn. SAGE Publications Inc., California (2018)
34. Bradford, P., Porciello, M., Balkon, N., Backus, D.: The blackboard learning system: the be all and end all in educational instruction? J. Educ. Technol. Syst. **35**(3), 301–314 (2007)

35. Porter, M.E.: The five competitive forces that shape strategy. Harv. Bus. Rev. **86**(1), 78–93 (2008)
36. Csik, M., Frankenberger, K., Gassmann, O.: 55 pattern Cards. FT Press (2014)
37. Accelerate your product management career. https://productschool.com/. Accessed 29 Jan 2024
38. Brooke, J.: SUS: A Quick and Dirty Usability Scale, 1st edn. Taylor & Francis, United Kingdom (1996)
39. Master the practice of innovation. https://www.strategyzer.com/. Accessed 29 Jan 2024
40. 5 ways to interpret a SUS score. https://measuringu.com/interpret-sus-score/. Accessed 11 Nov 2023
41. Get Discord for any device. https://discord.com/download. Accessed 16 Jan 2024
42. Vlachogianni, P., Tselios, N.: Perceived usability evaluation of educational technology using the system usability scale (SUS): a systematic review. J. Res. Technol. Educ. **54**(3), 392–409 (2022). https://doi.org/10.1080/15391523.2020.1867938
43. Kortum, P.T., Bangor, A.: Usability ratings for everyday products measured with the system usability scale. Int. J. Hum. Comput. Interact. **29**(2), 67–76 (2013). https://doi.org/10.1080/10447318.2012.681221
44. Brooke, J.: SUS: a retrospective. J. Usability Stud. **8**, 29–40 (2013)
45. Euchner, J., Ganguly, A.: Business model innovation in practice. Res. Technol. Manag. **57**(6), 33–39 (2014). https://doi.org/10.5437/08956308X5706013

Research Status and Trends of Virtual Simulation Technology in Clothing Design

Zichan Wang(✉)

National Academy of Chinese Theatre Arts, Peking 400 Wanquan Temple, Beijing, China
17810259385@163.com

Abstract. The application of virtual simulation technology in clothing design has become increasingly common in recent years. However, only some studies have conducted systematic bibliometric research on this topic. This study utilizes bibliometric techniques and integrates VOSviewer and CiteSpace to construct a knowledge map. It screens 241 papers in this field from the Web of Science core database from 1997 to 2023. This study aims to understand the evolutionary trajectory of research hotspots and the latest research progress in virtual clothing from 1997 to 2023. Finally, the study will discuss future research directions and challenges in this field.

Keywords: Computer Graphics and Virtual Trip · Design Methods and Techniques · Design Thinking · Virtual Simulation Technology · Clothing Design

1 Introduction

Virtual simulation technology is a product of the development of the Internet era. In recent years, with the continuous improvement of computer technology and graphic processing capabilities, virtual simulation clothing research is getting more and more attention. The application of virtual simulation technology is extensive, including apparel modeling, historical heritage restoration, pattern making, teaching, and so on. The use of three-dimensional virtual technology in apparel design can improve design efficiency. Various virtual prototyping techniques have been developed to innovate the apparel industry. For each step of the apparel design process, one can find specialized tools [1]. The use of three-dimensional virtual technology in clothing design can improve design efficiency. The textile and clothing industry is moving towards digitalization and has many development spaces. Many scholars have adopted virtual simulation technology in clothing research and, at the same time, produced many results in the literature. However, current scholars need a more comprehensive grasp of the research field of virtual simulation technology in apparel design. In order to promote the research and development of the proposed simulation technology in apparel design, it is indispensable to analyze the relevant literature. So, it is necessary to summarize further, analyze the current status and trends of virtual simulation research in clothing design, and then provide some reference for subsequent research.

F. F.-H. Nah and K. L. Siau (Eds.): HCII 2024, LNCS 14721, pp. 115–129, 2024.
https://doi.org/10.1007/978-3-031-61318-0_9

The point worth emphasizing is that, On the one hand, the literature output is very diverse and involves knowledge from many disciplinary fields. The changes and development trends of research hotspots in this field cannot be effectively sorted out, analyzed, and summarized solely through traditional literature review methods. On the other hand, quantitative analysis of all kinds of literature with the help of the bibliometric method can help to discover potential patterns and information in a large amount of literature data. Therefore, in this study, we use the literature related to Virtual Simulation in Clothing Design (VS-CD) in the WOS database as the data source and analyze the retrieved data through scientific bibliometrics to visualize and analyze the knowledge structure to provide scholars in this field with reference and an overall overview. Therefore, this paper will develop an analysis of the current state of VS-CD research by addressing the following research questions through bibliometrics, providing a comprehensive overview of the VS-CD:

1. Which significant scholars and institutions are working on VS-CD topics? What are their collaborative relationships?
2. What are the hot research topics about VS-CD? What are the trends in the evolution of research content?
3. Which journals/conferences are the most contributing sources of literature in the field of VS-CD research, and what are their research areas?
4. Sorting out these issues will help scholars to keep track of the findings and latest trends in VS-CD as a complement to previous qualitative work and provide a corresponding knowledge base for further research in the field.

2 Research Design

2.1 Data Sources

Data Sources Web of Science is the world's most comprehensive English language database. The literature in Web of Science's core collection has been rigorously scrutinized by peer review through published journals. It is, therefore, considered more disciplined than other more representative databases [2]. Through this platform, researchers can find bibliographic data from over 12,000 scientific journals in the natural sciences, social sciences, arts, and humanities [3]. Therefore, in this paper, we chose the core database of the Web of Science as the primary source of research. The search strategy was set as TS = ((Virtual Simulation OR Virtual OR 3D virtual-reality) AND (Costume Design OR Costume Design OR garment Design OR Clothing Design OR Fashion Design OR apparel Design OR dress Design OR Clothes Design)), selecting SSCI, SCI-Expanded, A&HCI, CPCI-S as the search sources, exported the retrieved documents to txt files in the format of "Full Record and Cited References," and eliminated the biased and interfering articles so that a total of 241 documents were obtained in the end.

2.2 Research Methodology

Bibliometrics and information visualization methods are the main research methods in this paper. Bibliometrics is a multifaceted endeavor that includes structural, dynamic, evaluative, and predictive scientometrics. It constantly evolves and has been widely used to analyze scholarly communication patterns. Several software packages have been used for bibliometric analysis, each with different capabilities and limitations [4]. In this paper, CiteSpace (V1.6.18) and VOSviewer (V6.1.R3) software are used to analyze the keyword co-occurrence of the retrieved literature as well as the citation analysis of the literature and to draw the corresponding scientific knowledge maps.CiteSpace and VOSviewer can effectively establish the mapping relationship between the knowledge units of the literature and display the macroscopic structure of the knowledge through visual information. VOSviewer has advanced graphical representation capability, which is suitable for large-scale data to locate the focus and hotspot of the research topic. CiteSpace can statistically analyze the authors, keywords, citations, and other information of the literature related to a particular knowledge field and present the analysis results through visualization, which can help researchers understand the development trend of a knowledge field from the overall perspective and it has been widely used in the analysis of bibliometrics in recent years.

3 Results

3.1 Trend Analysis of Annual Outputs of VS-CD Literature

The number of published academic literature can visualize the development process of the research field, and the statistics of literature published in the research field of virtual simulation clothing from 1997 to 2023 are shown in Fig. 1. The distribution of literature output in Fig. 1 shows that the research on virtual simulation clothing is getting more and more attention in the academic world. The first literature in the WOS database was published in 1997, and the literature output was at most seven articles/year until 2007, when the growth rate was slower. 1991–2007 was the beginning stage of virtual simulation clothing research. 2008–2021 is the development period of the research, which is the stage of the development of virtual simulation clothing. The period of 2008–2021 is the development period of research, in which the overall fluctuation of literature output is significant. However, the overall literature output shows a growing trend, and the highest number of publications reaches 19 articles per year. With the new Crown pneumonia epidemic in the world, fashion designers cannot meet face-to-face with their real-world clients. In this situation, there is an unprecedented demand for virtual assembly and virtual dynamic presentation of garments [5]. In 2022, the literature output in this area increased dramatically, reaching 32 articles/year, and VS-CD research entered a boom phase.

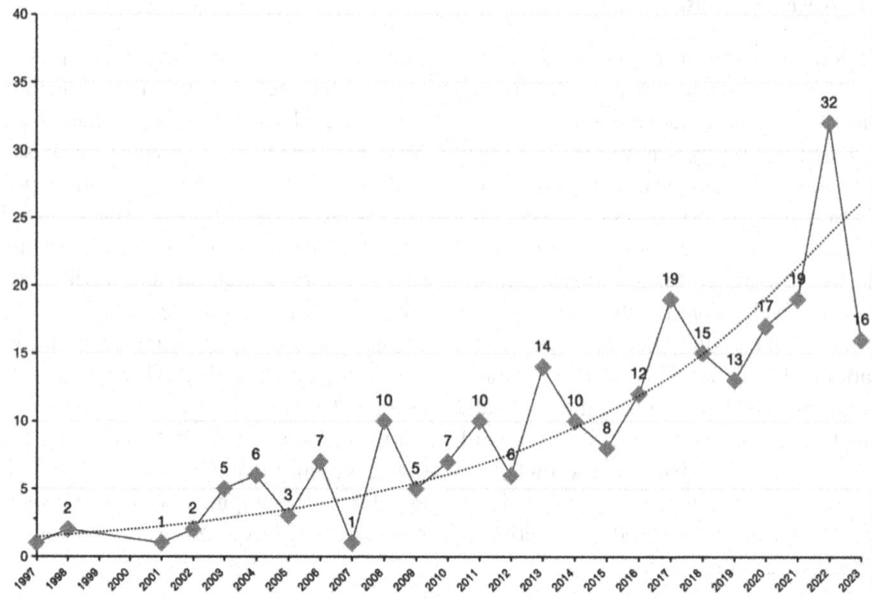

Fig. 1. A distribution map of the annual publication volume of VS-CD.

3.2 Research Hotspots

The literature keywords summarize the author's research results, usually containing the research object, perspective, and methodology. Co-occurrence analysis is essential in bibliometric methods, and high-frequency co-occurring keywords reflect the research hotspots of VS-CD for a long time. In this study, 241 literature records were selected for subsequent analysis. The WOS data containing 917 keywords were imported into VOSviewer, and the co-occurrence frequency was set to be more than 2. We obtained the keyword co-occurrence clustering of VS-CD in Fig. 2. Among them, the keyword co-occurrence mapping of the foreign studies had 188 items and 1,113 links. The keywords with the same color in the figure are the same clusters, forming a total of 5 major clusters, which are #1 - theoretical approach and cultural communication, #2 - technology implementation, #3 - technology development, #4 - user experience and fashion design, #5 - user research and healthcare design, Table 1 provides detailed information on the VS-CD research clusters, and the following sections summarize the specific research in these clusters.

Cluster #1- Theoretical Approach and Cultural Communication, contains 66 members, including computer-aided design, cultural heritage, garment construction method, generation, pattern-making, reconstruction, and reverse engineering. The cluster consists of many themes that deal with the basic theory of virtual simulation technology, research methodology, and cultural communication in the digital age, among other topics. The basic theory and research methodology of virtual simulation technology are the

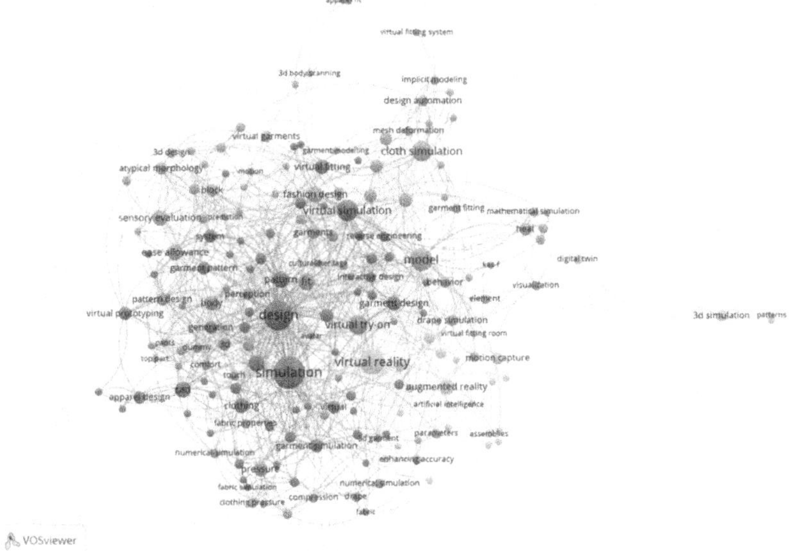

Fig. 2. Keyword co-occurrence cluster diagram in VS-CD literature. (Color figure online)

cornerstone of VS-CD field research and the fundamentals of virtual simulation technology, and the research of this cluster is an essential step to realizing VS-CD from the conceptual to the practical level. This cluster also includes research on cultural communication in the digital era, such as the digital reconstruction of cultural heritage [6, 7] and digital Inheritance of traditional costumes [7]. Digital replicas enrich museum collections, multimedia exhibitions, and online presentations.

Cluster #2- Technology Implementation, contains 44 members, collision detection, condensation, coupled diffusion, drape simulation, enhancing accuracy, parameterization, skeleton-driven deformation, physically based modeling, and so on. This clustering is mainly concerned with implementing virtual reality technology in apparel and discusses the implementation methods and processes of computer vision technology and assisted design systems. For example, an efficient collision detection algorithm is proposed for human dynamic simulation of 3D garments to achieve real-time and virtual simulation effects [8]. A survey of currently used mechanical parameters and methods for evaluating simulation accuracy in apparel CAD systems is conducted to explore simulation-based parameter estimation methods to improve the accuracy and fidelity of apparel CAD systems for advances in virtual apparel design and production. Additionally, the clustering includes how the development of virtual systems can be advanced to provide designers with practical tools for designing and engineering garments.

Cluster #3- Technology Development, contains 33 members, including body scan, digital clothing pressure, dummy, fabrics, fit, pattern design, perception, virtual try-on, comfort, and finite element method. This clustering discusses the development of virtual models and systems to explore how the efficiency and effectiveness of the decision-making process in the fashion industry can be improved through advances in virtual simulation technology. For example, new methods or tools were developed to measure

Table 1. Specific information on VS-CD's 5 research clusters.

Cluster	Cluster Label	Keywords
1 (red)	Theoretical approach and cultural communication	computer-aided design, cultural-heritage, garment construction method, generation, pattern-making, reconstruction, reverse engineering
2 (green)	Technology implementation	collision detection, condensation, coupled diffusion, drape simulation, enhancing accuracy, parameterization, skeleton-driven deformation, physically based modeling
3 (blue)	Technology development	body scan, digital clothing pressure, dummy, fabrics, fit, pattern design, perception, virtual try-on, comfort, finite element method
4 (yellow)	User experience and fashion design,	artificial intelligence, augmented reality, customization, computer graphics, human-computer interaction, motion capture, virtual fitting room, simulation interfaces
5 (violet)	User research and healthcare design,	atypical morphology, collaborative design, ease allowance, fashion design, disabled people, products, sensory evaluation

digital garment pressure to assess dynamic wearing comfort [9, 10]. A more accurate pressure prediction model was developed to reduce material waste and energy consumption in garment manufacturing [11]. In addition, this cluster discusses the development of methodologies for apparel design, such as proposing a new interactive design methodology for sustainable fashion using machine learning techniques [12]. The research in this cluster aims to contribute to the advancement of virtual apparel design and production and to facilitate the development of personalized apparel customization.

Cluster #4- User Experience and Fashion Design, contains 28 members and includes artificial intelligence, augmented reality, customization, computer graphics, human-computer interaction, motion capture, virtual fitting room, simulation interfaces, and so on. This clustering started late, and from the high-frequency keywords, the clustering reflects the research direction of applying virtual simulation of clothing in the field of interaction design in order to enhance the user experience. The research studies the usability and user experience of virtual fitting rooms from the user's point of view. It emphasizes the intelligence of the fashion industry's overall supply chain and the retail terminals' services. Attempts were made, for example, to enhance user satisfaction by developing a more comprehensive virtual dressing room (interface that meets user satisfaction, accurate fitting of virtual garments, realistic virtual garments, reliable and low-cost, and user-friendly purchasing process) [13].

Cluster #5-User Research and Healthcare Design, contains 17 members, including atypical morphology, collaborative design, ease allowance, fashion design, disabled people, products, sensory evaluation, and so on. This cluster presents a very inclusive perspective and is more concerned with special needs groups such as Wheelchair-Bound people [14], scoliosis people [15], and so on, such as product customization or finding the best apparel design solutions to meet individual's unique needs through new technologies [16–18]. This clustering study starts from the users' needs and focuses on the users' emotional experience and satisfaction testing and evaluation through better design output for their individual needs.

3.3 Trends in Hotspot Outputs

In order to further study the cutting-edge themes and development trends of VS-CD research, the average occurrence time of keywords was statistically analyzed separately to obtain Fig. 3, which visualizes the evolution of the themes of VS-CD as well as the development trends of keywords for each period within the search scope [19]—the larger the keyword node, the higher the frequency of occurrence. For example, design, virtual reality, simulation, garments, virtual garments, garments design, clothes simulation, and garments simulation appear highly frequently. However, because these words are subject terms included in the search formula, it is not meaningful to analyze them individually. Other high-frequency keywords reflecting the field of clothing design include computer-aided design, interactive design, augmented reality, sensory evaluation, ease allowance, patterns, garment design, virtual fitting, collision detection, numerical simulation, garment patterns, and so on. Due to the high frequency of these keywords, they are likely the research hotspots in this field at a particular stage. From Fig. 3, it can be found that the keywords that appear later than 2017 on average in the WOS data are mainly collaborative design, disabled people, virtual fitting, atypical morphology, enhancing accuracy,

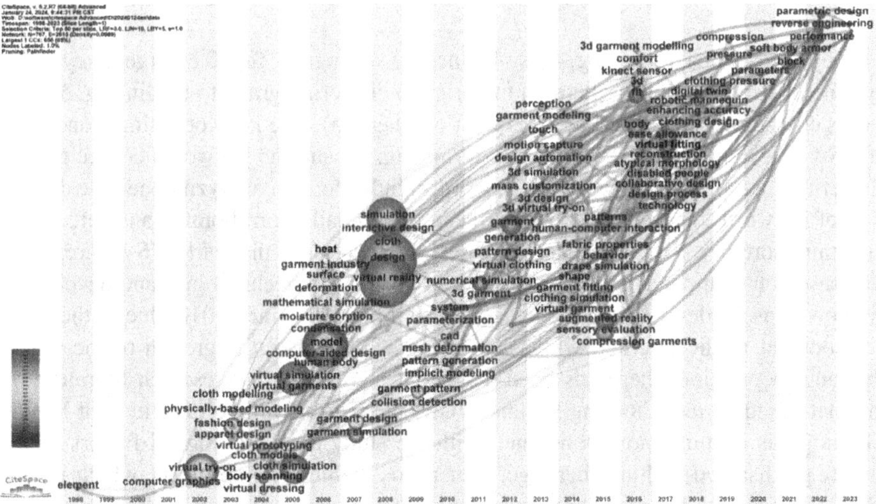

Fig. 3. Keywords time zone view of VS-CD.

reconstruction, digital twin, customization, and so on. The high-frequency keywords that have appeared recently are parametric design, reverse engineering, soft body armor, sustainable fashion, ancient costume, and so on.

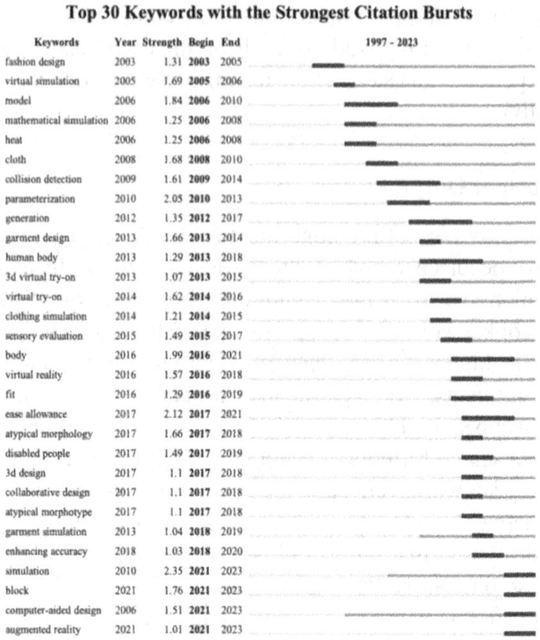

Top 30 Keywords with the Strongest Citation Bursts

Keywords	Year	Strength	Begin	End	1997 - 2023
fashion design	2003	1.31	**2003**	2005	
virtual simulation	2005	1.69	**2005**	2006	
model	2006	1.84	**2006**	2010	
mathematical simulation	2006	1.25	**2006**	2008	
heat	2006	1.25	**2006**	2008	
cloth	2008	1.68	**2008**	2010	
collision detection	2009	1.61	**2009**	2014	
parameterization	2010	2.05	**2010**	2013	
generation	2012	1.35	**2012**	2017	
garment design	2013	1.66	**2013**	2014	
human body	2013	1.29	**2013**	2018	
3d virtual try-on	2013	1.07	**2013**	2015	
virtual try-on	2014	1.62	**2014**	2016	
clothing simulation	2014	1.21	**2014**	2015	
sensory evaluation	2015	1.49	**2015**	2017	
body	2016	1.99	**2016**	2021	
virtual reality	2016	1.57	**2016**	2018	
fit	2016	1.29	**2016**	2019	
ease allowance	2017	2.12	**2017**	2021	
atypical morphology	2017	1.66	**2017**	2018	
disabled people	2017	1.49	**2017**	2019	
3d design	2017	1.1	**2017**	2018	
collaborative design	2017	1.1	**2017**	2018	
atypical morphotype	2017	1.1	**2017**	2018	
garment simulation	2013	1.04	**2018**	2019	
enhancing accuracy	2018	1.03	**2018**	2020	
simulation	2010	2.35	**2021**	2023	
block	2021	1.76	**2021**	2023	
computer-aided design	2006	1.51	**2021**	2023	
augmented reality	2021	1.01	**2021**	2023	

Fig. 4. Keywords Burst Term.

Keywords Burst Term (Fig. 4) is obtained by sorting the Top30 emergent keywords by time and superimposing them on the original clustering graph to obtain Fig. 5 (keywords co-occurrence clustering superposition diagram). The keywords time zone view of VS-CD (Fig. 3), the keywords burst term (Fig. 4), and the keywords co-occurrence clustering superposition diagram (Fig. 5) are all indicators for analyzing the time dimension of keywords, and the three of them can be mutually corroborated and referenced to obtain more objective and accurate results. It can be seen that early VS-CD research was based on Computer graphics, such as proposing new technologies and developing systems to assist designers in the apparel design process. Since 2010, due to the rapid development of the Internet, which led to the beginning of the growth of the online shopping population, there has been a gradual increase in the research literature on Internet-based Virtual Clothing Rooms (VCRs) and virtual fittings. The field of Virtual Clothing has become more concerned with the study of psychological factors of the user (e.g., disabled, cultural heritage, user needs, habits, and comfort), which is also in accordance with the results obtained from the time-zone diagram in Fig. 3. A comprehensive analysis of the time zone map and high-density emergent words shows that future research in the international academic community focuses on reverse engineering, computer-aided design, and integration of artificial intelligence, which has a solid cross-disciplinary integration characteristic.

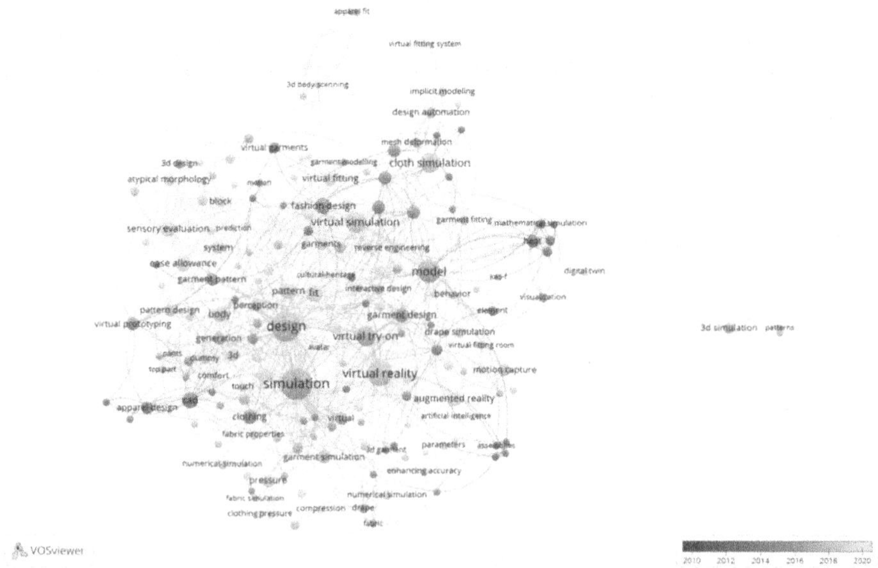

Fig. 5. Keywords co-occurrence clustering superposition diagram.

3.4 Distribution of VS-CD Literature Sources

Using the VOSviewer (V6.1.R3) software allows it to analyze the number of citations and papers from the sources, which can help the searcher know which journals/conferences are the most contributing literature sources in the VS-CD research area. According to the statistics, 241 publications were published in the valid literature within the search. Table 2 lists the top five high-impact journals/conferences in descending order of total global citations. The number of citations is the sum of the total number of citations to the articles published within the journal search, while the average number of citations is the average number of times an article published in a journal is cited [20]. Journals with high citation counts play an essential role in the relevance of research and provide a large number of references for scholars in the field. From the number of documents in Table 1, computer-aided design has the highest number of citations in this field of research, with a total of 453 citations and 11 papers published. The following highest number of citations is the International Journal of Clothing Science and Technology, which published 25 papers with 230 citations. In third place was computers in industry, with four papers published and 155 citations. Textile Research Journal published 13 papers with a total of 150 citations. The fifth place is the computer graphics forum, which has 116 citations. There is also a relative concentration of journals publishing in this field, with the top 5 journals accounting for 20.24% of the publications.

These journals/conferences are from different research areas. According to the WOS disciplinary statistical analysis system, 241 documents are related to 32 significant disciplines, such as Art, Computer Science, Materials Science, Engineering, Automation Control Systems, Science Technology, Other Topics, and others, which are significant research areas for VS-CD. Provide theoretical basis and methodological tools for VS-CD research. By analyzing the journals from which the literature is sourced, the VS-CD key research topics have solid interdisciplinary characteristics involving knowledge from multiple subject areas. In addition, Table 2 shows that the literature published in these journals has high citation counts, indicating that they provide essential reference content with significant impact in the field.

Table 2. Distribution of VS-CD literature source (top 5)

Ranking	Literature source	publications	citations	Average number of citations
1	computer-aided design	11	454	41.2727
2	international journal of clothing science and technology	26	234	9.2
3	computers in industry	4	155	38.75
4	textile research journal	13	150	11.5385
5	computer graphics forum	2	116	58

3.5 High-Impact Institutions Analysis

The amount of literature output of research institutions in the field of intelligent textiles and clothing plays a certain reference role in judging the influence of the institutions. A total of 261 international research institutions have conducted research related to VS-CD from 1997 to 2023, and the co-occurrence analysis of research institutions was conducted through the CiteSpace tool to obtain the core institutions in the field of VS-CD, as shown in Table 3. The top five core institutions are all universities. In terms of the total number of citations, the top five research institutions are korea adv inst sci & technol (352 citations), hong kong polytech univ (302 citations), donghua univ (293 citations), in order, ensait (292 citations) and univ geneval (228 citations). Regarding publications, donghua univ ranked first with 22 articles, and Hong Kong Polytech University ranked second with 17 articles. They are followed by Soochow univ (16 citations), ensait (12 citations), and xian polytech univ (12 citations).

Table 3. High-Impact Research Institutions (top 5)

Ranking	institutional	publications	citations	average number of citations
1	korea adv inst sci & technol	4	352	88
2	hong kong polytech univ	17	302	17.7647
3	donghua univ	22	293	13.3182
4	ensait	12	292	24.3333
5	univ geneva	5	228	45.6

3.6 High-Impact Author Analysis

Through author co-citation, we can grasp the mainstream direction of the research field through the high-quality literature published by core authors and collaborative research topics. A total of 651 authors within the search contributed to the research area of VS-CD. Table 3 lists the top five authors with the highest scores in terms of literature production and their respective total citations. According to the statistics, the top author in terms of the number of publications, Liu Kaixuan, with 19 publications and totaling 307 citations, focuses his research in the field of intelligent textiles and apparel on the digital restoration of historical garments, and his research area is materials science and computer science. The second-ranked author, Zeng, Xianyi, with 16 publications and 244 citations, focuses on engineering materials science and computer science. Bruniaux, Pascal has 16 publications and 231 citations. The fourth and fifth-ranked authors are magnenat-thalmann, n and Hong Yan, who published 4and 12 papers with 215 and 192 citations, respectively. The total citations of the above authors in the search scope are also among the highest in the world, and these high-impact authors are regarded as pioneers and essential contributors to the research of VS-CD. According to the research themes of the core authors, computer-aided design, garment simulation, and reverse engineering research are the mainstream directions in the field (Table 4).

Table 4. TOP5 High-Impact authors (top 5)

Ranking	Author	publications	citations	average number of citations
1	liu, kaixuan	19	307	16.1579
2	zeng, xianyi	16	244	15.25
3	bruniaux, pascal	16	231	14.4375
4	magnenat-thalmann, n	4	215	53.75
5	hong, yan	12	192	16

Author co-occurrence analysis was performed using CiteSpace software to obtain a core author collaboration mapping in the field of VS-CD, as shown in Fig. 5. In the mapping, the size of the nodes indicates the amount of literature published by the authors, and the connecting lines between the nodes indicate the cooperation between the authors. Overall, the core authors are relatively closely connected, and the collaborative research of the remaining authors is mainly carried out within a small team (Fig. 6).

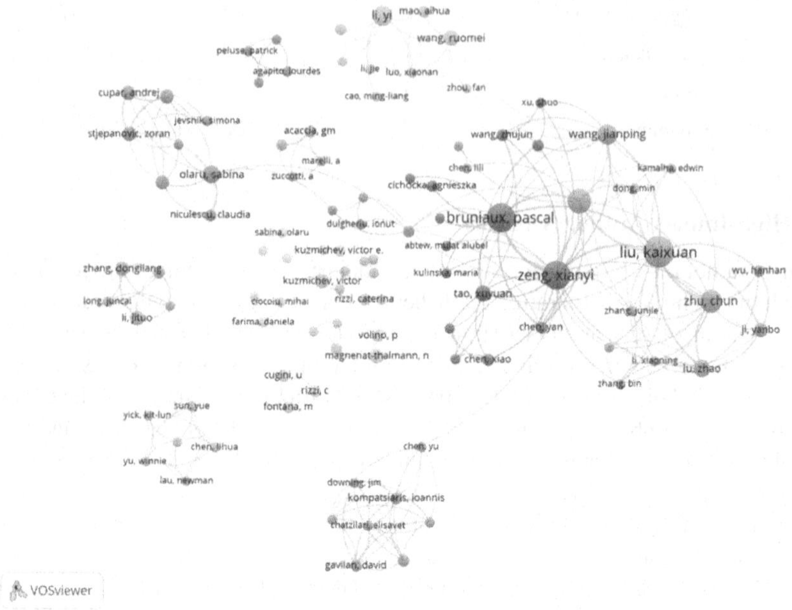

Fig. 6. Authors cooperation network.

4 Discussion

1. The research results show that the application of virtual simulation in apparel design has received more and more attention in recent years. The output of VS-CD research literature has fluctuated significantly in the past five years. However, the overall trend has been upward, reflecting the fact that there is still a significant research space in virtual simulation technology in clothing design.
2. Studying the performance of journals and institutions by counting the distribution of VS-CD literature sources enables the identification of journals and institutions with a very high academic impact in the field of virtual clothing design research. Scholars can be guided on which journals or partner institutions to choose for their submissions.
3. Liu Kaixuan, Zeng Xianyi, bruniaux, Pascal, magnenat-thalmann, n, and Hong Yan are the core authors in the international community; the author found that scholars with a high number of publications have established good cooperation with other scholars, and the rest of the authors' collaborative research is mainly carried out within small

teams. The remaining authors' collaborative research is mainly carried out within small-scale teams. In recent years, interdisciplinary and cross-unit collaborations have begun to increase, which is not only conducive to the output and depth of research but also helps to promote the progress of scientific research. Counting the core authors in the field of VS-CD can provide guidance to researchers in choosing which scholars to follow and co-author, thus facilitating communication and collaboration among researchers in the community.

4. Through keyword clustering, VS-CD research is comprehensive and diversified, and the research hotspots can be categorized into five main research clusters, which are #1 theoretical approach and cultural communication, #2 technology implementation, #3 technology development, #4 user experience and fashion design, #5 user research and healthcare design. The research on VS-CD in the international academic community emphasizes user research and presents a solid cross-disciplinary integration characteristic.
5. By analyzing the frequency of keywords, we can know the research areas popularized by VS-CD in different periods and the future research trends, which will help researchers grasp the dynamics of the research field more quickly. In the early period, the research was extensive and focused mainly on the static simulation of clothing and new technology research. In recent years, the high-frequency keywords are collaborative design, enhancing accuracy, disabled people, parametric design, reverse engineering, cultural heritage, and so on. These are also essential research combinations in the field of virtual simulation in clothing design. The keywords time zone view of VS-CD, keywords burst term, and the keywords co-occurrence clustering superposition diagram can be predicted that the future research focus of the international academic community will be concentrated on computer-aided design, integration of artificial intelligence, augmented reality, and reverse engineering, and so on;

Overall, there is still a significant research space in the field of VS-CD, and scholars in the field still have a long way to go. It is necessary to strengthen exchanges and cooperation among institutions and increase scientific research's strength, depth, and breadth. Finally, the field of VS-CD is characterized by both natural disciplines and social sciences, so scholars should have a broad research vision, strengthen interdisciplinary exchanges among scholars, have more contact with interdisciplinary knowledge, focus on the integration of interdisciplinary research, and give full play to the strengths of various disciplines, to provide more research ideas and research methods for the study of VS-CD.

5 Conclusion

Based on a dataset of 241 documents obtained from WoS for core studies on VS-CD from 1997 to 2023, this study conducted a rigorous and comprehensive bibliometric analysis using CiteSpace with VOSviewer software. The bibliometric performance analysis is the main focus of journals, institutions, authors, keywords, and co-word network analysis, with its main contributions: 1. Identify the significant institutions and journals in the research field and their research areas 2. Identify the core authors in the research field and the core authors' collaborative relationships. 3. Identify the key research hotspots and trends in the field and propose potential research directions and opportunities.

By analyzing the current status of the research on virtual simulation technology in apparel design and the trend, we can grasp the current situation and hotspots more comprehensively and objectively and provide direct references for subsequent scholars' research. By analyzing the research status and trends of virtual simulation technology in apparel design, we can more comprehensively and objectively grasp the current situation and hotspots and provide direct references for the research of subsequent scholars. This study will help to promote further the development of virtual simulation technology in apparel design.

Despite the extensive bibliometric analysis, this study has some limitations. Only the WoS database was used as the primary source of research in this study, with searches limited to SSCI, SCI-Expanded, A&HCI, and CPCI-S. Therefore, both the coverage and the number of public are limited. Secondly, this study includes only English papers, which may underestimate the trend and research conducted in other languages [21].

Disclosure of Interests. The authors have no competing interests to declare that are relevant to the content of this article.

References

1. Vitali, A., Rizzi, C.: Acquisition of customer's tailor measurements for 3D clothing design using virtual reality devices. Virtual Phys. Prototyping **13**(3), 131–145 (2018)
2. Mongeon, P., Paul-Hus, A.: The journal coverage of web of science and Scopus: a comparative analysis. Scientometrics **106**, 213–228 (2016)
3. Tan, H., et al.: User experience & usability of driving: a bibliometric analysis of 2000–2019. Int. J. Hum.-Comput. Interact. **37**(4), 297–307 (2021)
4. Fahimnia, B., Sarkis, J., Davarzani, H.: Green supply chain management: a review and bibliometric analysis. Int. J. Prod. Econ. **162**, 101–114 (2015)
5. Cao, C., Gao, Y.: Research on national costume design based on virtual reality technology. Math. Probl. Eng. **2022**, 7503167 (2022)
6. Liu, K.X., et al.: Archaeology and virtual simulation restoration of costumes in the Han Xizai banquet painting. Autex Res. J. **23**(2), 238–252 (2023)
7. Wijnhoven, M.A., Moskvin, A.: Digital replication and reconstruction of mail armour. J. Cult. Herit. **45**, 221–233 (2020)
8. Liu, J.Z., Li, J.T., Lu, G.D.: Deformation similarity clustering based collision detection in clothing simulation. Int. J. Clothing Sci. Technol. **26**(5), 395–411 (2014)
9. Cheng, Z., Kuzmichev, V., Adolphe, D.: A digital replica of male compression underwear. Text. Res. J. **90**(7–8), 877–895 (2020)
10. Liu, K.X., et al.: Optimization design of cycling clothes' patterns based on digital clothing pressures. Fibers Polym. **17**(9), 1522–1529 (2016)
11. Cheng, Z., et al.: Research on the accuracy of clothing simulation development: the influence of human body part characteristics on virtual indicators. Appl. Sci.-Basel **13**(22), 12257 (2023)
12. Wang, Z.J., et al.: Design of customized garments towards sustainable fashion using 3D digital simulation and machine learning-supported human-product interactions. Int. J. Comput. Intell. Syst. **16**(1), 16 (2023)

13. Holte, M.B.: The virtual dressing room: a perspective on recent developments. In: Shumaker, R. (ed.) VAMR 2013. LNCS, vol. 8022, pp. 241–250. Springer, Heidelberg (2013). https://doi.org/10.1007/978-3-642-39420-1_26
14. Brooks, A.L., Brooks, E.: Towards an inclusive virtual dressing room for wheelchair-bound customers. In: International Conference on Collaboration Technologies and Systems (CTS). Minneapolis, MN (2014)
15. Mosleh, S., et al.: Developments of adapted clothing for physically disabled people with scoliosis using 3D geometrical model. Appl. Sci.-Basel **11**(22), 10655 (2021)
16. Hong, Y., et al.: Interactive virtual try-on based three-dimensional garment block design for disabled people of scoliosis type. Text. Res. J. **87**(10), 1261–1274 (2017)
17. Hong, Y., et al.: Visual-simulation-based personalized garment block design method for physically disabled people with scoliosis (PDPS). Autex Res. J. **18**(1), 35–45 (2018)
18. Hong, Y., et al.: Virtual reality-based collaborative design method for designing customized garment for disabled people with scoliosis. Int. J. Clothing Sci. Technol. **29**(2), 226–237 (2017)
19. Chen, C., Dubin, R., Kim, M.C.: Emerging trends and new developments in regenerative medicine: a scientometric update (2000–2014). Expert Opin. Biol. Ther. **14**(9), 1295–1317 (2014)
20. Wang, J., et al.: Research hotspots and trends of social robot interaction design: a bibliometric analysis. Sensors **23**(23), 9369 (2023)
21. Tran, B.X., et al.: The current research landscape of the application of artificial intelligence in managing cerebrovascular and heart diseases: a bibliometric and content analysis. Int. J. Environ. Res. Public Health **16**(15), 2699 (2019)

Improving User Experience and Service Efficiency

An Investigation of Readability, User Engagement, and Popularity of E-Government Websites in Saudi Arabia

Obead Alhadreti(✉)

Umm Al-Qura University, Mecca, Kingdom of Saudi Arabia
oghadreti@uqu.edu.sa

Abstract. This paper presents the results of a study that aimed to examine the readability, user engagement and popularity of e-government websites in the Kingdom of Saudi Arabia using three evaluation tools, namely the readability calculator, Similarweb and CheckPageRank. The results show that the majority of the targeted websites fell short in the three dimensions. The readability analysis showed that the mean score of the FRE was 48.62 and the FKGL was at ninth-grade level, which indicates a difficult reading level. Only 30% of the websites had a score below or equal to 7 on the SMOG index. The results also reveal that all of the websites had a bounce rate above 30%, indicating that users are not engaging positively with the websites evaluated. The results of CheckPageRank showed that 85% of the websites had scores below or equal to 5, which suggests a lower page ranking than expected. Interestingly, there is a statistically significant relationship between the readability index scores and the bounce rate, implying that users engage more with the websites that have better reading levels. The study's findings suggest that the web readability, user engagement and popularity of e-government websites in Saudi Arabia should be enhanced to make them more usable for all their potential users.

Keywords: Website Evaluation · Readability · User Engagement · Popularity · E-government

1 Introduction

E-government have made accessing government services and information easier than ever before. It can be defined as "the delivery of government services over the internet in general and the Web in particular" [30]. The government departments which have embraced digital technology have greater efficiency and improve a country's international competitiveness significantly. Therefore, many countries are positioning themselves to take advantage of these novel opportunities to help them transform complex bureaucracies into responsive, innovative, citizen-focused operations [20].

In Saudi Arabia, the government has an on-going Digital Transformation Programme for modernising the infrastructure of information and communications technology (ICT) in the country. This is one of the key goals of the Saudi Vision 2030, which aims to

F. F.-H. Nah and K. L. Siau (Eds.): HCII 2024, LNCS 14721, pp. 133–148, 2024.
https://doi.org/10.1007/978-3-031-61318-0_10

develop e-government as a means of improving its economy. It seeks to take advantage of the fourth industrial revolution [31] to facilitate the growth of a flourishing digital society and turning the Kingdom into a globally competitive ICT hub. Vision 2030 has 13 Vision Realisation Programmes and 96 strategic objectives, which are intended to improve the life of its citizen. As part of this, several initiatives have been taken by the Kingdom's Ministry of Communications and Information Technology (MCIT) to embed digital technology in the infrastructure of government. The Saudi National Portal currently offers over 2500 government services. The Kingdom has achieved several milestones with its programme of digital transformation to serve citizen and attract foreign investment [19].

For an e-government website to be successful, it needs to be well-organized, accessible, easily comprehended, engaging and able to address users' specific needs [2]. However, little is known about the current state of the readability, user engagement and popularity of e-government websites in the Kingdom of Saudi Arabia. This study contributes by filling this knowledge gap and raises many implications for success in e-governance. The rest of this paper is structured into the following sections. Firstly, a review of the current literature in the field of readability, user engagement, and popularity of government websites will be presented. The chosen methodological approach is then discussed; the data analysis is examined and the results presented. A brief conclusion is given, along with suggestions for future work.

2 Related Work

2.1 Readability Evaluations of E-Government Websites

Readability has been defined by Klare [14] as "the ease with which a reader can understand a document due to the style of writing". Idler [10] proposed eight guidelines for improving web readability, on which developers and designers should focus. These eight guidelines are shown in Table 1.

Table 1. Eight guidelines [10].

ID	Guidelines
1	Choose fonts wisely
2	Font size and line spacing are important
3	Use high contrast
4	Keep the lines short
5	Keep paragraphs short
6	Get straight to the point
7	Don't use jargon
8	Use lists, images and highlights

E-government website readability has been assessed by many studies. In Turkey, Akgül et al. [2] used six indices to investigate the content readability of e-government websites at local (247 sites) and state (77 sites) levels. The websites complied with generally accepted standards for their accessibility, performance, heuristic usability and mobile readiness, but had made many mistakes in terms of accessibility, usability, readability and quality. Morato et al. [18] used an automatic process to assess the readability of e-government websites in Spain concerned with administrative procedures. The analysis considered the various linguistic characteristics which are thought to provide clearer explanations of the relevant resources and revealed that the web pages assessed were difficult to comprehend. The study went on to propose the development of an official Spanish-language corpus of e-government websites.

An Indian study looked at the site-ranking, accessibility and readability of the 20 top-ranked government websites in the country, using six indices to measure readability, which showed scores in the range of acceptability [11]. Misra et al. [17] undertook a readability analysis of 17 educational websites focused on healthcare, using a tool called Readability Studio Professional. The results showed that the information provided on the sites exceeds the reading level of the average American. Another study analysed the readability of 121 patient-orientated website articles related to spinal issues [28]. The study recommended that the readability of texts aimed at patients should be at a level below that of sixth grade, the recommended readability level for adult patients in the United States. Kumar et al. [15] analysed the readability of patient education materials (PEM) in 2016. They also compared content readability of online PEM between those with institutions with and without fellowships, where the average Flesh Kincaid grade was 13.8 for the former and 10.8 for the latter. Furthermore, the readability score was higher than the recommended grade of six or lower. As far as this author is aware, no Saudi government websites have been subjected to readability evaluation.

2.2 User Engagement Evaluations of E-Government Websites

User engagement is a key concept in designing user-centered websites and applications. It refers to the "quality of the user experience that emphasizes the positive aspects of the interaction, and in particular the phenomena associated with being captivated by technology" [3]. This definition derives from the idea of successful technologies being engaged with rather than simply being used [3].

A number of studies has analysed how users engage with websites of government agencies. In one study [1], the Alexa online tool was used to assess Indian e-government websites, giving a rank and a bounce rate for each site. The study found that approximately 44% of the evaluated websites had a bounce rate lower than 40%, 28% were average, and 32% above average. Furthermore, only 22% had bounce rates within an acceptable range and 78% failed to engage their users. A Malaysian study [5] explored evaluation measures for information-based municipal websites to assess their user engagement and credibility, revealing that there is work still to be done to make the country's e-government portal engaging and usable. Another study has shown local government websites to have low usage and that information quality is a key factor in the level of utilisation of government websites [7]. A review article [29] found that usability was the critical factor in website use across a range of studies. Website evaluations have usually found at least one accessibility or usability problem, such as speed and broken links or page-not-found errors [16]. Other studies have shown that broken links reduce the credibility of a website [9]. O'Brien and Cairns [22] developed a framework, known as the user engagement scale (UES), which measures perceived usability and user-involvement, novelty, aesthetics, attention focus and "endurability" (here meaning overall success in use and likelihood to return or recommend the site to others).

The process of determining the effectiveness of e-government website content has also suggested that features likely to promote citizen engagement might include frequent podcasts, with updates initiatives and opportunities for engagement in the local area, public consultation calendars, maps showing planned events, comprehensive information on how local government functions, and links to social media feeds [6]. To this author's knowledge, Saudi government websites have not yet been evaluated for user engagement.

2.3 Popularity Evaluations of E-Government Websites

The popularity of websites refers to the ranking of the websites [12]. The Alexa tool was used to find the global rank of Indian e-government websites in India in [1], giving an average ranking of 599,025.77. 40% (66 of the 164) websites studied, had an Alexa ranking lower than 100,000, and the remaining 60% than 100,000, which is taken as a global threshold ranking. This inferior ranking shows that service delivery by the Indian government's web portals has a low popularity in the population. The highest ranking was achieved by the website of the income tax department (444th) and the lowest (6,625,034th) by the online application system used by medical colleges.

Jati and Dominic [13] assessed e-government website quality in five Asian countries using online web diagnostics tools, showing that these websites were neglecting quality and performance criteria. According to the popularity of links used from the Google search engine, the five countries were in the following order: Hong Kong, Singapore, Malaysia, Japan and Korea. Another study [11] compared the two forms of ranking available from the Alexa tool, i.e. National Ranking and Global Ranking, for the 20 top-ranked government websites in India. The results showed that the national and global rankings are highly correlated (0.947).

[23] performed a comprehensive analysis across 582 African websites providing e-government services, looking at website types, kind of services and features offered, and how well-developed e-government services were. In addition, the study calculated e-government indexes and e-government rankings. The results showed that there has been progress, but that much work is left to do in order to address North-South inequalities as well as disparities between regions. A cross-sectional study [12] of Arabic websites conveying information about breast cancer found that most performed poorly for all measures, including ranking.

2.4 Aims and Research Questions

As mentioned earlier, e-government websites in Saudi Arabia have received little attention in relation to their readability, user engagement, and popularity. This study, therefore, aims to investigate these three factors for Saudi e-government websites. The study also seeks to examine the existence and nature of any association between the measures of the three dimensions involved in the evaluation. As the first study of its kind, it is believed that the findings will assist decision-makers and top management in the government sector in formulating policies for success in e-government. This study addresses the following research questions.

Research Question 1 (RQ1). What is the current readability level for e-government websites in Saudi Arabia?
Research Question 2 (RQ2). What is the current user engagement level for e-government websites in Saudi Arabia?
Research Question 3 (RQ3). What is the current popularity level of e-government websites in Saudi Arabia?
Research Question 4 (RQ4). Is there any association between the readability, user engagement and popularity measures?

3 Methods

3.1 Website Selection

This study employs a cross-sectional analysis approach [12] to evaluate government websites in the Kingdom of Saudi Arabia. The targeted websites were identified via the Saudi Unified National Platform[1]. There are twenty-three ministry websites listed on the platform. However, due to security restrictions, three websites cannot be accessed by automated tools. Accordingly, only testable websites that were available in Arabic were considered. This means that the study sample is composed of twenty websites. The readability, user engagement and popularity of the homepage of each website was assessed using automated tools. The homepage was selected due to its importance to the website as a whole, since it is a gateway to all the services and information provided by the website [27]. Data collection about the websites began on 1st of October 2022 and took approximately one week to complete. The whole data collection took place within the one week to avoid gaps arising through sites or their pages being unavailable or subject to major modifications. The ministries' names were kept anonymised for reasons of confidentiality.

3.2 Readability Evaluation

The readability of the website materials was assessed using an online readability calculator tool developed by Online Utility[2]. This tool analyzes the text using different common, well-known formulae. Whilst it was designed to analyse text in English, the tool can be used with other languages, which is promoted on the website of the Online Utility. In addition, many other studies have used the tool with Arabic texts and proved its validity for such use [12].

There are many reading level algorithms available with this tool that can be used for determining how readable the content is and also give a useful indication of whether the content created is at the right level for the intended audience. Such formulae include

[1] https://www.my.gov.sa/wps/portal/snp/main.
[2] http://www.online-utility.org/english/readability_test_and_improve.jsp.

the Gunning Fog Index (GFI), the Coleman Liau Index (CLI), the Flesch Reading Ease (FRE), the Flesch Kincaid Grade Level (FKGL), the Automated Readability Index (ARI) and the Simple Measure of Gobbledygook (SMOG) [10]. In the current investigation, FRE, FKGL and SMOG were used due to their suitability for the analysis of the Arabic text. The other indices are not suitable for the Arabic language since they count the number of letters, and the Arabic written language is comprised of words made up of letters that are linked to each other, not like in English [12].

1. Flesch Reading Ease (FRE)
FRE is a straightforward method for finding the understandability of a text. FRE is a 100-point scale, the higher the score the more easily understood a text is [11]. The scoring is shown in Table 2. The formula for computing FRE is in Eq. (1):

$$FRE = 206.835 - 1.015 * \left(\frac{words}{sentences}\right) - 84.6 * \left(\frac{syllables}{words}\right) \tag{1}$$

Following this readability formula, shorter words and sentences give better text, with an acceptable reading ease score between 60 and 70. Table 2 shows a summary of a text's understanding status according to this formula for FRE.

Table 2. FRE: Summary of understanding status of text [11].

Readability score	Understanding status
90–100	Very Easy
80–89	Easy
70–79	Fairly Easy
60–69	Standard
50–59	Fairly Difficult
30–49	Difficult
0–29	Very Confusing

2. Flesch Kincaid Grade Level (FKGL)
This formula is an updated version of the FRE formula. The FKGL formula is in Eq. (2):

$$FKGL = 0.39 * \left(\frac{words}{sentences}\right) + 11.8 * \left(\frac{syllables}{words}\right) - 15.59 \tag{2}$$

The FKGL is designed to give as an output an education grade level according to the system used in the USA. Thus, if a text scores 7, it is considered readable by someone who has been in education for seven years. Thus, contrary to FRE, text with a higher FKGL score is more difficult to understand. A score above 12 using this formula should be treated as being the same as grade twelve. A score of 5.0 can therefore be assumed to indicate readability at primary school level and a score of, for instance, 7.4 indicates readability by a typical seventh grade pupil [11].

3. Simple Measure of Gobbledygook (SMOG)
SMOG provides a formula that is considered suitable for readers of secondary school age and which gives an output in terms of the grade system used in US schools. Thus, as with FKGL, a text scoring 7.4 can be read by the typical seventh-grader [11]. The SMOG index formula is given as Eq. (3):

$$SMOG = 1.0430 * \left(\sqrt{(30 * \frac{complexwords}{sentences})} \right) + 3.1291 \tag{3}$$

SMOG is contrary to FRE, as the higher it is the more difficult the text is to understand. For the readability level to be satisfactory, the FRE has to be ≥80.0, whereas both FKGL and SMOG have to be <7 [12].

3.3 User Engagement Evaluation

To examine user engagement with the evaluated websites, the Similarweb[3] tool was applied, and three measures were considered, namely bounce rate, pages per visit, and average visit duration.

The bounce rate is the percentage of website users leaving the having only visited one page and without initiating any events of activity. If the bounce rate is low, engagement may be better. A higher bounce rate could come about from a website loading slowly or having a large number of broken links. The Similarweb tool claims that average bounce rate was 30.5% across all industries worldwide in 2021 [25].

Pages per visit is a measure of the amount of content (in terms of web pages) viewed on a single visit by each user, as an average calculated from total page views divided by total visitors [26].

Average visit duration is a measure of the average amount of time spent on a website by a user in one session, calculated as the time from accessing the first page to leaving the last page viewed [24].

3.4 Popularity Evaluation

The dimension of popularity assesses the ranking of the websites. To do so, an online tool called CheckPageRank[4] was used to analyse Google's PageRank (PR). The most-visited websites have a PR of ten, while the least-visited websites have a PR of zero [12]. Google PR is one of the methods Google uses to determine a page's relevance or importance. Important pages are more likely to appear at the top of the search results. Similarweb was also utilised to find the targeted websites' ranking at both the national and global levels.

---

[3] https://www.similarweb.com/.

[4] https://checkpagerank.net/index.php.

4 Results and Discussion

4.1 Readability of the Evaluated Websites

The results of the readability analysis showed that the mean of the FRE was 48.62 (See Table 3). None of the websites scored above 90. 10% of the websites studied scored between 80 and 89, and 25% scored between 70 and 79. The most difficult website to read was W3, with an FRE score of −39.63. According to the FKGL, only 20% (W2, W10, W11 and W14) of the websites studied had the recommended score level of sixth grade and below, while 80% of the websites had a score above the seventh grade (≥7). The mean of FKGL was at the ninth-grade level, which indicates a challenging level of reading. Again, W3 was the most difficult website. As for the SMOG index score, 70% of the websites had a score above or equal to 7, and the mean score was 8.92, which also indicates a challenging reading level. Table 5 shows the detailed assessments of each website. Figure 1 also shows a representation of readability indices for each website. Overall, the results reveal that the evaluated websites are written at readability levels above the recommended reading levels. Therefore, ministries should carefully develop the content on their websites in a manner that is easy to read. These results are in line with [11], which indicated a poor readability of the government websites in India.

Table 3. Readability analysis of the websites.

	FRE	FKGL	SMOG
Mean	48.62	9.55	8.92
SD	33.31	5.53	3.69
Median	56.33	9.03	8.76
Minimum	−39.63	0.61	16.93
Maximum	88.46	20.32	2.65
<7 score	NA	20% (n = 4)	30% (n = 6)
≥7 score	NA	80% (n = 16)	70% (n = 14)
<80 score	90% (n = 18)	NA	NA
≥80 score	10% (n = 2)	NA	NA

NA Not Applicable.

4.2 User Engagement with the Evaluated Websites

Tables 4 and 6 present the engagement scores for the targeted websites. All the websites had a bounce rate that is above the worldwide average rate, which is 30% [25]. This may indicate that most users had a poor engagement with the sites, which may be attributed to the poor design of content. The average bounce rate was 45.16. The average number of pages browsed per visit was 5.65. The mean time spent on a visit was 4.61 min. Longer browsing time can indicate both positive or negative experiences. It may indicate that users are engaging positively with the site, or that they are basically lost [21]. However, in the current case, considering the readability and bounce rate results, it is more likely that the browsing time points to a negative experience. These results are in agreement with [5].

Table 4. Central tendency and variation statistics of the user engagement of the websites.

	% Bounce rate	Page per visit	Visit duration in minutes
Mean	45.16	5.65	4.61
SD	12.31	2.19	2.38
Median	42.16	5.03	4.17
Minimum	31.20	3.04	1.59
Maximum	67.77	10.82	11.00

Table 5. Readability Indices Score.

ID	Reading ease (FRE)*	Grade level (FKGL)	SMOG
W1	75.73	8.7	5.27
W2	76.73	3.46	4.84
W3	−39.63	20.32	15.25
W4	29.48	10.54	10.48
W5	54.90	9.36	9.13
W6	60.44	8.43	8.39
W7	67.59	7.19	9.81
W8	70.71	7.00	8.11
W9	65.87	8.65	8.04
W10	88.46	0.61	2.65
W11	73.61	1.29	5.86
W12	23.52	16.34	12.69
W13	55.54	11.79	10
W14	80.27	1.76	4.63
W15	−20.40	19.12	14.93
W16	50.10	10.68	9.72
W17	19.32	16.56	13.28

(continued)

Table 5. (*continued*)

ID	Reading ease (FRE)*	Grade level (FKGL)	SMOG
W18	33.74	11.32	10.45
W19	75.73	7.18	5.06
W20	57.12	10.64	8

* 100–90 very easy, 89–80 easy, 79–70 fairly easy, 69–60 standard, 59–50 fairly difficult, 49–30 difficult, 29–0 very confusing.

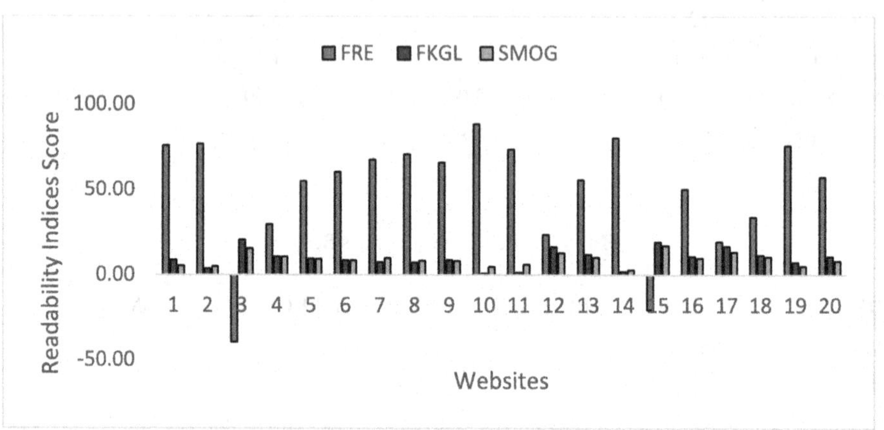

Fig. 1. Readability indices for the targeted websites.

Table 6. User engagement scores of the websites.

ID	% Bounce rate	Page per visit	Visit duration in minutes
W1	33.20	5.26	4.35
W2	31.20	10.82	10.13
W3	60.64	5.53	5.38
W4	55.16	4.99	4.22
W5	42.49	7.78	6.18
W6	40.81	8.68	11.00
W7	36.19	4.31	5.41
W8	34.07	8.5	5.37
W9	41.83	3.04	2.52

(*continued*)

Table 6. (*continued*)

ID	% Bounce rate	Page per visit	Visit duration in minutes
W10	31.31	3.13	3.54
W11	45.08	4.71	2.33
W12	62.80	4.12	3.19
W13	50.63	4.28	4.07
W14	32.96	3.63	2.20
W15	67.77	7.12	5.43
W16	56.59	5.07	3.49
W17	66.22	4.38	3.40
W18	42.79	8.42	4.11
W19	33.44	3.25	1.59
W20	37.96	5.9	4.28

4.3 Popularity of the Evaluated Websites

Tables 7 and 9 present the ranking for the evaluated websites. 85% of the websites had a Google ranking below or equal to 5, which suggests a lower than expected page ranking. Table 8 shows that three websites (W4, W6, W15) had rankings >5. The average ranking of the websites is 4.33. The average national and global ranking of the websites according to Similarweb are 2658.56 and 209,433 respectively, which also indicate a poor ranking. Only two websites (W6 and W15) were ranked among the top 20 websites on the national level.

Table 7. Central tendency and variation statistics of the ranking of the websites.

	Google ranking/10	National ranking	Global ranking
Mean	4.33	2658.56	209433.83
SD	0.69	7176.40	426374.89
Median	4.00	391.00	43726.50
Minimum	3.00	13.00	2307.00
Maximum	5.00	30954.00	1811483.00

4.4 Correlation Analysis

To explore the correlations between the different measures used in the current study, Spearman's correlation coefficient was utilised [8]. Table 8 shows that there is a statistically significant correlation between the three readability indices. All of the websites with FKGL or SMOG scores >7 had the lowest scores for FRE. Website W3 had the most difficult text content (FKGL score of 20.32, SMOG score of 15.25 and FRE score of –39.63), perhaps due to long texts with many lengthy sentences. In contrast, website W10 had the easiest text content (FKGL score of 0.61, SMOG score of 2.65 and FRE score of 88.46), mostly owing to the large number of short sentences, despite the text being long. Interestingly, there is a statistically significant relationship between the readability index scores and the bounce rate, suggesting that users engage more with the websites that have better reading level used (see Fig. 4, Fig. 5 and Fig. 6). Furthermore, there is a statistically significant relationship between pages visited and the time spent in the session, suggesting that the users who spent more time on a website visited significantly more pages (Figs. 2 and 3).

Table 8. Correlation analysis between the different measures used in the current study.

	FRE	FKGL	SMOG	% Bounce rate	Page per visit	Visit duration in minutes	Google ranking
FRE	1	−.906**	−.919**	−.892**	−.094	−.049	−.002
FKGL	−.906**	1	.923**	.847**	.040	.178	−.159
SMOG	−.919**	.923**	1	.898**	.093	.083	−.070
% Bounce rate	−.847**	.830**	.878**	1	−.075	−.085	.004
Page per visit	−.094	.040	.093	−.075	1	**.808	−.168
Visit duration in minutes	−.049	.178	.083	−.085	**.808	1	.036
Google ranking	−.002	−.002	−.070	.004	−.168	.036	1

Table 9. The ranking of websites

ID	Google ranking/10	National ranking	Global ranking
W1	4	1901	252.369
W2	4	703	2307.00
W3	5	193	27591.00
W4	6	106	19020.00
W5	3	347	38752.00
W6	7	13	19560.00

(continued)

Table 9. (*continued*)

ID	Google ranking/10	National ranking	Global ranking
W7	5	538	80407.00
W8	4	3774	429255.00
W9	4	238	48701.00
W10	4	2683	330215.00
W11	5	191	34623.00
W12	3	30954	1811483.00
W13	5	143	11066.00
W14	5	1481	107255.00
W15	7	19	3440.00
W16	5	127	21722.00
W17	4	435	72509.00
W18	4	4008	459534.00
W19	4	1464	185593.00
W20	4	3314	398385.00

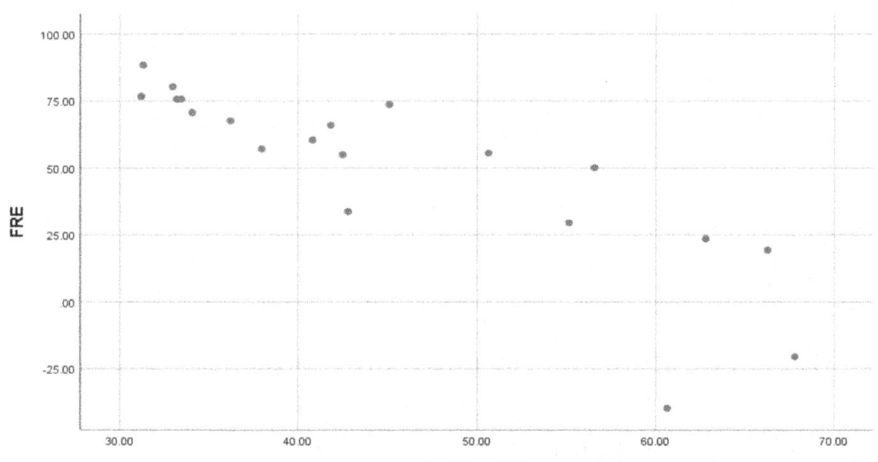

Fig. 2. Representation of the strong correlation between FRE and bounce rate

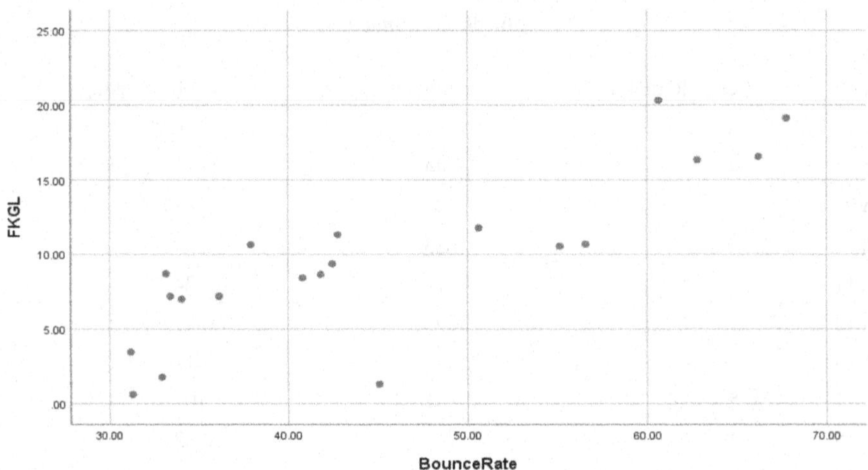

Fig. 3. Representation of the strong correlation between FKGL and bounce rate

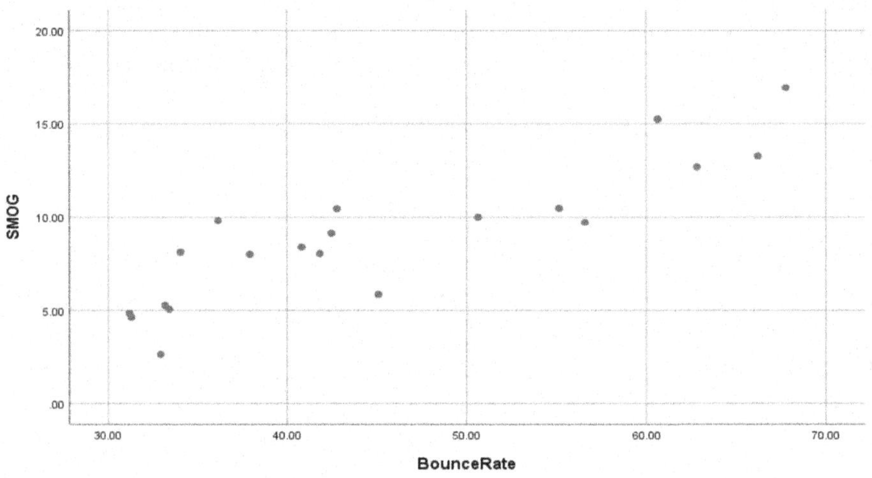

Fig. 4. Representation of the strong correlation between SMOG and bounce rate

5 Conclusions and Future Work

The aim of this study was to assess the readability, user engagement and popularity of e-government websites in the Kingdom of Saudi Arabia using three evaluation tools, namely the readability calculator, Similarweb and CheckPageRank. The results overall showed that the websites evaluated were unsatisfactory in all three dimensions. The readability analysis showed the mean score for FRE was 48.62 and the FKGL was at ninth grade level, which is considered a difficult reading level. The SMOG score for the websites showed only 30% at 7 or below. Every website showed a bounce rate over 30%, showing a lack of user engagement with them. The CheckPageRank results revealed that

85% of the sites scored equal to or below 5, i.e. a low page ranking. Overall, the findings of this study show a need to improve the readability, user engagement and page ranking of Saudi Arabia's e-government websites, so that they are usable by all potential users.

The study presented in this paper has some limitations that could be improved in future work. First, the study relied on automated tools to conduct the investigation. If the findings are to be conclusive, an element of human judgement is an important aspect of the evaluation procedure. Second, the study only considered the original language when examining the government websites. If the experience of users who are limited to a specific language is to be understood, work should be undertaken to assess the equivalent pages in Arabic and English. Third, only the homepages of the government websites were considered in the current study.

References

1. Agrawal, G., Kumar, D., Singh, M.: Assessing the usability, accessibility, and mobile readiness of e-government websites: a case study in India. Univ. Access Inf. Soc. **21**(3), 737–748 (2022)
2. Akgül, Y.: The accessibility, usability, quality and readability of Turkish state and local government websites an exploratory study. Int. J. Electron. Gov. Res. (IJEGR) **15**(1), 62–81 (2019)
3. Akgül, Y.: Evaluating the performance of websites from a public value, usability, and readability perspectives: a review of Turkish national government websites. Universal Access in the Information Society, pp.1–16. (2022)
4. Attfield, S., Kazai, G., Lalmas, M., Piwowarski, B.: Towards a science of user engagement (position paper). In: WSDM Workshop on User Modelling for Web Applications, pp. 9–12 (2011)
5. Bahry, F.D., Masrom, M., Masrek, M.N.: Website evaluation measures, website user engagement and website credibility for municipal website. ARPN J. Eng. Appl. Sci. **10**, 18228–38 (2015)
6. Bucci, A.W., Hulford, L., MacDonald, A., Rothwell, J.: Citizen Engagement: a catalyst for effective local government. Dalhousie J. Interdisciplinary Manag. **7**, 11 (2015)
7. Detlor, B., Hupfer, M.E., Ruhi, U., Zhao, L.: Information quality and community municipal portal use. Gov. Inf. Q. **30**(1), 23–32 (2013)
8. Dewberry, C.: Statistical Methods for Organizational Research: Theory and Practice. Routledge, New York (2004)
9. Dominic, P.D., Jati, H.: A comparison of Asian airlines websites quality: using a non-parametric test. Int. J. Bus. Innovation Res. **5**(5), 599–623 (2011)
10. Idler, S.: https://www.getfeedback.com/resources/ux/8-guidelines-for-better-readability-on-the-web/. Accessed 27 Jun 2022
11. Ismail, A., Kuppusamy, K.S., Kumar, A., Ojha, P.K.: Connect the dots: accessibility, readability and site ranking–an investigation with reference to top ranked websites of Government of India. J. King Saud Univ.-Comput. Inf. Sci. **31**(4), 528–540 (2019)
12. Jasem, Z., AlMeraj, Z., Alhuwail, D.: Evaluating breast cancer websites targeting Arabic speakers: empirical investigation of popularity, availability, accessibility, readability, and quality. BMC Med. Inform. Decis. Mak. **22**(1), 1–5 (2022)
13. Jati, H., Dominic, D.D.: Quality evaluation of e-government website using web diagnostic tools: Asian case. In: 2009 International Conference on Information Management and Engineering, pp. 85–89. IEEE (2009)
14. Klare, G.: The Measurement of Readability. Iowa State Uni- versity Press, Ames, Iowa (1963)

15. Kumar, G., Howard, S.K., Kou, A., Kim, T.E., Butwick, A.J., Mariano, E.R.: Availability and readability of online patient education materials regarding regional anesthesia techniques for perioperative pain management. Pain Med. **18**(10), 2027–2032 (2017)
16. Latif, M.H., Masrek, M.N.: Accessibility evaluation on Malaysian e-government websites. J. E-Gov. Stud. Best Pract. **2010**, 11 (2010)
17. Misra, P., Agarwal, N., Kasabwala, K., Hansberry, D.R., Setzen, M., Eloy, J.A.: Readability analysis of healthcare-oriented education resources from the American academy of facial plastic and reconstructive surgery. Laryngoscope **123**(1), 90–96 (2013)
18. Morato, J., Iglesias, A., Campillo, A., Sanchez-Cuadrado, S.: Automated readability assessment for Spanish e-government information. J. Inf. Syst. Eng. Manag. **6**(2), em0137 (2021)
19. Muzafar, S., Jhanjhi, N.Z.: Success stories of ICT implementation in Saudi Arabia. In: Employing Recent Technologies for Improved Digital Governance, pp. 151–163. IGI Global (2020)
20. Nadrah, R., Gambour, Y., Kurdi, R., Almansouri, R.: E-government service in Saudi Arabia. PalArch's J. Archaeol. Egypt/Egyptol. **18**(16), 21–9 (2021)
21. NNGroup. How to Interpret User Time Spent and Page Views. https://www.nngroup.com/videos/interpret-time-spent-page-views/. Accessed 25 Jul 2022
22. O'Brien, H., Cairns, P.: An empirical evaluation of the user engagement scale (UES) in online news environments. Inf. Process. Manag. **51**(4), 413–427 (2015)
23. Rorissa, A., Demissie, D.: An analysis of African e-Government service websites. Gov. Inf. Q. **27**(2), 161–169 (2010)
24. Similarweb. Average Visit Duration. https://support.similarweb.com/hc/en-us/articles/115000501485-Average-Visit-Duration. Accessed 20 Apr 2022
25. Similarweb. Bounce Rate Explained: Everything You Need to Achieve Growth. https://www.similarweb.com/corp/blog/research/market-research/bounce-rate/. Accessed 25 Apr 2022
26. Similarweb. Page Per visit. https://support.similarweb.com/hc/en-us/articles/360000802405-Pages-Per-Visit-the-web/. Accessed 20 Apr 2022
27. Vigo, M., Brajnik, G.: Automatic web accessibility metrics: where we are and where we can go. Interact. Comput. **23**(2), 137–55 (2011)
28. Vives, M., Young, L., Sabharwal, S.: Readability of spine-related patient education materials from subspecialty organization and spine practitioner websites. Spine **34**(25), 2826–2831 (2009)
29. Wang, F.: Explaining the low utilization of government websites: using a grounded theory approach. Gov. Inf. Q. **31**(4), 610–621 (2014)
30. West, D.M.: E-Government and the transformation of service delivery and citizen 813 attitudes. Public Adm. Rev. **64**(1), 15–27 (2004)
31. Xu, M., David, J.M., Kim, S.H.: The fourth industrial revolution: opportunities and challenges. Int. J. Finance. Res. **9**(2), 90–95 (2018)

The Implementation of Advanced AIS and the Accounting Data Quality: The Case of Jordanian SMEs

Esraa Esam Alharasis[1](✉) and Abeer F. Alkhwaldi[2]

[1] International Accounting Standards, Department of Accounting, College of Business, Mutah University, Karak, Jordan
esraa_alharasis@mutah.edu.jo

[2] Management Information Systems, Department of Management Information Systems, College of Business, Mutah University, Karak, Jordan

Abstract. This paper examines the effect of implementing advanced Accounting Information Systems (AIS) on Accounting Data Quality (DQ) in light of the Technology-Organization-Environment (TOE) theory. As organisations become more reliant on AIS to manage their financial data, it becomes essential to evaluate whether advanced AIS technologies contribute to greater DQ. This study seeks to investigate this relationship empirically by analysing data from 17,849 small and medium-sized (SMEs) Jordanian businesses. On the basis of the data collected via an online survey, the structural model was developed. The analysis of 412 completed surveys demonstrates that, among technological factors, "compatibility and relative advantage" have a significant impact on AIS implementation, "security and privacy" did not, however. AIS implementation is significantly influenced by all four organisational variables: top management support, firm size, IT-infrastructure, and employee IT-competency. Addressing the environmental constructs, it was confirmed that "Clients AIS Complexity" has a significant effect on AIS implementation. All hypothesised ties regarding the impact of AIS implementation on Accounting DQ were confirmed. The research concludes that AIS implementation is essential for organisations to maintain effective operation that can sustain productivity.

Keywords: Accounting information systems · technology-organization-environment theory · accounting data quality · small and medium-sized businesses · Middle East · Jordan

1 Introduction

In today's environment, AIS is an indispensable instrument for managers looking to maintain a competitive edge in the face of rapid technological innovation, increasing awareness, and tough expectations from consumers and business owners [1]. The purpose of this study is to assess the effect of AIS on Accounting DQ through a review of the empirical literature. The capacity of organisations to design and use computerised

F. F.-H. Nah and K. L. Siau (Eds.): HCII 2024, LNCS 14721, pp. 149–173, 2024.
https://doi.org/10.1007/978-3-031-61318-0_11

systems to track and record financial transactions has had the greatest influence on accounting in terms of management decision making, internal controls, and financial report quality [2]. Managers rely on AIS to stay ahead of the competition in the face of greater awareness, demanding consumers and business owners, and rapid technical improvement [3].

In order to aid and direct the decision-making process inside an organisation, an AIS may be defined as an electronic system that is computer-based and used to gather, store, process, and communicate accounting and accounting data through financial statements [4]. Since they provide the groundwork for all information systems to function, computers serve as the nerve centre of accounting data. In order for an AIS to function, the computer system that is going to be utilised must have the proper software programme installed. Accounting DQ is a pivotal moment in every organization's journey to success. To guarantee the efficacy of AIS on DQ and financial performance, several AIS have been implemented and utilised [5]. At this time, the majority of companies are allocating larger portions of their expenditures on information systems. Information cost constraints are also generated by competitiveness and economic conditions. To help individuals, government organisations, and parastatals do their jobs better, information systems are often built utilising IT [6].

Information systems are becoming an essential tool for accountants in today's fast-paced corporate world. Accounting nowadays is highly dependent on sophisticated information systems to offer real-time financial insights, improve accuracy and quality, and eliminate the need for human record-keeping and ledger entries [7, 8]. This article provides insight into the factors motives the adoption of AIS and its advantages and the development of this vital crossroads of technology and finance. Traditionally, accounting was a paper-based activity, typified by handwritten ledgers and laborious calculations. This method accomplished what it set out to do, but it was laborious, prone to mistakes, and could not provide fast access to accounting data [9]. As a result of the revolution brought about by the proliferation of computers and other digital technologies, AIS came into existence. In order to manage accounting data efficiently, AIS provides a complete framework that combines accounting principles with information technology. It comprises a range of operations, including data entry, processing, storage, and reporting [10]. Due to developments in technology, AIS has undergone substantial evolution over the years. Modern versions include Enterprise Resource Planning (ERP) software, cloud computing, and other high-tech software [11].

In order for stakeholders, investors, regulators, and companies to make informed decisions, high-quality accounting data is essential. Modern AIS systems may streamline financial processes, reduce human error, and enhance data quality. Thus, the purpose of this research is to determine whether the use of more sophisticated AIS results in a considerably higher level of Accounting DQ. As a result, the primary motivation for the current study is to provide empirical and theoretical evidence on the effect of implementing AIS on Accounting Data Quality (DQ) using the Technology-Organization-Environment (TOE) model in the context of Jordanian SMEs. This paper focuses on small and medium-sized enterprises (SMEs), which account for more than 97% of Jordan's businesses [12]. The optimal application of AIS by SMEs indicates a high level of competitiveness and successful adaptation to changing environments, which improves a company's dynamic nature (even a small company must incorporate AIS). That is, administrative management has advanced in the fields of accounting and finance. AIS allows organisations to use advanced statistical software to forecast future profits or assess the risk of specific operations. All of these benefits should be possible to extend to SMEs because they have been developed and tested in larger companies [3].

The earlier literature focused on AIS adoption, implementation possibilities, and users' perspectives on this adaptation [13, 14]. So far, no research has focused on the DQ in relation to AIS implementation, particularly in developing countries such as Jordan. Based on the literature review, we can differentiate the research objective between AIS technology and DQ in SMEs. Because of the importance of real-time reporting in reducing the gap between actual conditions and accounting reports, some researchers have focused on technology that can be used to develop AIS. The topic of AIS success has only gradually emerged in recent literature, with a few attempts to investigate the impact of AIS on DQ. However, there is still a significant gap to be drained, particularly in the culture of analytical decision-making, which contributes to the system's success in developing nations [1]. The DQ factors are incorporated into this study's model to fill this knowledge gap.

The goal of this study is to conduct the first empirical investigation into this relationship by analysing data from Jordanian SMEs. It is in response to recent call for additional research on AIS and Accounting DQ [15], which emphasised the importance of empirical findings. The researcher emphasizes the lack of such empirical evidence in the context of develop country, Middle Eastern (ME) region in particular. Hence, this study responds to that call. As a result, this analysis adds to the body of AIS knowledge in several ways. First, it creates an updated empirical model that measures the empirical evidence of AIS adoption on Accounting DQ using the TOE factors. Second, focusing on the context of ME region, like Jordan, a developing country with a high proportion of SMEs. Third, providing updated timeline analysis and results on TOE factors in AIS and their impact on Accounting DQ. Fourth, this study is the first of its kind to combine the TOE theory with the Contingency and Activity Theories in order to add a new contribution to the existing body of literature, which may be useful in settings with a low prevalence of the AIS framework [15, 16]. Fifth, given the rapidly changing nature of the global economy as a result of globalisation, this study is viewed as a preliminary lesson that can be applied in other ME contexts to implement AIS applications within their accounting framework [17].

Data collected through an online survey were used to develop the structural model. There are 412 surveys that indicate that the implementation of AIS is influenced by "compatibility and relative advantage" among technological factors. The effect of "security and privacy" factor was not confirmed. The support of top management, the size of the company, the IT infrastructure, and the IT competence of employees all have an impact on the implementation of AIS. It was demonstrated through environmental constructs that "Clients AIS Complexity" has an effect on AIS implementation. It was determined that each and every hypothesis concerning the impact of AIS implementation on Accounting DQ was correct. Because there are so few works that deal with the relationship between the application and use of AIS and performance and productivity indicators in SMEs in developing countries, this research offers a significant contribution to the field of accounting, both for practitioners and academics [18]. It is now possible to conduct additional research on this subject as a result of this analysis.

The paper is organised in the following way. A quick summary of AIS is provided in Sect. 2. In Sect. 3, the study's theoretical underpinnings are laid out. The theoretical structure is described in Sect. 3. The review of relevant literature and formulation of working hypotheses are detailed in Sect. 4. The study's methodology and data are presented in Sect. 5. Sections 6 and 7 contain the results and the discussion. Section 8 discusses the analysis implications to theory and practice. Section 9 concludes the paper and outlines the limitations, and potential directions for further study.

2 Accounting Information Systems Overview

A complete framework for an organization's financial and accounting data collection, processing, storage, and dissemination, an AIS integrates people, procedures, and technology. The main goal of an AIS is to help businesses with issues including financial reporting, regulatory compliance, and making decisions based on up-to-date, accurate information. The AIS is comprised of fundamental components. First, data entry and collection. Information systems enable the automated collection of financial data from a variety of sources, including point-of-sale terminals, bank statements, and invoices. Afterwards, this information can be entered into a database for further examination. According to [19], this makes it less likely that errors could be made during the process of data entry and ensures that the data could be accurate. Secondly, the processing of data. Following the completion of the data collection process, the AIS processes the information by utilising a number of different modules, including payroll, general ledger, accounts payable, and accounts receivable. Automation makes it possible to perform calculations, accruals, and reconciliations in a short amount of time [20]. This reduces the amount of manual labour that is required and the number of errors that can occur [3]. In the third place, data storage. Modern AIS makes use of encrypted databases in order to store monetary information. Consequently, this not only ensures that the data is accurate, but it also makes it easy to retrieve and examine the data whenever it is necessary [1]. The ability of the AIS to generate comprehensive financial reports and to carry out in-depth data analysis is perhaps one of the most valuable aspects of the system. The fourth aspect is reporting and analysis. Accountants have easy access to balance sheets, income statements, and cash flow statements, which enables them to make decisions that are based on accurate information [16].

The field of AIS is one that stands to gain a great deal from the employment of information systems. Since automation reduces the likelihood of errors caused by humans, it ensures that monetary data is entered and processed correctly. This is the most important benefit of automation. First, maintaining the integrity of the financial statements and improves the degree to which they comply with the standards of accounting [1]. Information systems make accounting procedures more efficient by accelerating the data entry, processing, and reporting of financial information. Second, this brings us to the effectiveness in accounting systems. Because of this, timely financial decisions and insights are produced as a result of the consequence of this. Putting in place an information system might require an initial investment; however, the cost reductions that are realised over the lifetime of the system are substantial. Third, AIS purposed to the term of cost reductions. According to [21], the overall cost-effectiveness of a company can be improved by lowering the costs of labour, lowering the error rates, and improving the efficiency of the business and its operations. Fifth, in order to protect financial data from unauthorised access and cyber threats, a modern AIS typically incorporates stringent security measures. This results in an increase in both data security and compliance. Sixth, accessibility. Cloud-based accounting systems allow for remote access to financial data, which makes it easier for teams that are located in different locations to collaborate with one another and provides real-time insights to stakeholders. Internet access is required in order to use accounting software that is hosted in the cloud. In the sixth place, compliance. Seventh, AIS plays a role in ensuring that legal requirements and accounting standards are adhered to by carrying out compliance checks and developing audit trails. One of the ways in which accounting information systems contribute to the maintenance of compliance is as described above [3].

3 Theory Foundation

3.1 TOE Theory

The research conducted by [22] and [23] demonstrates that the TOE framework offers a useful starting point for examining the utilisation of AIS. From the perspective of the TOE framework, there are three distinct types of characteristics that have an effect on the manner in which businesses utilise technology. To begin, the technical background serves to provide a description of the characteristics that are thought to be associated with innovation. Observability, relative advantage, complexity, compatibility, and trialability are some of the aspects that fall under this category. In accordance with the findings of a meta-analysis research project that was carried out by [22], the characteristics that have been taken into consideration in this study are the ones that are more frequently relevant and positively significant. These characteristics include compatibility and relative advantage. The second aspect of the "organisational context" is the quantity of unused resources that are available within the organisation. The third factor that is most frequently found to be essential in the utilisation and acceptance of innovations is the commitment of the owner or manager, as well as the readiness of the organisation being considered. During the course of this investigation, these aspects are going to be taken into consideration additionally. [22] state that the third factor is the "environmental context," which is defined as "the arena in which a firm conducts its industry and its business, competitors, and dealings with government." In other words, the environmental context is the environment in which a company operates. The dissemination of Innovation Theory (DOI), which was developed by [24], is consistent with this paradigm. When it comes to the factors that drive the dissemination of technology, DOI places a strong emphasis on the technological qualities and organisational characteristics of an organisation.

Through the utilisation of a different research approach that takes into account the impact of the utilisation of technology, the TOE framework has been further expanded. According to the Resource-based view (RBV), which states that businesses generate value and influence by integrating a variety of resources that are either economically difficult to copy or desirable to other businesses [25], the methodology is based on the rationale of the RBV. Furthermore, the ability of an organisation to capitalise on an invention is a more significant factor in determining the resource effect than the innovation itself [26]. This is according to research conducted by [26, 27]. To put it another way, the impact of innovation is determined by the extent to which innovation is used to support critical operations of the company's value chain. According to [28], the likelihood of a company developing a distinctive effect as a result of its invention increases in proportion to the extent to which the innovation is adopted by the company.

Authors [29–31] developed a body of research that focuses on the factors that lead up to innovation adoption as well as the results of its implementation. This methodology has resulted in a body of research that focuses on both of these aspects. Having said that, when taking into consideration AIS, it appears that a relatively small number of publications have concentrated their attention on this particular subject. Subsequently, the present study concentrates on the implementation and consequences of AIS, thereby bridging the gap in the current corpus of knowledge by establishing a correlation between the TOE conceptual framework and Accounting DQ. According to [32], the TOE is the primary conceptual framework that guided the majority of the earlier research that was conducted with the purpose of gaining an understanding of the factors that led to the implementation of IS/AIS. From this point forward, the RBV was being utilised in order to conduct an analysis of the effects of utilising IS in light of earlier studies.

3.2 Contingency Theory

This suggests that when developing an AIS, an organisation should adopt a flexible approach that accounts for the specific environment and organisational structure it encounters. Furthermore, the AIS must have the capability to adapt to the specific decisions that are being deliberated upon. In a different way of putting it, the AIS needs to be developed within the framework of an adaptive architecture. In the accounting literature, [33] was the first to specifically focus on the contingency view of AIS (A Contingency Framework for the Design of AIS). According to the findings of these researchers, the fundamental structure that must be utilised when approaching AIS from a contingent viewpoint was determined. [33] came to the realisation that environmental unpredictability is a primary motivation for the implementation of management accounting systems in organisations that have achieved success. Additionally, they observed that when decision makers perceive an increase in environmental uncertainty, they have a tendency to seek more information from external sources, non-financial sources, and ex-ante sources in addition to information from internal sources, financial sources, and ex-post sources. This is due to the fact that predictive accuracy of ex-ante information is higher than that of ex-post information. In the fundamental contingency framework, the variables of the environment and the factors that are specific to the institution are what make up the framework. These two categories of factors each have an impact on the competitive strategy that is employed. The extent of the information, the timeliness of the information, and the amount of aggregation that is done are the factors that determine whether or not the competitive strategy is successful. All in all, the effectiveness of AIS has a direct influence on the effectiveness of Accounting DQ and performance, and this influence is proportional.

3.3 Activity Theory

Activity theory is a conceptual framework that integrates human employment and technology, prioritising the well-being and contentment of users and employees as they go about their daily tasks and professional lives. The goal of action theorists is to develop work practices that are not only conducive to the development of skills and knowledge, but also enrich the work experience [34]. In order to accomplish this, they disprove the idea that there is "one best way" to complete a task and instead give preference to processes that are determined by the user instead. Those who subscribe to the activity theory argue that the degree to which a new piece of technology satisfies these objectives within the context of the user's own job is a significant factor in determining whether or not it will be accepted. For the most part, activity theory is consistent with the more general humanistic goals as well as the methodologies of the socio-technical approach. This is the case for the most part. According to [35], socio-technical systems thinking is distinguished from other types of systems thinking by the fact that it places an emphasis on the outcomes of organisational processes. This is the bare minimum that can be said to differentiate it from other types of systems thinking.

AIS and Accounting DQ can be better understood and researched with the help of a comprehensive framework that is provided by the integration of TOE theory, Contingency theory, and Activity theory. The TOE theory investigates the dynamic relationship that exists between technological factors, organisational characteristics, and an external environment. The theory of contingency places an emphasis on the necessity of aligning organisational structures and technologies in order to achieve desired levels of performance [3, 36]. Activity theory, on the other hand, is concerned with the ways in which human activities interact with technological processes and corporate procedures. It is possible to identify the technological factors that influence the adoption and effectiveness of AIS through the use of TOE theory in the context of AIS and accounting DQ research. On the other hand, Contingency theory investigates how organisational structures and contexts influence the design and implementation of AIS. The activity theory lends support to these viewpoints by putting an emphasis on the part that human activities play in the utilisation of AIS and in contributing to the quality of accounting data. A comprehensive understanding of the factors that influence the effectiveness of AIS and the quality of accounting information can be obtained through the use of this integrated approach, which enables researchers to investigate the dynamic relationships that exist between technology, organisations, and the environment [37].

This study aims to examine the effects and application (post-adoption phases) of AIS from an organisational viewpoint. Prior research on this subject has taken a two-pronged approach: first, it has examined the elements that influence decisions about the use of innovations; second, it has sought to understand both the causes and effects of these decisions. To account for the environment and organisational structure that an organisation faces, the contingency framework suggests developing an AIS in a flexible way. The components of the contingency framework include both the institutions and the aspects specific to the environment. The strategy employed by the competitors is impacted by these elements. The quantity, timeliness, and breadth of the aggregated data decide the viability of the competitive strategy. The performance and quality of financial reporting are potentially impacted by the effectiveness of the AIS. Accounting information consumers or workers' long-term welfare is the focus of activity theory, a way of understanding human labour and technology. Because of this, it can be linked to the Contingency Theory, another framework for comprehending technological and human endeavours.

All in all, this article addresses some of the most important aspects of advanced AIS, such as the integration of real-time data, automation, and data validation. To address the elements that affect AIS application choice and the impact of AIS on Accounting DQ, the conceptual framework of the current study integrates the TOE framework, the most popular theories in the realm of IS/IT [24], as well as the Contingency and the Activity Theories. In addition, the theories were modified in order to analyse the use of AIS and its impact on SMEs. This study explores the variables impacting the adoption of AIS by SMEs from diverse contexts, developing nations, Jordan in particular. It also investigates the influence of AIS usage on DQ. Figure 1, which provides a conceptual framework for the study. The model takes into account three different kinds of contexts for its elements: first, technologically related factors; second, organizationally related aspects; and third, environmentally associated ones [3].

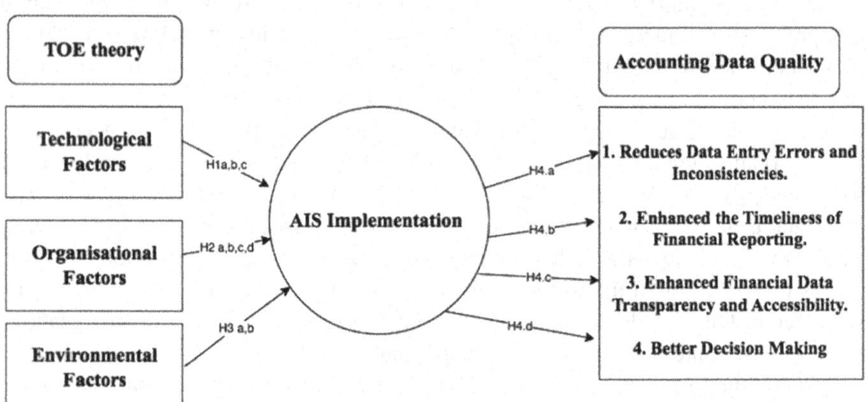

Fig. 1. Conceptual framework: Final proposed model

4 Hypotheses Development

4.1 Technological Factors

"Technological advancements" enhance the capabilities of the AIS. Strong encryption, access controls, and audit trails are useful and essential for protecting sensitive accounting information. Furthermore, compliance capabilities are a common feature of modern AIS, which helps businesses meet the standards set by regulatory agencies. Technical aspects reveal characteristics of the technology that could be used [38]. Innovative technologies, according to [24], have traits that might influence their uses. As an example, one definition of relative advantage is the degree to which a novel invention is believed to offer more advantages than its competitors. What sets AIS apart from other types of IT innovations is its ability to collect, store, and process accounting and financial data for the purpose of internal management decision-making [39]. When compared to other forms of IT innovation, AIS stands out due to its many other unique features. Both [40] and [41] state that relative advantage is the most important factor in information systems and the use of IT. The overall adoption of IS/IT is positively affected by relative advantage, according to [42–44].

Accounting data cannot be accessed or altered by unauthorised parties due to the implementation of security measures. This ensures the accuracy and reliability of the data stored in the AIS [45]. Secure and private AIS use may increase trust from customers, clients, and other stakeholders [9]. It shows that the company cares about keeping customers' private financial data safe. Security measures must be put in place to ensure that sensitive accounting information cannot be accessed by unauthorised parties [1, 46]. In order to keep financial records and transactions secret, this is essential. The security of users' private information and data is of the utmost importance in an effective AIS [47]. Organisations can make wise financial/business decisions and protect their assets when they contribute to data integrity, regulatory compliance, trust, and overall DQ. According to [3], neglecting security and privacy concerns can lead to data breaches, financial fraud, legal issues, and damage to one's reputation. Research by [48] and [49], among others, found that the "security risk" concept is directly related to the consequences and spread of technology.

The degree to which a novel technology is thought to be consistent with the established norms and values is known as compatibility [24]. In the context of AIS, compatibility refers to the extent to which existing processes align with the required procedures for implementing and utilising AIS. Technological incompatibility can hinder the utilisation of technologies and hinder the pace of innovation [50]. Therefore, the use of a technology necessitates the acquisition of new skills and procedures in order to effectively employ and execute it. Consistent with findings from [23, 31, 51], compatibility is considered the factor that exerts the greatest influence on innovation. According to [11], compatibility is thought to be the most important factor in motivation to use technology. [11] assert that implementing AIS is congruent with the current business environment and the organisational culture. This information is in accordance with the organisations. The level of advancement in organisations' AIS systems is one potential explanation. These organisations possess an infrastructure for IT that is already compatible with AIS, thus enabling them to utilise it. The existing IT/IS systems are compatible with and appropriate for use with the IT-infrastructure, as demonstrated by [47]. Drawing from the theoretical evidence mentioned earlier, the subsequent hypotheses have been formulated:

- Hypothesis (H1): Technological factors will positively influence firms' decision to implement AIS.
- *Hypothesis (1.a): Relative advantage will positively influence firms' decision to implement AIS.*
- *Hypothesis (1.b): Security and privacy will positively influence firms' decision to implement AIS.*

- *Hypothesis (1.c): Compatibility will positively influence firms' decision to implement AIS.*

4.2 Organizational Factors

Organisational factors, as stated by [52], refer to the characteristics of an organisation that influence its utilisation of technology. Internal factors are the elements within a company that directly influence its resource management policy and working environment [22, 53]. These factors are necessary for the company to accomplish its objectives and fulfil its missions. [51] assert that the level of preparedness within an organisation significantly impacts the adoption and utilisation of information systems and technology in SMEs. [54] found that numerous businesses lack the necessary expertise and financial resources to effectively implement AIS, especially those that involve complex technology. [2, 23, 31, 55] and [56] collectively determined that organisational preparedness is a crucial factor for the effective integration of technological advancements. Parallel to this, [41] concluded that companies with more readiness were more likely to use information technology and information systems efficiently. As posited by [22], top management support is an additional factor that could potentially be pertinent. The role that management plays in encouraging and facilitating the implementation of technology within businesses is the subject of this particular aspect. The findings of a number of studies indicate that this component has a significant influence on the utilisation of technology [31, 32]. Additionally, the findings of these studies indicate that this component plays a role in the utilisation of technology. Consequently, this article investigated government assistance as a possible indicator of AIS adoption among SMEs in Jordan [5].

"Owner/Manager Commitment" (OMC) is the degree of dedication, involvement, and support provided by management regarding the implementation and planning of technological systems within an organisation that guarantee staff members' use of those technologies [53]. The concept of OMC can be categorised into three distinct components: commitment, active engagement, and support. Ensuring the commitment of the owner or manager to effectively utilise all available resources is crucial for the successful implementation of AIS and overcoming the inherent resistance associated with the technology [57]. Given that decision-making typically rests with the owner or manager, it is crucial to ensure their dedication to effectively utilise all available resources [1, 9]. Prior research has shown that the effectiveness of IS/IT is greatly influenced by the level of dedication exhibited by the owner/manager [23]. Furthermore, research has demonstrated that the commitment of owners/managers has a favourable impact on the success of information systems/technology within SMEs, thereby making the commitment of the business's owner a vital factor in the utilisation of AIS [43, 52].

Higher readiness levels have been linked to a higher likelihood of AIS system use and implementation for businesses [1, 9]. The term "organisational readiness/IT infrastructure" pertains to the extent to which an organisation is prepared to efficiently implement and utilise AIS systems. This encompasses the IT infrastructure, IT resources, and IT leadership support of the organisation. Furthermore, it signifies that the strategic goals and objectives of an organisation are in harmony with the implementation of AIS. Typically, businesses that prioritise data-driven decision-making and recognise the importance of using information and analytics capabilities to gain a competitive edge tend to demonstrate higher levels of preparedness. By aligning the necessary information technology resources and support, the implementation of AIS is facilitated, thereby enhancing the project's probability of success [1]. Additionally, it was agreed that "employees' IT technical ability" had a substantial impact on AIS implementation [2]. The presence of IT professionals fosters a positive environment within the organisation on which the accounting department's personnel can depend in the event that they encounter any difficulties while utilising AIS. Drawing from the theoretical evidence mentioned earlier, the subsequent hypotheses have been formulated:

- Hypothesis (H2): Organizational factors will positively influence firms' decision to implement AIS.
- Hypothesis (2.a): Top management support will positively influence firms' decision to implement AIS.
- Hypothesis (2.b): Firm size will positively influence firms' decision to implement AIS.
- Hypothesis (2.c): IT-infrastructure will positively influence firms' decision to implement AIS.
- Hypothesis (2.d): Employees' IT competency (Skills and training) will positively influence firms' decision to implement AIS.

4.3 Environmental Factors

In accordance with the findings of [58], environmental factors comprise those external elements which are beyond the jurisdiction of SME management. One of the environmental factors is "competitive pressure," which denotes the degree of pressure that small and medium-sized enterprises (SMEs) encounter from their industry rivals. Organisations might be capable of exploiting novel approaches to surpass competitors, modify competition regulations, influence industry configuration, and consequently transform the competitive environment through the implementation of novel technologies [55]. This investigation could potentially be applied to AIS. Existing research indicates that a company experiences heightened pressure to adopt technology when an increasing number of its competitors in the same industry do so. As a result, the company feels compelled to do the same in order to maintain its competitive edge [59]. Consequently, the pressure exerted by competitors is a substantial factor that influences the adoption of AIS by SMEs.

As stated by [45], the concept of "competitive pressure" has an impact on the widespread use of information technology and information systems. It is not uncommon for organisations to face challenges in acquiring the resources they require, such as having limited budgets among other obstacles. Organisations may choose to prioritise investments in areas that are closely linked to their main objectives when faced with a shortage of resources. Under these conditions, "competitive pressure" might not be an adequate justification for allocating resources to the implementation of AIS, which can be expensive in terms of skilled personnel and the IT infrastructure in addition to the system itself.

The level of complexity of a client organization's accounting system is called its "clients AIS complexity" [22]. This determinant comprises the degree of implementation of advanced computerised reporting systems by the clients, in addition to the complexity, difficulty, and nature of the transactions that the AIS processes at their respective organisations. The nature of the tasks, the industry, and the business environment are additional factors that contribute to the complexity of utilising technology (such as AIS). Accountants in different companies may have to deal with complicated transactions that need the use of advanced and difficult AIS to be processed [5]. Drawing from the theoretical evidence mentioned earlier, the subsequent hypotheses have been formulated:

- Hypothesis (H3): Environmental factors will positively influence firms' decision to implement AIS.
- Hypothesis (3.a): Competitive pressure will positively influence firms' decision to implement AIS.
- Hypothesis (3.b): Clients AIS complexity will positively influence firms' decision to implement AIS.

4.4 The Implementation AIS on Accounting Data Quality

Automatic identification systems are used in conjunction with other activities, and the term "AIS usage" refers to the extent to which these systems are utilised. As a matter of fact, the ultimate goal of each and every company that implements AIS is to improve the overall efficiency of that particular company. Companies that make effective use of AIS are more likely to report positive impacts on the performance of the company, as demonstrated [3]. This is in contrast to companies that make limited use of AIS, which are less likely to report successful outcomes. The TOE suggests that the impact of a technology is contingent on the extent to which it is utilised to carry out commercial operations [31]. This is in line with the previous statement. In light of this, in order for a company to acknowledge the impact that it has on the effectiveness of AIS, the company must first put the system into application. A positive correlation between the utilisation of AIS and the impact of AIS is asserted by the authors of this study, who make use of the RBV theory to support their claim. According to [60], it is hypothesised that the organisation would be more likely to generate impacts that are valuable, difficult for competitors to imitate, and sustainable over the course of time if the scope and depth of its use of AIS are increased. Given the importance of information in improving both individual and organisational performance, [60] discovered that the quality of financial and accounting information was a critical factor in IS/IT success. It provides an overview of the system outputs by analysing them across a number of dimensions, including readability, timeliness, accuracy, and relevance. Drawing from the theoretical evidence mentioned earlier, the subsequent hypotheses have been formulated:

- Hypothesis (H4): The implementation of advanced AIS positively correlates with improved Accounting data quality.
- Hypothesis (H4a): The implementation of advanced AIS reduces data entry errors and inconsistencies.
- Hypothesis (H4b): The implementation of advanced AIS enhances the timeliness of financial reporting.
- Hypothesis (H4c): The implementation of advanced AIS enhances the financial data transparency and accessibility.

- Hypothesis (H4d): The implementation of advanced AIS positively correlates with better decision-making processes, resulting.

5 Research Methodology

5.1 Participants and Data Collection

Following the quantitative tradition within IT adoption research, this study employed a survey-based approach to validate the developed conceptual framework. Utilizing data from the Jordan Chamber of Industry (2017), which registers 17,849 Jordanian SMEs, researchers distributed the online questionnaire to a convenient sample of potential and existing AIS users within 700 SMEs' accounting departments [61]. Participation was voluntary, and respondents were recruited through mailing lists, personal emails, and social media groups in August 2023. Qualified participants were individuals with knowledge of AIS services. After receiving information about the study's objectives and the right to withdraw at any stage, respondents were invited to complete the online survey, sharing their beliefs and opinions regarding AIS implementation.

Prior to full-scale data collection, the study conducted a pretest with 12 participants knowledgeable in AIS services. This targeted group included both practitioners and academic researchers. Based on their feedback, minor modifications were made to the survey instrument to optimize its clarity, conciseness, and suitability for measuring the study's constructs. Recognizing the potential for language nuances, the originally English questionnaire was translated into Arabic, the native language of the target population. To ensure fidelity, the translation process was reviewed by four experts: two academics from Jordanian public universities and two linguist teachers. Recognizing the importance of accessible language, the research employed a back-translation method, a common practice in international research [62].

Table 1 depicts the demographic characteristics of the study sample. The gender distribution reveals a male majority at 61.2%, with females comprising 38.8%. Age-wise, the largest demographic segment falls within the (30–39) year-old range (44.4%). Regarding educational background, undergraduate degrees dominate, held by approximately two-thirds of the respondents (69.7%). Additionally, a significant majority (81.3%) possess five or more years of professional experience.

Table 1. Demographic characteristics of study participants

Demographic category	Description	Percentage	Frequency
Gender	Male	61.2%	252
	Female	38.8%	160
Age	20–29	27.43%	113
	30–39	44.41%	183
	40–49	23.31%	96
	<50	4.85%	20

(continued)

Table 1. *(continued)*

Demographic category	Description	Percentage	Frequency
Education background	High school	0.7%	3
	Diploma	7.5%	31
	Undergraduate	69.7%	287
	Masters	21.6%	89
	Doctorate	0.5%	2
Work Experience	<5 years	18.7%	77
	5–10 years	25.2%	104
	11–15 years	30.6%	126
	>15 years	25.5%	105

5.2 Questionnaire Development

The online survey instrument comprised three distinct sections. The first section presented an illustrative overview of Accounting Information Systems (AIS) services. Completion and subsequent affirmation of understanding served as a screening mechanism for participant eligibility. The second section collected demographic information on gender, age, educational background, and work experience. The third section introduced and measured the study's variables through established scales adapted from relevant prior research on TOE framework and IT/IS acceptance and use [3, 11, 45]. A seven-point Likert scale ranging from "1: strongly disagree" to "7: strongly agree" was employed to quantify the constructs under investigation.

6 Data Analysis and Results

6.1 Descriptive Statistics

Descriptive statistics for the study variables within the research framework evidenced that all mean values exceeded 3.998, indicating a strong tendency among respondents to endorse the investigated constructs. Cronbach's alpha coefficients (0.826–0.963) show that robust internal consistency estimates exist for all factors, which further supports this observation.

6.2 Measurement Model

This study employed a two-step analytical approach to investigate the interrelationships within the proposed theoretical framework. Following [63], the study first assessed the measurement model to ensure the validity and reliability of the survey instrument. This involved confirmatory factor analysis (CFA) conducted using AMOS 25.0 [64]. Maximum likelihood estimation and variance-covariance matrices were utilized for parameter estimation and model analyses [63]. Subsequently, the hypothesized relationships between framework constructs were examined under the light of the structural model. [63] emphasize the importance of assessing model fit through various indices, balancing absolute, incremental, and parsimony considerations. Accordingly, this study employed χ^2/df alongside absolute indices like "root mean square error of approximation" (RMSEA), "standardized root mean square residual" (SRMR), "goodness-of-fit index" (GFI), and "adjusted goodness-of-fit index" (AGFI), complemented by incremental indices like "normed fit index" (NFI) and "comparative fit index" (CFI) [63]. Table 2 summarizes the chosen fit indices and their acceptable levels, demonstrating satisfactory model fit for both the measurement and structural models.

Table 2. Model fit indices

Fit Indices	Measurement model	Structural model	Acceptable Value
χ2/df	2.521	2.528	≤3.00
GFI	0.917	0.920	≥0.9000
AGFI	0.872	0.873	≥0.8000
CFI	0.915	0.914	≥0.9000
SRMR	0.045	0.047	≤0.05
NFI	0.910	0.906	≥0.9000
RMSEA	0.071	0.071	≤0.080

Table 3. "Construct reliabilities, convergent validity, and discriminant validity"

	CR	AVE	RA	SP	COM	TMS	FS	ITI	ITC	CP	AISC	AISI	DEE	TL	TA	DM
RA	0.912	0.774	**0.879**													
SP	0.883	0.826	0.428	**0.908**												
COM	0.874	0.788	0.763	0.528	**0.887**											
TMS	0.940	0.739	0.785	0.689	0.812	**0.859**										
FS	0.918	0.814	0.621	0.669	0.718	0.810	**0.902**									
ITI	0.923	0.825	0.498	0.417	0.529	0.621	0.445	**0.908**								
ITC	0.889	0.695	0.175	0.214	0.433	0.220	0.439	0.361	**0.833**							
CP	0.850	0.758	0.414	0.475	0.435	0.343	0.559	0.315	0.209	**0.870**						
AISC	0.903	0.615	−0.529	0.326	0.202	0.561	0.122	0.546	0.541	0.311	**0.784**					
AISI	0.924	0.729	0.569	0.295	0.487	0.333	0.303	0.336	0.321	0.369	0.501	**0.853**				
DEE	0.895	0.816	-0.489	0.266	0.221	0.369	0.116	0.356	0.502	0.362	0.113	0.451	**0.903**			
TL	0.882	0.789	0.511	0.351	0.236	0.402	0.197	0.344	0.326	0.318	0.215	0.329	0.462	**0.888**		
TA	0.903	0.688	0.403	0.422	0.302	0.236	0.217	0.232	0.245	0.298	0.260	0.450	0.417	0.288	**0.829**	
DM	0.911	0.875	0.428	0.368	0.318	0.211	0.229	0.218	0.118	0.290	0.305	0.396	0.446	0.296	0.362	**0.935**

** "relative advantage (RA); security and privacy (SP); Compatibility (COM); Top management support (TMS); Firm size (FS); IT-infrastructure (ITI); Employees' IT competency (ITC); Competitive pressure (CP); Clients AIS complexity (AISC); AIS implementation (AISI); data entry errors (DEE); timeliness (TL); Transparency and accessibility (TA); decision-making (DM)"
** "factor correlation matrix with the square root of AVE on the diagonal line (shown as bold)"

Building upon [63], this study assessed construct reliability and validity through composite reliability (CR) and average variance extracted (AVE). [63] define acceptable discriminant validity as the average construct AVE exceeding its respective correlations with other constructs. Furthermore, they suggest adequate reliability thresholds of CR > 0.7, AVE > 0.5, and CR > AVE. As Table 3 reveals, all constructs exceeded these thresholds, indicating sufficient reliability and convergent validity. Specifically, AVEs ranged from 0.615 to 0.826, and CRs surpassed both AVEs and the 0.7 threshold. Additionally, the square root of each AVE exceeded its corresponding correlations, demonstrating that all measured variables exhibited satisfactory discriminant validity.

6.3 Structural Model

Table 4. Structural model (hypotheses testing)

Hypotheses	Path	Path coefficient (β)	Results
H1.a	RA >> AISI	0.146^{**}	Supported
H1.b	SP >> AISI	−0.013	**Not supported**
H1.c	COM >> AISI	0.118^{*}	Supported
H2.a	TMS >> AISI	0.124^{*}	Supported
H2.b	FS >> AISI	0.139^{**}	supported
H2.c	ITI >> AISI	0.113^{*}	Supported
H2.d	ITC >> AISI	0.178^{**}	Supported
H3.a	CP >> AISI	0.011	**Not supported**
H3.b	AISC >> AISI	0.182^{**}	Supported
H4.a	AISI >> DEE	0.168^{**}	Supported
H4.b	AISI >> TL	0.171^{**}	Supported
H4.c	AISI >> TA	0.193^{**}	Supported
H4.d	AISI >> DM	0.129^{*}	Supported

Note: (* $p < 0.05$, ** $p < 0.01$)

As shown in Table 4, the analysis of the structural model was performed to test the study hypotheses regarding the adoption of AIS. The results revealed that among the technological factors, both "compatibility and relative advantage" (H1. b and H1.c) have a significant impact on AIS implementation. However, "security and privacy" (H1. a) did not. Whereas, all four organizational variables, i.e., top management support (H2. a), firm size (H2. b), IT infrastructure (H2.c), and employees' IT competency (H2. d), significantly influence AIS implementation. Regarding the environmental constructs, "Clients AIS complexity" was confirmed to have a significant impact on AIS implementation (H3.b). Lastly, concerning the effect of AIS implementation on "financial data quality", all the hypothesized relationships were validated (H4a, b, c and d).

7 Discussion

The outcomes of this study report show that "relative advantage" is a substantial facilitator of AIS implementation, suggesting that H1.a is supported. While this conclusion supports earlier studies on the adoption and deployment of modern technologies [45], it contradicts others e.g., [44]. This shows that organisations are more likely to adopt the AIS if the perceived advantages outweigh the added expenses and drawbacks associated with this adaptation.

Based on the "technological factors" given in this study, data analysis findings show that "security and privacy" have no significant impact on AIS implementation. However, the findings were unexpected given that previous research [10, 49] revealed that the "security risk" construct had a direct influence on technological implication and adoption. One possible explanation is because the SP construct is a multidimensional factor that relates to more than only AIS services [3], but in this study, SP is primarily focused on the whole implementation of AIS services. As a result, there is insufficient evidence to support the hypotheses. Despite the modest findings for this category, many organisations remain concerned about the "security and privacy" of institutional data, as well as the security of their clients' data. Because of the high speed and automated nature of transaction processing, the executives expressed concern about the risks related to "privacy and security". They discovered that data security and monitoring are the most significant risks associated with using AIS, and AIS data becomes a target for attacks. In particular, they are concerned that their data or that of their clients may be stolen while it is being sent via the Internet or local network systems. As a result, these issues raise system vulnerabilities and limit the number of users.

The results demonstrated that "compatibility" has a substantial impact on AIS implementation, indicating that H1. c. is supported. Statistical findings are similar with previous research [11]. This demonstrates that organisations think that employing AIS is consistent with their current business environment and culture. One probable explanation is the maturity of AIS in organisations. These organisations' existing IT infrastructures are AIS-compatible. Businesses' previous experience has indicated that the IT/IS systems in place are interoperable and well-suited to the existing IT infrastructure. This conclusion is consistent with earlier research [47].

"Top management support" has a favourable impact on AIS implementation, indicating that H2.a is supported. This outcome is consistent with prior studies [1]. In reality, senior management influences organisational decisions and priorities. When executives within those organisations support and advocate the introduction of AIS technology, it sends a message to the rest of the organisation that AIS is a strategic priority. This assistance helps to create a favourable atmosphere for AIS projects. Additionally, "top management support" frequently translates into resource allocation. Leaders who recognise the importance of AIS and support its implementation are more likely to devote the required financial, human, and technological resources to implementing and maintaining AIS programmes. This resource availability aids in the effective installation and integration of AIS systems inside the target organisations.

The investigation also found that "Firm Size" has a substantial and positive relationship with AIS implementation (H2. b), implying that AIS adoption occurs in larger organisations with greater resources. This contradicts [65], who discovered that this factor had no influence on novel IT/IS adoption, and [6], who confirm a negative relationship between size and implementation of novel innovations. Contradictory outcomes connected to "firm size" may be attributed to the kind of applied technologies, nature of the setting, type of firm size measure (e.g., yearly revenue, number of workers, number of customers/citizens, management complexity, and number of departments), and so on. The findings are consistent with previous studies [5, 8].

The results also show that "organisational readiness/IT infrastructure" has a significant and favourable impact on AIS implementation, implying that hypothesis (H2.c) is validated. This conclusion aligns with previous studies [1, 9]. This implies that organisations with more preparedness are more likely to successfully set up and install AIS systems. "Organisational readiness/IT infrastructure" refers to the organization's ability to efficiently adopt and operate AIS technologies, including its IT infrastructure, resources, and IT leadership support. It also denotes the compatibility of an organization's strategic goals and objectives with the implementation of AIS. In general, organisations that prioritise data-driven decision-making and recognise the need of using information and analytics capabilities for competitive advantage have greater degrees of preparedness. Such alignment ensures that the necessary IT resources and support are available for the AIS deployment, increasing the likelihood of success. Furthermore, "employees' IT competency" was found to have a considerable impact on AIS deployment (H2.d). The presence of IT specialists fosters a favourable environment inside the organisation on which the accounting department workers may rely when faced with difficulty when using AIS. The findings of this investigation support Abed's (2020) conclusions.

Although organisations in general work in a competitive environment and may feel pressure to implement AIS in order to remain competitive and meet the growing needs and expectations of clients, staff, and other stakeholders, this study discovered that "competitive pressure" does not have a statistically significant impact on AIS implementation. This finding contradicts H3. a. This is consistent with the findings of certain published research publications [8, 9], but conflicting with others e.g., [45], in which "competitive pressure" was shown to influence IT/IS adoption. Some organisations frequently face resource constraints, particularly restricted funding. When faced with resource limitation, organisations may prioritise investments in areas closely related to their primary purpose. In these circumstances, "competitive pressure" alone may not be sufficient to justify dedicating resources towards AIS installation, which can be costly in terms of system, IT infrastructure, and staff expertise.

The results we have obtained also show a considerable influence of "Clients' AIS complexity" on AIS implementation (H3. b). The term "clients AIS complexity" refers to how sophisticated the clients' accounting system is. This element is made up of the extent to which clients use a highly computerised reporting system, as well as the complexity, difficulty, and kind of transactions processed by the AIS in their organisations. Another factor contributing to the complexity of technology use is the nature of the business environment, industry, and duties (for example, AIS). For instance, accounting professionals in diverse organisations may encounter complicated transactions that need the use of advanced and complex AIS to handle. This conclusion is consistent with previous investigations [5].

Lastly, the statistical assessments in this study indicated that AIS deployment had a substantial influence on "financial data quality" (H4a, H4b, H4c, H4d), including fewer data entry mistakes, improved timeliness, increased financial data openness and accessibility, and a better decision-making process. In terms of the quality of financial and accounting information, it was verified to be an essential aspect in IS/IT success since information plays an important role in boosting individual and organisational performance in general. It summarizes the system's output properties in terms of timeliness, correctness, relevance, readability, and clarity [60].

8 Research Implications

The AIS research seeks substantial implications for ensuring the quality of accounting data in organizations. It has been stated that AIS automates fundamental accounting activities, hence reducing the likelihood of errors caused by humans and ensuring that data is consistent. For stakeholders, regulators, and decision-makers who use accounting data to evaluate the performance of the organization, this improved accuracy is absolutely necessary. Consequently, the goal of our investigation was to analyze the impact of AIS on accounting DQ by making use of the TOE theory. This study analyses Jordanian SME data to investigate this link for the first time. This is in response to [15] demand for more AIS and accounting DQ studies, which stressed empirical findings. The outcomes of this investigation carry several practical and theoretical implications. With regard to the theoretical one, this study is the first attempt of its kind to combine the TOE theory with the Contingency and the Activity theories in order to add an innovative contribution to the existing body of literature. This may be advantageous for environments that have a low incidence of AIS framework. This research was carried out with the intention of contributing a novel contribution to the existing body of literature, which offers the possibility of being beneficial to settings that have a low incidence of AIS framework. Entities may rely on our validated model to learn how AIS affects the quality of accounting data and what factors promote its broad adoption. To be more specific, the model helps entities concentrate on the potential implications of AIS utilisation that they might have neglected. When it comes to determining the value and efficacy of AIS for enterprises, as well as guiding decision-making surrounding AIS initiatives, this may be of great assistance.

Practically, our findings offered the theoretical framework for evaluating the capacity of AIS to enhance the quality of accounting data among Jordanian SMEs [66–69]. Given the limited number of studies that have been conducted on the subject of the link between the application and usage of AIS and performance and productivity indicators in SMEs in Jordan, this offers practitioners and academics in the field of accounting the opportunity to gain additional value. Likewise, this theoretical model has the potential to serve as a guide for future research endeavours in the sector that is connected to it. In particular, we were able to place an emphasis on the influence that Technological, Organisational, and Environmental elements have on accounting data quality in terms of mistakes that occur during data input, the timeliness of financial reporting, the openness and accessibility of accounting data, and the process of decision-making. Given that the research validates the positive association between the adoption of AIS and Accounting

DQ, this demonstrates that the TOE elements are key factors in the adoption of AIS by SMEs in Jordan [70–77]. Therefore, it is necessary for governments and accounting agencies to improve the awareness training initiatives and to develop an official policy suggestions and guidelines for the most effective practices of AIS approaches. Training must be obligatory for professions that include digital accounting, and users ought to be kept up to date on the most recent developments on a consistent basis.

9 Conclusion, Limitations and Future Research

The accounting profession has been completely transformed by the introduction of information technology, which have brought about several benefits in terms of accuracy, efficiency, cost savings, and safety. Incorporating technology into accounting procedures has resulted in a significant transformation in the manner in which firms handle their financial data. This has enabled decision-makers to obtain information that is both fast and trustworthy. As technology continues to grow, it is expected that AIS eventually become even more complex. This would, in turn, significantly improve financial management and contribute to the success of organisations in a world that is becoming increasingly data driven. The use of information systems in accounting is not only a choice for companies that seek to succeed in the digital age; rather, it is a need for those companies who are in this period of fast change.

In order to evaluate this link, this study examines data from 17,849 SMEs operate in the context of Jordan, a developing, ME nation. In order to construct the structural model, data from a web-based survey was utilised. After conducting 412 surveys, it was determined that "compatibility and relative advantage" are significant technological variables in the installation of AIS. Nevertheless, "security and privacy" does not hold true. Top management support, the scale of the organisation, the IT infrastructure, and the IT expertise of staff members are all factors that influence AIS deployment. There is a substantial influence that "Clients AIS Complexity" has on the deployment of AIS with regard to environmental structures. Every single association that was found between the installation of AIS and "financial data quality" was verified. The research comes to the conclusion that the implementation of AIS is essential for the achievement of effective operations and productivity.

Despite the study's positive contributions, and similar to prior investigations, the current research possesses limitations. First, this study is a cross-sectional examination that does not demonstrate the potential for changes in users' AIS behaviour over time, as it only captures a certain period. Hence, it is advisable to carry out longitudinal study. Furthermore, in the future, it might be beneficial to conduct further studies using alternative research methods, such as case studies, in order to get more profound insights. Second, despite the model's effectiveness in Jordan, further research may be required to confirm its applicability in other developing economies regions. In this way, future researchers could obtain additional evidences from various countries in order to investigate the effects of cultural variations on the research environment. Third, this analysis exclusively concentrated on SMEs. Subsequent investigations might utilize data from several operational sectors in the capital market. Fourth, this study also investigates the direct impacts of TOE factors on IS/IT impact in order to find any potential direct correlations between TOE variables and AIS usage. As a result, such relationships exist between AIS deployment and outcome. Future study should look at how TOE elements, in light of globalization and other areas of auditing practices, such as the efficiency and quality of digital auditing. Thus, recommendations for future study topics, such as investigating the long-term consequences of advanced AIS and comparing various AIS systems. Fifth and finally, since Covid-19 pandemic is regarded as a significant factor in the choice to implement AIS and in determining accounting DQ, further study is crucial in order to examine the influence of extraordinary circumstances on the adoption of AIS and Accounting DQ. The insights might be augmented by longitudinal study that compares the findings before and post the Covid-19 period.

Acknowledgments. The authors thank Mutah University, Karak, Jordan, for funding this article. Furthermore, the authors thank the HCI 2024 committee and anonymous reviewers for their time and effort in reviewing this work and for their helpful comments and suggestions, which improved the initial version of this paper.

Disclosure of Interests. The authors declare that they have no competing interests.

References

1. Al-Okaily, M., Alqudah, H.M., Al-Qudah, A.A. Alkhwaldi, A.F.: Examining the critical factors of computer-assisted audit tools and techniques adoption in the post-COVID-19 period: internal auditors perspective. VINE J. Inf. Knowl. Manag. Syst. (2022)
2. Abed, S.S.: Social commerce adoption using TOE framework: an empirical investigation of Saudi Arabian SMEs. Int. J. Inf. Manage. **53**, 102118 (2020)
3. Al-Okaily, M., Alkhwaldi, A.F., Abdulmuhsin, A.A., Alqudah, H., Al-Okaily, A.: Cloud-based accounting information systems usage and its impact on Jordanian SMEs' performance: the post-COVID-19 perspective. J. Financ. Reporting Acc. **21**(1), 126–155 (2023)
4. Tilahun, M.: A review on determinants of accounting information system adoption. Sci. J. Bus. Manag. **7**(1), 17–22 (2019)
5. Siew, E.G., Rosli, K., Yeow, P.H.: Organizational and environmental influences in the adoption of computer-assisted audit tools and techniques (CAATTs) by audit firms in Malaysia. Int. J. Account. Inf. Syst. **36**, 100445 (2020)
6. Dewett, T., Jones, G.: The role of information technology in the organization: a review, model, and assessment. J. Manag. **27**(3), 313–346 (2001)
7. Jamieson, D., et al.: Data for outcome payments or information for care? A sociotechnical analysis of the management information system in the implementation of a social impact bond. Public Money Manag. **40**(3), 213–224 (2020)
8. Lowe, D.J., Bierstaker, J.L., Janvrin, D.J., Jenkins, J.G.: Information technology in an audit context: have the big 4 lost their advantage? J. Inf. Syst. **32**(1), 87–107 (2017)
9. Maroufkhani, P., Tseng, M.L., Iranmanesh, M., Ismail, W.K., Khalid, H.: Big data analytics adoption: determinants and performances among small to medium-sized enterprises. Int. J. Inf. Manage. **54**, 102190 (2020)
10. Shin, D.-H.: User centric cloud service model in public sectors: policy implications of cloud services. Gov. Inf. Q. **30**(2), 194–203 (2013)
11. Rababah, K.A., Al-nassar, B.A., Al-Nsour, S.N.: Factors influencing the adoption of cloud computing in small and medium enterprises in Jordan. Int. J. Appl. Comput. **10**(3), 96–110 (2020)

12. Shqair, M.I., Altarazi, S.A.: Evaluating the status of SMEs in Jordan with respect to Industry 4.0: a pilot study. Logistics **6**(4), 69 (2022)
13. Al-Okaily, M.: Does AIS usage matter in SMEs performance? An empirical investigation under digital transformation revolution. Inf. Discov. Deliv. **52**(2), 125–137 (2023)
14. Ahmed, Z., Nathaniel, S.P., Shahbaz, M.: The criticality of information and communication technology and human capital in environmental sustainability: evidence from Latin American and Caribbean countries. J. Clean. Prod. **286**, 125529 (2021)
15. Al-Hattami, H.M.: Impact of AIS success on decision-making effectiveness among SMEs in less developed countries. Inf. Technol. Dev. 1–21 (2022)
16. Gunarathne, N., Lee, K.H., Hitigala Kaluarachchilage, P.K.: Tackling the integration challenge between environmental strategy and environmental management accounting. Acc. Auditing Accountability J. **36**(1), 63–95 (2023)
17. Namazi, M., Rezaei, G.: Modelling the role of strategic planning, strategic management accounting information system, and psychological factors on the budgetary slack. In: Accounting Forum, pp. 1–28 (2023)
18. Alharasis, E.E., Haddad, H., Shehadeh, M., Tarawneh, A.S.: Abnormal monitoring costs charged for auditing fair value model: evidence from Jordanian finance industry. Sustainability **14**(6), 3476 (2022)
19. Trieu, H.D., Nguyen, P.V., Tran, K.T., Vrontis, D., Ahmed, Z.: Organisational resilience, ambidexterity and performance: the roles of information technology competencies, digital transformation policies and paradoxical leadership. Int. J. Organ. Anal. (2023)
20. Alharasis, E.E., Alhadab, M., Alidarous, M., Jamaani, F., Alkhwaldi, A.F.: The impact of COVID-19 on the relationship between auditor industry specialization and audit fees: empirical evidence from Jordan. J. Financ. Reporting Accounting (2023a)
21. Al-Dmour, A., Zaidan, H., Al Natour, A.R.: The impact knowledge management processes on business performance via the role of accounting information quality as a mediating factor. VINE J. Inf. Knowl. Manag. Syst. **53**(3), 523–543 (2023)
22. Tornatzky, L., Fleischer, M.: The Process of Technology Innovation. Lexington, MA, Lexington Books (1990)
23. Alqudah, H., Lutfi, A., Al Qudah, M.Z., Alshira'h, A.F., Almaiah, M.A., Alrawad, M.: The impact of empowering internal auditors on the quality of electronic internal audits: a case of Jordanian listed services companies. Int. J. Inf. Manag. Data Insights **3**(2), 100183 (2023)
24. Rogers, E.M.: Diffusion of Innovations, 5th edn. Free Press, New York, NY (2003)
25. Peteraf, M.: The cornerstones of competitive advantage: a resource-based view. Strateg. Manag. J. **14**, 179–191 (1993)
26. Clemons, E.K., Row, M.C.: Sustaining IT advantage: the role of structural differences. Manag. Inf. Syst. Q. **15**(3), 275–292 (1991)
27. Ross, J.W., Beath, C.M., Goodhue, D.L.: Develop long-term competitiveness through IT assets. Sloan Manag. Rev. **38**(1), 31–42 (1996)
28. Zhu, K.: The complementarity of information technology infrastructure and e-commerce capability: a resource-based assessment of their business value. J. Manag. Inf. Syst. **21**(1), 175–211 (2004)
29. Salwani, M., Marthandan, G., Daud Norzaidi, M., Choy Chong, S.: E-commerce usage and business performance in the Malaysian tourism sector: empirical analysis. Inf. Manag. Comput. Secur. **17**(2), 166–185 (2009)
30. Picoto, W.N., Bélanger, F., Palma-dos-Reis, A.: A technology–organisation–environment (TOE)-based m-business value instrument. Int. J. Mobile Commun. **12**(1), 78–101 (2014)
31. Zhu, K., Kraemer, K.L.: Post-adoption variations in usage and value of e-business by organizations: cross-country evidence from the retail industry. Inf. Syst. Res. **16**(1), 61–84 (2005)

32. Aloulou, M., Grati, R., Al-Qudah, A.A., Al-Okaily, M.: Does FinTech adoption increase the diffusion rate of digital financial inclusion? A study of the banking industry sector. J. Financ. Reporting Accounting **22**(2) (2023)
33. Gordon, L.A., Miller, D.: A contingency framework for the design of accounting information systems. Acc. Organ. Soc. **1**(1), 59–69 (1976)
34. Bhatt, G.D.: An empirical examination of the effects of information systems integration on business process improvement. Int. J. Oper. Prod. Manag. **20**(11), 1331–1359 (2000)
35. Martin, J., Leben, J.: Strategic Information Planning Methodolofies, 2nd edn. Prentice-Hall, Englewood Cliffs, NJ (1989)
36. Alharasis, E.E., Haddad, H., Alhadab, M., Shehadeh, M., Hasan, E.F.: Integrating forensic accounting in education and practices to detect and prevent fraud and misstatement: case study of Jordanian public sector. J. Financ. Reporting Acc. **1**(1), 3 (2023b)
37. Kelton, A.S., Murthy, U.S.: Reimagining design science and behavioral science AIS research through a business activity lens. Int. J. Account. Inf. Syst. **50**, 100623 (2023)
38. Henderson, D., Sheetz, S.D., Trinkle, B.S.: The determinants of inter-organizational and internal in-house adoption of XBRL: a structural equation model. Int. J. Account. Inf. Syst. **13**(2), 109–140 (2012)
39. Khairi, M.S., Baridwan, Z.: An empirical study on organizational acceptance accounting information systems in Sharia banking. Int. J. Accounting Bus. Soc. **23**(1), 97–122 (2015)
40. Garg, A.K., Choeu, T.: The adoption of electronic commerce by small and medium enterprises in Pretoria East. Electron. J. Inf. Syst. Dev. Countries. **68**(1), 1–23 (2015)
41. Rahayu, R., Day, J.: Determinant factors of e-commerce adoption by SMEs in developing country: evidence from Indonesia. Procedia Soc. Behav. Sci. **195**, 142–150 (2015)
42. Ahmad, A., Maynard, S.B., Shanks, G.: A case analysis of information systems and security incident responses. Int. J. Inf. Manage. **35**(6), 717–723 (2015)
43. Al-Alawi, A.I., Al-Ali, F.M.: Factors affecting e-commerce adoption in SMEs in the GCC: an empirical study of Kuwait. Res. J. Inf. Technol. **7**(1), 1–21 (2015)
44. Yaseen, H., Al-Adwan, A.S., Nofal, M., Hmoud, H., Abujassar, R.S.: Factors influencing cloud computing adoption among SMEs: the jordanian context. Inf. Dev. **39**(2), 317–332 (2023)
45. Boonsiritomachai, W., McGrath, G.M., Burgess, S.: Exploring business intelligence and its depth of maturity in Thai SMEs. Cogent Bus. Manag. **3**(1), 1220663 (2016)
46. Alharasis, E.E.: Evaluation of ownership structure and audit-quality in the wake of the Covid-19 crisis: empirical evidence from Jordan. Int. J. Law Manag. **65**(6), 635–662 (2023)
47. Oliveira, T., Thomas, M., Espadanal, M.: Assessing the determinants of cloud computing adoption: an analysis of the manufacturing and services sectors. Inf. Manag. **51**(5), 497–510 (2014)
48. Shin, J., Shin, W.S., Lee, C.: An energy security management model using quality function deployment and system dynamics. Energy Policy **54**, 72–86 (2013)
49. Chang, H.H., Chen, S.W.: The impact of online store environment cues on purchase intention: trust and perceived risk as a mediator. Online Inf. Rev. **32**(6), 818–841 (2008)
50. Cho, S.H., Kim, J.W.: Analysis of residual stress in carbon steel weldment incorporating phase transformations. Sci. Technol. Weld. Joining **7**(4), 212–216 (2002)
51. Grandon, E.E. Pearson, J.M.: Electronic commerce adoption: an empirical study of small and medium US businesses. Inf. Manag. **42**(1), 197–216 (2004)
52. Maryeni, Y.Y., Govindaraju, R., Prihartono, B., Sudirman, I.: Technological and organizational factors influencing the e-commerce adoption by Indonesian SMEs (2012)
53. Thong, J.Y., Hong, S.J., Tam, K.Y.: The effects of post-adoption beliefs on the expectation-confirmation model for information technology continuance. Int. J. Hum. Comput. Stud. **64**(9), 799–810 (2006)

54. Mehrtens, J., Cragg, P.B., Mills, A.M.: A model of Internet adoption by SMEs. Inf. Manag. **39**(3), 165–176 (2001)
55. Chwelos, P., Benbasat, I., Dexter, A.S.: Empirical test of an EDI adoption model. Inf. Syst. Res. **12**(3), 304–321 (2001)
56. Nasiren, M.A., Abdullah, M.N., Asmoni, M.: Critical success factors on the BCM implementation in SMEs. J. Adv. Res. Bus. Manag. Stud. **3**(1), 105–122 (2016)
57. Grover, V., Goslar, M.: Toward an empirical taxonomy and model of evolution for telecommunications technologies. J. Inf. Technol. **8**(3), 167–176 (1993)
58. Jeyaraj, A.: Models of information technology use: meta-review and research directions. J. Comput. Inf. Syst. **63**(4), 809–824 (2023)
59. Chong, T.T.L., Lu, L., Ongena, S.: Does banking competition alleviate or worsen credit constraints faced by small-and medium-sized enterprises? Evidence from China. J. Bank. Finance **37**(9), 3412–3424 (2013)
60. Sabeh, H.N., Husin, M.H., Kee, D.M.H., Baharudin, A.S., Abdullah, R.: A systematic review of the DeLone and McLean model of information systems success in an E-learning context (2010–2020). IEEE Access **9**, 81210–81235 (2021)
61. JCI.: Jordan chamber of industry - industrial directory (2017). available at: www.jci.org.jo/
62. Saunders, M.N.K., Thornhill, A. Lewis, P.: Research Methods for Business Students, 8th ed., Pearson Education, Harlow, Essex (2019)
63. Hair, J.F., Black, W.C., Babin, B.J. Anderson, R.E.: Multivariate Data Analysis: Pearson New International Edition, Always Learning, Pearson Harlow, Essex (2014)
64. Byrne, B.M.: Structural Equation Modelling with AMOS: Basic Concepts, Applications, and Programming, Routledge, New York (2023)
65. Lefebvre, L.A., Lefebvre, E., Elia, E., Boeck, H.: Exploring B-to-B e-commerce adoption trajectories in manufacturing SMEs". Technovation **25**(12), 1443–1456 (2005)
66. Alharasis, E., Alidarous, M., Alkhwaldi, A.F., Haddad, H., Alramahi, N., Al-Shattarat, H.K.: Corporates' monitoring costs of fair value disclosures in pre-versus post-IFRS7 era: Jordanian financial business evidence. Cogent Bus. Manag. **10**(2), 2234141 (2023)
67. Salhab, H., Allahham, M., Abu-AlSondos, I., Frangieh, R., Alkhwaldi, A., Ali, B.: Inventory competition, artificial intelligence, and quality improvement decisions in supply chains with digital marketing. Uncertain Supply Chain Manag. **11**(4), 1915–1924 (2023)
68. Alharasis, E.E., Mustafa, F.: The effect of the Covid-19 epidemic on auditing quality and the reaction of family vs non-family businesses to Covid-19: the case of Jordan. J. Family Bus. Manag. (2023)
69. Alharasis, E.E., Prokofieva, M., Clark, C.: The moderating impact of auditor industry specialisation on the relationship between fair value disclosure and audit fees: empirical evidence from Jordan. Asian Rev. Accounting **31**(2), 227–255 (2022)
70. Alkhwaldi, A.F., Abdulmuhsin, A.A.: Crisis-centric distance learning model in Jordanian higher education sector: factors influencing the continuous use of distance learning platforms during COVID-19 pandemic. J. Int. Educ. Bus. **15**(2), 250–272 (2022)
71. Alkhwaldi, A.F., Aldhmour, F.M.: Beyond the bitcoin: analysis of challenges to implement blockchain in the Jordanian public sector. In: Convergence of Internet of Things and Blockchain Technologies (2021)
72. Alharasis, E.E., Alhadab, M., Alidarous, M., Jamaani, F., Alkhwaldi, A.F.: The impact of COVID-19 on the relationship between auditor industry specialization and audit fees: empirical evidence from Jordan. J. Financ. Reporting Accounting (2023)
73. Jamaani, F., Alidarous, M., Alharasis, E.: The combined impact of IFRS mandatory adoption and institutional quality on the IPO companies' underpricing. J. Financ. Reporting Accounting (2022)

74. Alkhwaldi, A. F.: Understanding learners' intention toward Metaverse in higher education institutions from a developing country perspective: UTAUT and ISS integrated model. Kybernetes (2023)
75. Alharasis, E.E., Alidarous, M., Jamaani, F.: Auditor industry expertise and external audit prices: empirical evidence from Amman Stock Exchange-listed companies. Asian J. Accounting Res. **8**(1), 94–107 (2023)
76. Alkhwaldi, A.F.: Investigating the social sustainability of immersive virtual technologies in higher educational institutions: students' perceptions toward metaverse technology. Sustainability **16**(2), 934 (2024)
77. Alharasis, E., Alkhwaldi, A., Hussainey, K.: Key audit matters and auditing quality in the era of COVID-19 pandemic: The case of Jordan. International Journal of Law and Management. In press (2024)

A Study on Speech Emotion Recognition in the Context of Voice User Experience

Annebeth Demaeght[1](✉), Josef Nerb[2], and Andrea Müller[1]

[1] Hochschule Offenburg, Badstrasse 24, 77652 Offenburg, Germany
annebeth.demaeght@hs-offenburg.de
[2] Pädagogische Hochschule Freiburg, Kunzenweg 21, 79117 Freiburg, Germany

Abstract. With the increasing popularity of voice user interfaces (VUIs), there is a growing interest in the evaluation of not only their usability, but also the quality of the user experience (UX). Previous research has shown that UX evaluation in human-machine interaction is significantly influenced by emotions. As a consequence, the measurement of emotions through the user's speech signal may enable a better measure of the voice user experience and thus allow for the optimization of human-computer interaction through VUIs.

With our study, we want to contribute to the research on speech emotion recognition in the context of voice user experience. We recorded 45 German participants while they were interacting with a voice assistant in a Wizard-of-Oz scenario. The interactions contained some typical user annoyances that might occur in voice-based human-computer interaction. Three analysis modules provided insight into the voice user experience of our participants: (1) a UX-questionnaire; (2) the UEQ+ scales for voice assistants; (3) speech emotion recognition with OpenVokaturi.

Keywords: Voice User Experience · Conversational User Experience · Voice Assistants · Speech Emotion Recognition

1 Introduction

As voice user interfaces (VUIs) become more widespread, there is a growing interest in evaluating not only their usability, but also the overall user experience (UX) [1]. Previous studies have shown that emotions play a central role in the evaluation of user experience in human-machine interaction [2]. As a result, assessing emotion through the user's speech signal could provide a better assessment of the voice user experience, facilitating the improvement of interactions between humans and voice-controlled systems.

Through our research, we aim to enhance the understanding of voice user experience and speech emotion recognition. We recorded 45 German participants while interacting with a voice assistant in a Wizard-of-Oz scenario. The interactions contained a number of typical user annoyances occurring in voice-based human-computer interaction, described in Demaeght et al. [3]. We used a UX-questionnaire based on Gast [4] and the UEQ+ scales for voice assistants [5] to evaluate the voice user experience (VUX) of our participants. Furthermore, we analyzed a sample of the participants' voice signal with OpenVokaturi [6], an open source software for speech emotion recognition.

F. F.-H. Nah and K. L. Siau (Eds.): HCII 2024, LNCS 14721, pp. 174–188, 2024.
https://doi.org/10.1007/978-3-031-61318-0_12

2 Related Work

2.1 Emotions

Theory of Emotions. In psychology, defining emotion is a highly controversial subject. There is no generally accepted definition of emotion in any of the disciplines that study emotion [7]. The number of proposed scientific definitions has become so large that even simply counting the number of definitions seems rather pointless, considering that Kleinginna and Kleinginna [8] found 92 different definitions of the term "emotion" [9].

For our study, the emotion theory approach and the Component Process Model (CPM) by Scherer [10] are of particular relevance. Scherer filled a gap in cognitive emotion theory by developing a component-process model (CPM) that describes the appraisal process. It views appraisal as a serial sequence of Stimulus Evaluation Checks (SECs), these are considered to be the central control element of emotions [11]:

1. Relevance check: The assessment of the personal relevance, novelty, and subjective pleasantness of an event.
2. Check of event implications: Dealing with the immediate or longer-term consequences and implications of the event.
3. Check of coping potential: Assessing how well one can cope with or adapt to the consequences of the event.
4. Check of normative significance: Examining the significance of the event in relation to one's own and society's norms and values. In the context of the component process model.

Scherer defines the term emotion as "[…] an episode of interrelated, synchronized changes in the states of all or most of the five organismic subsystems in response to the evaluation of an external or internal stimulus event as relevant to major concerns of the organism" [9].

Categorizing Emotions. The two main approaches to categorize emotions are the "dimensional" and "categorical" one [4]. The dimensional approach describes emotions in terms of different dimensions (valence, arousal, and dominance). Categorical approaches assume that there is a set of a limited number of basic emotions, e.g. Ekman identifies six basic emotions: anger, disgust, fear, sadness, surprise and joy [12].

Measuring Emotions. The methods for measuring emotions can be divided into three levels of measurement [13]: The first level is subjective measurement: The individual concerned provides information about his or her emotional state. Usually, survey methods are used to record the inner state, which can be divided into qualitative (e.g. guided interviews, open-ended questionnaires) and quantitative (e.g. standardized questionnaires) methods. The second level is to measure emotions physiologically. Physiological emotion measurement generally relies on non-invasive medical diagnostics to record changes triggered by the autonomic nervous system. Physiological indicators for emotions include heart rate, skin conductance or activation of certain brain regions. The third level is measuring motor changes to detect emotions, e.g.: Facial expressions, gestures and posture.

2.2 Speech Emotion Recognition

Data Sets of Emotional Speech. Data sets of emotional speech are an essential component of speech emotion recognition research. Existing databases for emotional speech can be divided into three categories [14–17].

1. Simulated: Simulated emotional speech data is usually recorded by professional voice actors. The data collection considered to be much easier than the other two methods. However, there is a discrepancy between acted and natural emotions [18]. A prominent German data set of simulated emotional speech is the emoDB Berlin [19]: Ten professional actors (five men and five women) each spoke ten sentences in seven different emotions (neutral, anger, fear, joy, sadness, disgust and boredom). The material is freely available online and has been used for numerous studies.
2. Provoked: When collecting provoked speech data, speakers are in a simulated emotional situation that can elicit different emotions. Provoked emotions are closer to natural emotions than simulated data. In 2002, Scherer and colleagues recorded and analyzed speech data from German- and English-speaking study participants in stressful situations [20]. The participants performed a logic test on a computer under two different conditions: 1) without interference or 2) simultaneously with an auditory monitoring task. During the test, a pop-up window appeared at various times asking the test subjects to pronounce a standardized phrase and a series of numbers (e.g. "This is task 345629"). The speech data was then analyzed and revealed that workload and, in some cases, stress had an effect on speech rate, energy, F0 and spectral parameters.
3. Natural: Spontaneous emotional speech data mostly comes from talk shows, call center recordings, radio broadcasts and similar sources. A German example is the "Vera am Mittag" database [21]. The database consists of 12 h of audio-visual recordings of the German TV talk show Vera am Mittag. The corpus contains spontaneous, emotional voice recordings of authentic discussions between the talk show guests. The audio-visual data were labeled with the dimensions: arousal, valence and dominance.

A list of prominent German-language databases, extracted from a more elaborate overview in Swain et al. [17], is summarized in Table 1.

Voice Features. In the literature, prosody, spectral characteristics and voice quality are often identified as indicators of the speaker's emotional state [14, 16, 17].

In linguistics, prosody stands for suprasegmental characteristics such as intonation, volume, accent and (speech) pauses. It can convey both linguistic (e.g. the difference between a question and a statement) and paralinguistic (e.g. the emotional state of the speaker) information [14]. The prosodic characteristics most commonly used in voice-based emotion measurement are fundamental frequency, loudness and duration [14, 24].

Voice quality is determined by the degree of closure of the glottis and the stiffness of the vocal folds. For example, incomplete closure and insufficient stiffness of the vocal folds leads to a breathy voice. A rough voice is caused by strong irregularities in the vocal fold vibrations, both in amplitude (shimmer) and frequency (jitter).

Table 1. A selective overview of data sets containing German emotional speech, extracted from a more elaborate overview in Swain et al. [17]

Data set	Category	Size	Emotions/Affect
Scherer et al. (2002) [20]	Provoked	100 native speakers	Stress and workload
Tato et al. (2002) [22]	Provoked	14 native speakers	Anger, boredom, joy, sadness, neutral
"FAU Aibo Emotion Corups" Batliner et al. (2004) [23]	Provoked	51 children	Anger, boredom, joy and surprise
"emoDB-Berlin" Burkhardt et al. (2005) [19]	Simulated	10 professional actors	Neutral, anger, fear, joy, sadness, disgust and boredom
"Vera am Mittag" Grimm et al. (2008) [21]	Natural	104 native speakers	Dimensional: Valence, arrousal, dominance

In order to increase the performance of voice-based emotion recognition systems, additional indicators can be used, such as visual signals, physiological signals and word recognition [14].

Speech Emotion Recognition Systems. Since Picard's pioneering work on affective computing [25], there has been great interest in automatic emotion recognition. Two examples of systems which are used in the field are the feature extraction and audio analysis tool OpenSMILE [26] and Vokaturi [6]. Vokaturi categorizes emotions into five types (anger, happiness, neutral, sadness, and fear) and presents the result as a weighted vector of probabilities associated with each emotion [27]. In our work we use the freely available open source version of Vokaturi: OpenVokaturi [6].

3 Study

3.1 Research Goal

The goal of this study was to develop and implement a procedure for collecting emotional speech data from users interacting with voice-controlled applications. Furthermore, the collected data was used to investigate how emotions occurring during the use of voice user interfaces can be measured empirically.

3.2 Population

The study was conducted with 45 German-speaking participants of whom 18 were male and 27 were female. 38 participants were between 18 and 25 years old, the remaining 7 were between 26 and 35 years old. All participants were students at the University of Applied Sciences in Offenburg, Germany.

The participants were asked which voice assistant they use most. The results show that most participants use Siri (15), followed by Google Assistant (7), Alexa (7), and other voice assistants (2). 14 participants indicated that they do not use any voice assistant.

Participants were also asked how often they use voice assistants: several times a day (5), about daily (7), several times a week (4), several times a month (5), about once a month (10), or never (14).

3.3 Setup

The study took place in the Customer Experience Tracking Lab of Offenburg University of Applied Sciences (Fig. 1). The CXT-lab enabled participants to engage with the voice application in a setting that mimics a natural living room environment, without the perceived control of the experimenter since there is a physical separation between the experimental area and the observation space [4].

A Røde NT-USB Mini microphone and the audio recording software Audacity were used to capture the participant's voice during the interaction. Amazon's smart speaker Echo (4th generation) was used for the output of the synthetic voice. The participants assumed they are talking to a voice assistant, but in reality, the voice output was simulated by a human (Wizard-of-Oz experiment). Via Bluetooth, the smart speaker was connected to the experimenter's computer in the adjoining room. The experimenter simulated the voice application with the tool Speechtester, a browser-based web application that allows texts to be read aloud by a computer voice [28].

Furthermore, two cameras were used for recording the experiment. One camera was placed in front of the participants, a second camera with an integrated microphone filmed the participants and captured the audio.

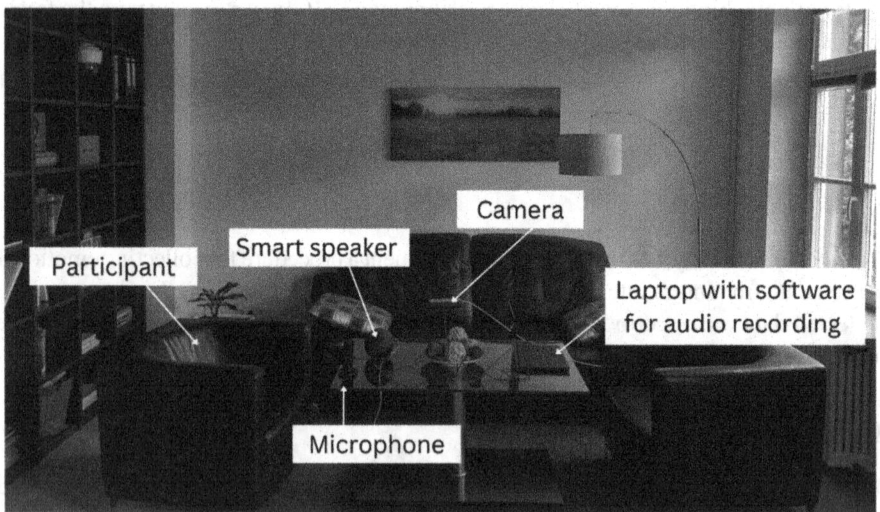

Fig. 1. Study setup

3.4 Procedure

The phases of the study procedure can be found in Table 2 and are described in more detail below. The procedure is mainly based on the work of Gast [4].

Table 2. Study procedure

Procedure	Participant alone
Arrival and consent with privacy policy	No
Background story	No
Explanation of the tasks	No
Processing the tasks	Yes
Processing the UX questionnaire	Yes
De-briefing	No
Farewell	No

Arrival and Consent with Privacy Policy. After a personal welcome and a brief introduction to the Customer Experience Tracking Lab, the participants were informed about the study's privacy policy and asked for their consent.

Background Story. The test manager described the background story and explained the procedure for the subsequent testing. The participants were told that they would contribute to the user experience optimization of a voice application for the canteen at Offenburg University of Applied Sciences.

Explanation of the Tasks. In the next step, the test manager explained the tasks to the participants. Since the tasks were processed without the presence of the test manager, it had to be ensured that participants had understood the content and desired sequence of the tasks in order to avoid distortions based on misinterpretation of the assignments within the tasks. Thus, each task was briefly discussed and checked for comprehension questions. The task descriptions were first explained orally and then given to the participants in written text. The task description was kept as short as possible and as long as necessary [4].

Processing the Tasks. All 45 participants completed three tasks with the voice application in German. An example of each of the three interactions can be found in Fig. 1. Task 1 served as a "warm-up" to familiarize the participants with the application and to put them at ease. The participants were asked to find out tomorrow's canteen menu and write it down on a sheet of paper. There were no irritations in this task.

In the second task, the participants were asked to use the voice application to pre-order the "Spaghetti Bolognese" lunch for pick-up at the canteen on the Gengenbach campus at 1 pm. While processing this task, the participants were confronted with the irritation of not being understood acoustically: They had to repeat their student number three times until the system recognized the voice input and they were able to pre-order.

Furthermore, participants could not transfer all the relevant information in one prompt. For example, if participants would say "I would like to pre-order Spaghetti Bolognese for pick-up at the Gengenbach campus at 1 pm", the system would ask for each piece of information separately: "Which dish do you want to order?", "Do you want to eat it at the restaurant or take-away?", "Where do you want to pick it up?", "When do you want to pick it up?". In addition, participants had to confirm each of their entries, which made the conversation unnecessary long.

In the third task, participants were asked to use the voice application to find out if the student restaurant offers coffee with lactose free milk and if so, how much it costs. While processing this task, the participants were confronted with the irritation of being misunderstood, which would lead to the system being aborted and the participant having to start again from the beginning. Furthermore, the interaction was unnecessary long as the voice application would follow a predefined conversational flow.

The irritations built into the interactions are based on a survey and literature study described in detail by Demaeght et al. [3] (Fig. 2).

Processing the UX-Questionnaires. When the third dialog was completed, the test manager briefly entered the experimental room to ask the participants to complete the UX-Questionnaire. During the completion of the online questionnaire, the participants were again left alone in the experimental room so that they could work undisturbed.

Two questionnaires were used to survey the subjective user experience of the participants. On the one hand, a questionnaire developed on the Table 3 Questionnaire (UEQ) [5].

Seven question items were asked using a six-point Likert scale with the poles "1 = No, not at all" and "6 = Yes, very much". The question "Would you have discontinued the application at any point?" was asked with the answer options "Yes" or "No". The questions were formulated in accordance with Gast [4], who based his questionnaire on the DIN standards 9241–210 and 9241–110. In addition, there were two free text fields where the participants could indicate what annoyed them and what they liked about the application. Includes all question items with the respective scales, poles and UX-relation on the basis of which Gast [4] derived the questions.

The second part of the questionnaire was based on the UEQ+, a modular extension of the User Experience Questionnaire (UEQ). It contains a larger list of UX scales that can be combined by UX-researchers to build a questionnaire adapted to their specific research question [29]. We used the scales which were developed specifically for voice interactions: "Response behavior", "Response quality" and "Comprehensibility" [5], they are listed in Table 4. Each scale consists of four bipolar adjective pair items. Users rate them using a seven-point scale ranging from −3 to + 3.

De-briefing. After completing the questionnaire, the experimenter entered the experimental room and began a brief debriefing. The participants were informed about the set-up and objectives of the testing.

Farewell. At the end of each individual testing session, the participants were thanked for their participation and bid farewell.

First interaction

User	Open the Mensa app.
Assistant	Welcome to the Mensa. How can I help you?
User	What's tomorrow's meal of the day?
Assistant	Tomorrow we're serving chicken with mushrooms. Would you also like to know the vegetarian option?
User	Yes.
Assistant	The vegetarian dish is potatoes filled with cream cheese.
User	Thank you.
Assistant	You're welcome. Enjoy your meal.

Second interaction

User	Open the Mensa app.
Assistant	Welcome to the Mensa. How can I help you?
User	I would like to pick up the Spaghetti Bolognese meal of the day at 1 p.m. at the Gengenbach campus.
Assistant	You would like to pre-order the meal of the day. Is that correct?
User	Yes.
Assistant	There are two dishes of the day: spaghetti Bolognese or pancakes with mushrooms. Which one would you like?
User	Spaghetti Bolognese.
Assistant	You want spaghetti Bolognese. Is that right?
User	Yes.
Assistant	Please give me your student number first.
User	1, 2, 3, 4, 5, 6.
Assistant	I'm afraid I didn't understand that. Please repeat the number again.
User	1, 2, 3, 4, 5, 6.
Assistant	Your number is 1 2 2 4 5 6. Is that correct?
User	No.
Assistant	Please enter the number again.
User	1, 2, 3, 4, 5, 6.
Assistant	Your number is 1, 2, 3, 4, 5, 6. Is that correct?
User	Yes.
Assistant	Where would you like to pick up the food? At the Gengenbach campus or the Offenburg campus?
User	At the Gengenbach campus.
Assistant	You would like to pick up the meal at the Gengenbach campus. Is that correct?
User	Yes.
Assistant	When would you like to pick up your meal?
User	At 1 pm.
Assistant	You would like to pick up your meal at 1 pm. Is that correct?
User	Yes.
Assistant	Would you like to take the dish away or eat it on site?
User	Take away.
Assistant	You would like to take the dish with you. Is that correct?
User	Yes.
Assistant	Your meal is pre-ordered. Your pick-up number is 102.
User	Thank you.
Assistant	Enjoy your meal.

Third interaction

User	Open the Mensa app.
Assistant	Welcome to the Mensa. How can I help you?
User	Does the Mensa offer coffee with lactose-free milk?
Assistant	You would like to order a coffee. Is that correct?
User	No.
Assistant	Please give me your student number first.
User	I don't want to order coffee.
Assistant	I'm afraid I didn't understand that. Please repeat it again.
User	I don't want to order coffee.
Assistant	Your student number consists only of numbers. Please enter the numbers.
User	1, 2, 3, 4, 5, 6
Assistant	Your student number is 1, 2, 3, 4, 5, 6. Is that correct?
User	Yes.
Assistant	Where would you like to pick up your coffee?
User	I don't want to order coffee.
Assistant	You don't want to order coffee? Then please say "cancel now".
User	Cancel now.
Assistant	Order canceled. How can I help you?
User	Can I get coffee with lactose-free milk at the Mensa?
Assistant	Unfortunately, I didn't understand that. I can help you pre-order your meal or give you information about our offer.
User	Information about your offer.
Assistant	Would you like to know more about what we offer? Is that correct?
User	Yes.
Assistant	Our offer is very varied and consists of main courses, salads, snacks and beverages. What would you like to know more about?
User	Beverages.
Assistant	We have a wide range of soft drinks or hot beverages. What would you like to know more about?
User	Hot beverages.
Assistant	We offer the following hot beverages: Espresso, cappuccino, latte, hot chocolate and coffee with lactose-free milk
User	How much does coffee with lactose-free milk cost?
Assistant	Coffee with lactose-free milk costs 2 euros 50.
User	Thank you.
Assistant	You're welcome.

Fig. 2. Example dialogs for all three interactions

Table 3. Items of the UX questionnaire based on Gast [4]

Question	Scale	Poles	UX-Relation
Would you recommend the application to others?	6-point Likert scale	1 = No, not at all 6 = Yes, very much	Recommendation as a UX indicator
Would you use the application again?	6-point Likert scale	1 = No, not at all 6 = Yes, very much	Reuse as a UX indicator
Were you annoyed by the errors in the application?	6-point Likert scale	1 = No, not at all 6 = Yes, very much	Concrete addressing of negative emotions as a UX-influencing factor
Did you enjoy using the application?	6-point Likert scale	1 = No, not at all 6 = Yes, very much	Joy of use as an element of UX
Do you find the application user-friendly?	6-point Likert scale	1 = No, not at all 6 = Yes, very much	Usability as an element of UX
Do you find the application pleasant to use?	6-point Likert scale	1 = No, not at all 6 = Yes, very much	Element of software ergonomics and part of UX
Do you find the application useful?	6-point Likert scale	1 = No, not at all 6 = Yes, very much	Element of software ergonomics and part of UX
Would you have aborted the application at any point?		Yes No	Termination as a UX-indicator

Table 4. The UEQ+ scales for voice assistants [5]

Scale	Item Left	Item Right
Response behavior	artificial	natural
	unpleasant	pleasant
	unlikeable	likeable
	boring	entertaining
Response quality	inappropriate	suitable
	useless	useful
	not helpful	helpful
	unintelligent	intelligent
Comprehensibility	complicated	simple
	inaccurate	accurate
	ambiguous	unambiguous
	enigmatic	explainable

3.5 Data Preparation for Analysis

First, the recordings needed to be prepared for analysis. An initial inspection showed that the recordings of one participant had to be excluded. The participant had pointed out to the experimenter that she assumed that it was not the system but her limited language skills that were responsible for the difficulties in the interaction (she was not a native speaker).

After the first inspection, the audio data of each participant was cut so that only the interactions with the assistant remained, resulting in 44 audio files. Then, each interaction (1, 2 and 3) was filed separately, resulting in 152 audio files. Next, the voice of the assistant was removed. Finally, certain utterances that were particularly interesting for the analysis (e.g. the repetitions in the second interaction) were filed as snippets (Fig. 3).

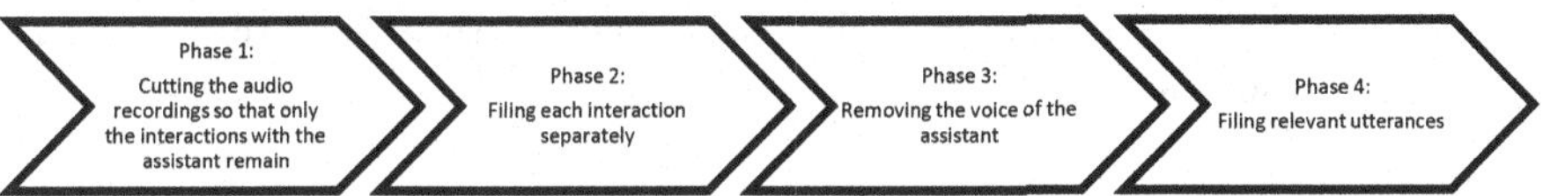

Fig. 3. Preparing the data for analysis

3.6 Results of the Study

(1) Analysis module 1: UX-questionnaire. The results of the UX survey are shown in Table 5. Most items receive an average score of 4 or higher. However, two items stand out. The question "Were you annoyed by the errors in the application?" has a mean score of 3. This question specifically addresses negative emotions as a UX-influencing factor. Furthermore, 23 out of 43 respondents (one participant did not answer this question) indicated that they would have terminated the application if they'd not been using it in a laboratory sitting. The latter is a significant indicator of user experience and allows us to compare the voice signal of the participants who were so annoyed, that they would have terminated the application with the participants who would have continued using it.

Table 5. The results of the UX-Questionnaire

Question	M	SD	n
Would you recommend the application to others?	4.05	1.02	44
Would you use the application again?	3.77	1.31	44
Were you annoyed by the errors in the application?	3.00	1.31	44
Did you enjoy using the application?	4.52	1.03	44
Do you find the application user-friendly?	4.02	1.10	44

(continued)

Table 5. *(continued)*

Question	M	SD	n
Do you find the application pleasant to use?	4.05	1.23	44
Do you find the application useful?	4.25	1.21	44
Question	**No**	**Yes**	**n**
Would you have aborted the application at any point?	20	23	43

(2) Analysis module 2: UEQ + scales for voice assistants. The results of the UEQ+ questionnaire, containing the mean value and standard deviation per item as well as Cronbach's Alpha as a measure for the consistency of the scales are shown in Table 6. All items, except for the word pair "artificial - natural", show positive mean values between 0.25 and 1.66.

Table 6. The results of the UEQ+ questionnaire for voice assistants

Scale	Item Left	Item Right	M	SD	n	α
Response behavior	artificial	natural	–0.80	1.47	44	0.76
	unpleasant	pleasant	1.02	1.22	44	
	unlikeable	likeable	1.16	1.20	44	
	boring	entertaining	0.25	1.13	44	
Response quality	inappropriate	suitable	1.39	1.19	44	0.83
	useless	useful	1.66	0.93	44	
	not helpful	helpful	1.52	0.92	44	
	unintelligent	intelligent	0.91	1.02	44	
Comprehensibility	complicated	simple	0.70	1.49	44	0.89
	inaccurate	accurate	0.45	1.29	44	
	ambiguous	unambiguous	0.58	1.47	43	
	enigmatic	explainable	1.23	1.29	43	

(3) Analysis module 3: Vokaturi. In the second interaction, participants had to utter the number series "1, 2, 3, 4, 5, 6" three times as the systems did not understand or misunderstood their input. This is illustrated in the following extract of the interaction:

Assistant Please give me your student number first.
User 1, 2, 3, 4, 5, 6.
Assistant I'm afraid I didn't understand that. Please repeat the number again.
User 1, 2, 3, 4, 5, 6.
Assistant Your number is 1, 2, 2, 4, 5, 6. Is that correct?
User No.

Assistant Please enter the number again.
User 1, 2, 3, 4, 5, 6.
Assistant Your number is 1, 2, 3, 4, 5, 6. Is that correct?
User Yes.

When analyzing the speech data with Vokaturi for this paper, we focused on this part of the interactions. We filed each utterance of the number series and then ran them through OpenVokaturi [6]. Table 7 shows the average results over all data. We see that the software detects an overall increase of the emotion "anger": From 6.4% anger in the first utterance, over 10.6% in the second utterance to 11.3% in the third utterance.

Table 7. Mean results of OpenVokaturi for repetitions of the utterance "1, 2, 3, 4, 5, 6" over all participants

	Neutrality	Happiness	Sadness	Anger	Fear
1st utterance	0.35645	0.22204	0.25037	0.06398	0.10717
2nd utterance	0.27528	0.24824	0.24582	0.10609	0.12458
3rd utterance	0.31022	0.23640	0.21760	0.11333	0.12244

We then compared the results of the participants who answered the question "Would you have aborted the application at any point?" with "yes" and thus indicated a negative user experience, with the participants who answered that they would not have terminated the application at any point.

The results in Table 8 show that OpenVokaturi detected more anger in the voice of the group who indicated a negative user experience, than in the voice of the other group. However, both groups show an increase of the emotion anger when comparing the first utterance with the third. Furthermore, the system detected a more neutral emotional state in the voice of the second group. When looking at the emotion happiness, it is noticeable that the system recognized more happiness in the first group. A reason for this could be that the arrousal in the voice was assigned a positive valence by the system.

Table 8. Mean results of OpenVokaturi for the 1st, 2nd and 3rd repetitions of the utterance "1, 2, 3, 4, 5, 6". (N = neutral, H = happiness, S = sadness, A = anger, F = fear)

	Group 1 (n = 23) Would have terminated the application (Negative UX)					Group 2 (n = 19) Would not have terminated the application (Neutral/Positive UX)				
	N	H	S	A	F	N	H	S	A	F
1st	0.25	0.26	0.28	0.10	0.11	0.46	0.19	0.23	0.02	0.10
2nd	0.19	0.26	0.25	0.14	0.16	0.38	0.25	0.23	0.06	0.08
3rd	0.22	0.27	0.26	0.13	0.11	0.41	0.20	0.17	0.09	0.13

4 Conclusion and Recommendations

We conducted a study involving 45 German participants engaging with a voice assistant within a Wizard-of-Oz scenario. The interactions included various common user irritations that typically arise in voice-based human-computer interaction. Three analysis modules provided insights into the voice user experience of our participants.

1. **Analysis Module 1: UX-Questionnaire.** Despite the built-in irritations, the UX-questionnaire showed a neutral to positive user experience. However, 23 out of 43 respondents would have abandoned the application, which is an indication of a negative user experience.
2. **Analysis Module 2: UEQ+.** The results of the UEQ+ questionnaire for voice applications show that particularly the response behavior of the assistant was perceived as unnatural. All other items had positive mean values.

   How to achieve natural and seamless interaction is a challenge many conversational designers face. The literature [30, 31] provides some key recommendations for implementing a more natural conversational flow:

   - Comprehend the needs of the users: User research enables the development of empathetic insights into the target audience and allows designers to find the right tone, style, and vocabulary.
   - Apply optimal design principles: Consider speech length and pauses, limit choices, maintain context, take latencies into account, consider social cues, provide the user orientation.
   - Perform user testings: Testing voice applications contributes significantly to the quality of the final product. Prototypes of the application can be tested in the developer team, with the customer, or ideally with test subjects from the target group. The goal is to continuously optimize the spoken texts to make it as easy and natural as possible for users to interact with the application.

3. **Analysis Module 3: OpenVokaturi.** We analyzed the participants' voice signal with OpenVokaturi [6], an open source software for speech emotion recognition, focusing on the voice of the users having to repeat the number series "1, 2, 3, 4, 5, 6" three times. OpenVokaturi detected slightly more anger in the data of the participants who had indicated a negative user experience. However, these preliminary results must be treated with caution as further data analyses need to be conducted.

References

1. Seaborn, K., Urakami, J.: Measuring voice UX quantitatively. In: Kitamura, Y., Quigley, A., Isbister, K., Igarashi, T. (eds.) Extended Abstracts of the 2021 CHI Conference on Human Factors in Computing Systems, Article 416, pp. 1–8. Association for Computing Machinery, New York, USA (2021). https://doi.org/10.1145/3411763.3451712
2. Thüring, M., Mahlke, S.: Usability, aesthetics and emotions in human–technology interaction. Int. J. Psychol. **42**(4), 253–264 (2007). https://doi.org/10.1080/00207590701396674

3. Demaeght, A., Nerb, J., Müller, A.: A survey-based study to identify user annoyances of German voice assistant users. In: Fui-Hoon Nah, F., Siau, K. (eds) HCI in Business, Government and Organizations. HCII 2022. LNCS, vol. 13327, pp. 261–271. Springer, Cham (2022). https://doi.org/10.1007/978-3-031-05544-7_20
4. Gast, O.: User Experience im E-Commerce. Messung von Emotionen bei der Nutzung interaktiver Anwendungen. Springer Gabler, Wiesbaden (2018). https://doi.org/10.1007/978-3-658-22484-4
5. Klein, A.M., Hinderks, A., Schrepp, M., Thomaschewski, J.: Measuring user experience quality of voice assistants. In: Rocha, A., Escobar Peréz, B., Garcia Peñalvo, F., del Mar Miras, M., Gonçalves, R. (eds.) In: 15th Iberian Conference on Information Systems and Technologies (CISTI), pp. 1–4. IEEE, Sevilla (2020). https://doi.org/10.23919/CISTI49556.2020.9140966
6. Vokaturi Homepage. https://vokaturi.com/. Accessed 26 Jan 2024
7. Mulligan, K., Scherer, K.R.: Toward a working definition of emotion. Emot. Rev. **4**(4), 345–357 (2012). https://doi.org/10.1177/1754073912445818
8. Kleinginna, P.R., Kleinginna, A.M.: A categorized list of emotion definitions, with suggestions for a consensual definition. Motiv. Emot. **5**(4), 345–379 (1981). https://doi.org/10.1007/BF00992553
9. Scherer, K.R.: What are emotions? And how can they be measured? Soc. Sci. Inf. **44**(4), 695–729 (2005). https://doi.org/10.1177/0539018405058216
10. Scherer, K.R.: On the nature and function of emotion: a component process approach. In: Scherer, K. R., Ekman, P. (eds.) Approaches to Emotion, Psychology Press, New York (1984). https://doi.org/10.4324/9781315798806
11. Brosch, T., Scherer, K.R.: Plädoyer für das Komponenten-Prozess-Modell als theoretische Grundlage der experimentellen Emotionsforschung. In: Janke, W., Schmitt-Daffy, M., Debus, G. (eds.): Experimentelle Emotionspsychologie: Methodische Ansätze, Probleme und Ergebnisse, pp. 193–204, Pabst, Lengerich (2008)
12. Ekman, P.: An argument for basis emotions. Cogn. Emot. **6**(3/4), 169–200 (1992)
13. Vogel, I.: Emotionen im Kommunikationskontext. In: Six, U., Gleich, U., Gimmler, R. (eds.): Kommunikationspsychologie Medienpsychologie, pp. 135–157, Weinheim (2007)
14. Akçay, M.B., Oğuz, K.: Speech emotion recognition: Emotional models, databases, features, preprocessing methods, supporting modalities, and classifiers. Speech Commun. **116**, 56–76 (2020). https://doi.org/10.1016/j.specom.2019.12.001
15. Basu, S., Chakraborty, J., Bag, A., Aftabuddin, M.D.: A review on emotion recognition using speech. In: 2017 International Conference on Inventive Communication and Computational Technologies, pp. 109–114, IEEE (2017). https://doi.org/10.1109/ICICCT.2017.7975169
16. Schuller, B.W.: Speech emotion recognition: two decades in a nutshell, benchmarks, and ongoing trends. Commun. ACM **61**(5), 90–99 (2018). https://doi.org/10.1145/3129340
17. Swain, M., Routray, A., Kabisatpathy, P.: Databases, features and classifiers for speech emotion recognition: a review. Int. J. Speech Technol. **21**(1), 93–120 (2018). https://doi.org/10.1007/s10772-018-9491-z
18. Vogt, T., Andre, E.: Comparing feature sets for acted and spontaneous speech in view of automatic emotion recognition. In: 2005 IEEE International Conference on Multimedia and Expo, pp. 474–477. IEEE (2005). https://doi.org/10.1109/ICME.2005.1521463
19. Burkhardt, F., Paeschke, A., Rolfes, M., Sendlmeier, W.F., Weiss, B.: A database of German emotional speech. In: Proceedings Interspeech 2005, pp. 1517–1520 (2005). https://doi.org/10.21437/Interspeech.2005-446
20. Scherer, K.R., Grandjean, D., Johnstone, T., Klasmeyer, G., Bänziger, T.: Acoustic correlates of task load and stress. In: Proceedings of the 7th International Conference on Spoken Language Processing, pp. 2017–2020 (2002). https://doi.org/10.21437/ICSLP.2002-554

21. Grimm, M., Kroschel, K., Narayanan, S.: The Vera am Mittag German audio-visual emotional speech database. In: 2008 IEEE International Conference on Multimedia and Expo, pp. 865–868 (2008). https://doi.org/10.1109/ICME.2008.4607572
22. Tato, R., Santos, R., Kompe, R., Pardo, J.M.: Emotional space improves emotion recognition. In: Proceedings of the 7th International Conference on Spoken Language Processing, pp. 2029–2032 (2002). https://doi.org/10.21437/ICSLP.2002-557
23. Batliner, A., et al.: 'You Stupid Tin Box' - Children Interacting with the AIBO Robot: A Cross-linguistic Emotional Speech Corpus. In: Lino, M. T., Xavier, M. F., Ferreira, F., Costa, R., Silva, R. (eds.) Proceedings of the Fourth International Conference on Language Resources and Evaluation (LREC'04), Lisbon, Portugal: European Language Resources Association (ELRA) (2004)
24. Mary, L.: Extraction of Prosody for Automatic Speaker, Language, Emotion and Speech Recognition. Springer, Cham (2018). https://doi.org/10.1007/978-3-319-91171-7
25. Picard, R.W.: Affective Computing. MIT Press, Cambridge (1997). https://doi.org/10.7551/mitpress/1140.001.0001
26. openSMILE 3.0 – audEERING Homepage. https://www.audeering.com/de/research/opensmile/. Accessed 26 Jan 2024
27. Datta, D., Jiang, W., Vogel, C., Ahmad, K.: Speech emotion recognition systems: a cross-language, inter-racial, and cross-gender comparison. In: Arai, K. (ed.) Advances in Information and Communication: Proceedings of the 2023 Future of Information and Communication Conference (FICC), pp. 375–390. Springer Nature, Cham (2023). https://doi.org/10.1007/978-3-031-28076-4_28
28. Labs Speechtester Homepage. http://speechtest.169labs.com/. Accessed 26 Jan 2024
29. Schrepp, M.: Measuring user experience with modular questionnaires. In: 2021 International Conference on Advanced Computer Science and Information Systems (ICACSIS), pp. 1–6 (2021). https://doi.org/10.1109/ICACSIS53237.2021.9631321
30. Pearl, C.: Designing Voice User Interfaces. O'Reilly Media, Sebastopol (2016)
31. Kahle, T., Meißner, D.: All About Voice. Konzeption, Design und Vermarktung von Anwendungen für digitale Sprachassistenten. Haufe Group, Freiburg (2020)

China's Evidence for the Determinants of Green Business Environment from a Dynamic fsQCA

Hang Jiang, Yongle Wang, Jiangqiu Wu, and Beini Zhuang(✉)

School of Business Administration, Jimei University, Xiamen 361021, China
200361000053@jmu.edu.cn

Abstract. The rapid growth of the Chinese mainland's economy was attributed to its reforms and opening-up policies. Establishing a conducive business environment has been identified as one factor contributing to the country's rapid growth. This study used dynamic fuzzy-set qualitative comparative analysis (fsQCA) to analyze panel data from 30 Chinese provinces between 2011 and 2022 to shed light on the antecedent conditions that influence green business environment. The findings reveal that two distinct configurations also result in great green business environment. Finally, based on our findings, specific policy suggestions are offered.

Keywords: Green Business Environment · Dynamic fsQCA · Configuration

1 Introduction

The rapid growth of the Chinese mainland's economy was attributed to its reforms and opening-up policies. According to official data from the National Bureau of Statistics, the GDP of the mainland of China has grown from 0.37 trillion yuan in 1978 to 121.02 trillion yuan in 2022, an increase of more than 300 times. Also, the total imports and exports and foreign direct investment reached 42.07 trillion yuan and 146.50 billion USD in 2022, respectively. Establishing a conducive business environment has been identified as one factor contributing to the country's rapid growth. The central government proposed "Forming a new system of opening up and improving a legalized, international, and convenient business environment" in October 2015 to optimize the business environment. Afterward, during an economic conference in July 2017, the government emphasized the importance of creating a conducive business environment.

Although the extensive development model has brought rapid economic growth, problems such as the disappearance of low-cost competitive advantage and the decline of demographic dividends, the deterioration of resources and the environment, excessive energy consumption, and limited innovation motivation have begun to emerge. Especially, the increasing environmental concerns have prompted many countries to focus on the link between economic development and sustainability development. As a result, China has proposed a dual carbon target, with carbon peaking before 2030 and carbon neutrality by 2060.

F. F.-H. Nah and K. L. Siau (Eds.): HCII 2024, LNCS 14721, pp. 189–199, 2024.
https://doi.org/10.1007/978-3-031-61318-0_13

The rise of the digital economy has steadily become the main driver of China's excellent economic development in which digital technology is still developing. It has also had a significant impact on the optimization of China's business environment. In addition, some academics have suggested creating a digital corporate environment. The expansion of the digital economy steadily creates a favorable business environment, and the business environment's optimization simultaneously fosters the growth of the digital economy (Pei & Hou, 2023).

The business environment evaluation indices established by many international organizations have achieved widespread acclaim after years of development. However, the indicator systems currently in common use are not integrated with sustainable development. As a result, this paper adds green development indicators on the basis of the original index system to construct a green business environment evaluation index. In addition, the indicators for evaluate digital economy were used to represent the technological development in business environment evaluation index. Then, the proposed evaluation index will be applied to evaluate the provincial green business environment in China. In addition, to analyze the combination effect exerted by determinants on green business environment, and seek the path to realize the green business environment, the panel fuzzy-set qualitative comparative analysis (fsQCA) was used.

The original fsQCA is based on set theory, and it uses Boolean minimization, fuzzy-set theory, and combinatorial logic to identify the combinations of case conditions that might be necessary or sufficient to result in an outcome. As a result, fsQCA employs an inductive methodology to identify the configurational relationship between conditions and outcomes. As a technique, fsQCA has also undergone development, particularly in regard to its proper application to panel data. The dynamic fsQCA, considering the temporal effect and cross-sectional effect simultaneously, recognizes the intrinsic panel data structure and suggests a new set of generic descriptive metrics for assessing set-theoretic relationship for such panel data.

This paper, referencing to business environment index of the World Bank, Institute for Management Development, and the Economist Intelligence Unit, etc., took into account the Chinese mainland's current business environment and combined the characteristics of sustainable development to establish a new green business environment evaluation index, which consists of the Economic Environment, Government Environment, Social Environment, Digital Environment, and Green Environment. In this case, the proposed evaluation index will be applied to score the green business environment for 30 provinces (with the exception of Tibet, Taiwan, Hong Kong, and Macao) in China. After that, the dynamic fsQCA is applied to find out the possible causal relationships that lead to high green business environment.

2 Literature Review

Creating an eco-friendly business environment is a necessary step toward high-quality economic development, which is a reflection, synopsis, and decision of social progress. To achieve global sustainable development, it is imperative to strike a balance between the environment and the economy and guarantee intergenerational justice. Sustainable development and environmental preservation must be prioritized when creating long-term growth strategies and achieving material benefits. The 17 sustainable development goals in the United Nations' 2030 Agenda for Sustainable Development span a wide variety of environmental and economic issues and urge action from all countries to preserve the environment while fostering economic growth.

In order to evaluate the business environment around the world, few global institutes established evaluation indexes, including World Bank Group (World Bank Group, 2020), World Economic Forum (World Economic Forum, 2020), Economist Intelligence Unit (Economist Intelligence Unit, 2014), and International Institute for Management Development (International Institute for Management Development, 2020), and also released evaluation reports, respectively.

The State Council published the Regulations for Optimization of Business Environment in October 2019 with the goal of enhancing the business environment. The regulations became formally operative on January 1, 2020. The Chinese-specific business environment evaluation index was made clearer by administrative regulations. In May 2020, the Opinions on Accelerating the Improvement of the Socialist Market Economic System in the New Era issued by the State Council clearly stated that the Regulations for Optimization of Business Environment should be implemented and carried out across the whole country, the regulation with market-oriented, legalized, and internationalized business environment should be accelerated. A first-rate business environment that is internationalized, governed by law, and market-oriented is proposed in the Communist Party of China's report for its 20th National Congress in 2022 (People's Daily Press, 2022).

Every province and city have carried out thorough assessments in response to the request for improving the business environment. Guangdong Province, the first province to do both theoretical and empirical study on the business environment, began exploring the business environment at the provincial level in 2012 (Liu & Wei, 2020). Notably, Shanghai was the first city to pilot the nationwide business environment reform in 2018 and gained a great deal of beneficial experience that may be replicated and promoted. In addition to the business environment assessments conducted by provincial and municipal governments, independent organizations also keep an eye on the business climate. In 2017, 2018, and 2020, the Guangzhou Institute of Greater Bay Area published a report on China's municipal business environment (Gu et al., 2020). The PwC, China Media Group, Chinese Academy of Social Sciences Information Research Center, and Wanbo New Economic Research Institute are among the third-party institutions that lead the business environment index evaluation system. Since 2018, the system has demonstrated a blowout development.

Referring to different business environment evaluation indexes aforementioned, combining with the Sustainable Development Goals Theory, the Green Economic Theory, Environment Social Governance Theory, and the Government Governance Theory, this paper established a green business environment evaluation index. There was a total of 27 indicators in the index, which comprised 5 dimensions: economic environment, government environment, social environment, digital environment, and green environment. The indicators and references for the green business environment evaluation index are compiled in Table 1..

Table 1. Green business environment index.

Dimensions	No.	Indicator	References
Economic Environment	X11	Economic Development	Huang and Li (2017) Jiang and Wang (2020) Shi et al. (2016) Sachs et al. (2021) Sachs et al. (2022)
	X12	International Trade	
	X13	Foreign Investment	
	X14	Enterprise Digitization	
	X15	Financing Capacity	
	X16	Transport Efficiency	
Government Environment	X21	Government Revenue Scale	Shang et al. (2020) Duan (2021) Sachs et al. (2021) Sachs et al. (2022)
	X22	Government Balance	
	X23	Tax	
	X24	Land Cost	
Social Environment	X31	Population	Huang and Li (2017) Jiang and Wang (2020) Scarlat et al. (2015) Shang et al. (2020) Wanner (2015)
	X32	Inflation	
	X33	Disposable Income	
	X34	Employment	
	X35	Social Security Level	
	X36	Wage Level	
Digital Environment	X41	Internet penetration	Pei and Hou (2023)
	X42	Internet-related employees	
	X43	Total telecommunications business	
	X44	Mobile internet users	
	X45	Digital inclusive finance development index	
Green Environment	X51	Power Consumption	Yi Cai (2022)
	X52	Environmental Protection Expenditure	
	X53	Waste Disposal	
	X54	Air Pollution	
	X55	Living Environment	

3 Methodology

3.1 Principal Component Analysis

In the domains of economics, management, and the geosciences, principal component analysis (PCA) has been extensively employed due to its capacity to reduce loss of information when compressing high-dimensional data into low dimensions (Gui et al., 2021; Wu & Li, 2021). In order to create a new combination with fewer and unrelated indicators or variables, PCA applied the concept of dimension reduction to many complicated and related variables or index combinations. This study considered five dimensions of the green business environment with specified indicators. PCA was done for each dimension's index in order to concentrate on key issues. The following steps were part of the PCA procedure. The z-score approach is used for variable normalization, as shown below.

$$Z_{ij} = \frac{X_{ij} - \overline{X}_j}{\sigma_j} \tag{1}$$

Then, determining whether the variables are appropriate for the PCA by applying the Bartlett's test of sphericity and the Kaiser-Meyer-Olkin (KMO) index. This study used the KMO index to quantify sample adequacy and Bartlett's test of sphericity ($p < 0.001$) to check the inter-correlation between variables in order to evaluate the validity of PCA. Subsequently, the number of principal components is computed using the eigenvalues and variance contributions, and expression for the linear combination of principle components is found. Lastly, the composite index is computed using the eigenvalues of the primary components as weights. This study used the PCA approach to reduce the dimension of 29 indicators in order to compute the comprehensive score of the green business environment in provinces and municipalities.

3.2 Dynamic fsQCA

To analyze the combination effect exerted by determinants on green business environment, the fuzzy-set qualitative comparative analysis (fsQCA) was used. In the book titled Redesigning Social Inquiry: Fuzzy Sets and Beyond, Ragin (2008) provided a thorough explanation of fsQCA, along with information on its associated set-theoretic approach. Its use as a method is growing in the fields of business and the social sciences (Beynon et al., 2020; Medina-Molina et al., 2022). The analysis is based on set theory, and it uses Boolean minimization, fuzzy-set theory, and combinatorial logic to identify the combinations of case conditions that might be necessary or sufficient to result in an outcome (Kent & Olsen, 2008). As a result, fsQCA employs an inductive methodology to identify the configurational relationships between conditions and outcome (Schneider & Wagemann, 2010).

As a technique, fsQCA has also undergone development, particularly in regards to its proper application to panel data (Garcia-Castro & Ariño, 2016). The dynamic fsQCA, considering temporal effect and cross-sectional effect simultaneously, recognizes the intrinsic panel data structure and suggests a new set of generic descriptive metrics for assessing set-theoretic relationships for such panel data. Although the dynamic fsQCA deviated from the central ideas of consistency and coverage, Garcia-Castro and Ariño (2016) proposed guidelines for assessing how stable the consistencies and coverage are across cases (within consistency and within coverage) and over time (between consistency and between coverage). Consequently, three alternative forms of consistency are proposed by dynamic fsQCA: pooled consistency (POCONS), between consistency (BECONS), and within consistency (WICONS). Specifically, BECONS evaluates the cross-sectional consistency for each year, WICONS measures the consistency of the relations across time for each case, and POCONS assesses the consistency of each causal combination (Guedes et al., 2016). In addition, distance is the key to be taken into account in the dynamic fsQCA. The distances between BECONS and POCONS represent how stability a consistency has held over time. As a results, the smaller the distance, the more stable the consistency. If it is high, it is important to assess the temporal effect on the panel. The distances between the WICONS and POCONS were also computed to assess how the WICONS vary between cases (Pineiro-Chousa et al., 2023).

According to the prior studies, the basic steps in dynamic fsQCA are shown in Fig. 1. Calibration is the first step after data collection, which determine the degree of membership for conditions and the outcome in the set they represent. In the following, a truth table should be generated, which is a data matrix for necessity and sufficiency analyses. Any condition that should present or absent in order to achieve the outcome can be found in the necessity analysis, with a consistency criterion over 0.9 (Wagemann, 2012). Thereafter, the sufficiency analysis examines every possible logical combination of causal conditions that could result in the outcome, which uses raw consistency benchmark of sufficiency analysis above 0.8 accompanied by a benchmark for proportional reduction in inconsistency (PRI) score of over 0.65 (Misangyi & Acharya, 2014). Finally, the causal configuration analysis can be drawn according to the complex, parsimonious, and intermediate solutions.

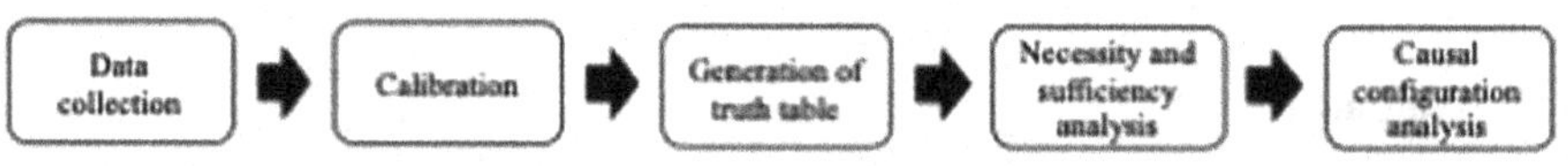

Fig. 1. Procedures for dynamic fsQCA

4 Empirical Study

4.1 Research Design and Data Collect

The green business environment evaluation index is made up of various dimensions, including the economic environment, government environment, social environment, digital environment, and green environment. In order to more accurately analyze the influence, principal component analysis (PCA) is used to determine the scores for the green business environment's five aspects. Following that, the multi-factors impact is investigated using the dynamic fsQCA.

This paper used panel data from 30 provinces in China (excluding Tibet, Taiwan, Hong Kong, and Macao) between 2011 and 2022. All data of selected variable come from China Statistical Yearbook, and Provincial Statistical Yearbooks. Aforementioned, PCA is used to evaluate the provincial green business environment. Consequently, Table 2. displays the descriptive statistics for all the variables used. In bellowing table, GBE, ECO, GOV, SOC, DIG, GRE represent the green business environment, economic environment, government environment, social environment, digital environment, and green environment, respectively.

Table 2. Descriptive statistics

Variables	N	Mean	Std.	Min	Max
GBE	360	45.00	13.61	25.94	95.65
ECO	360	45.00	13.18	4.23	85.73
GOV	360	45.00	13.37	15.01	94.98
SOC	360	45.00	13.95	13.05	85.45
DIG	360	45.00	17.07	21.56	114.90
GRE	360	45.00	13.57	16.65	98.71

4.2 Dynamic QCA Analysis

Data Calibration. Following the basic steps outlined above, the original data should be calibrated to a fuzzy-set membership degree ranged from 0 to 1 that represents the membership of a variable (Ragin, 2008). As described by Woodside (2013), full membership, crossover (neither in nor out), and full non-membership are denoted by 0.95, 0.5, and 0.05, respectively. Table 3. displays the calibration values and fuzzy value descriptive statistics for each condition and outcome.

Table 3. Calibration values

	Calibration values		
	95%	50%	5%
Outcome			
GBE	75.65	40.50	30.07
Conditions			
ECO	69.40	44.45	24.26
GOV	74.66	41.62	30.96
SOC	68.28	44.43	22.34
DIG	83.10	40.36	26.47
GRE	72.15	43.27	27.24

Necessity and Sufficiency Analyses. According to the analysis of the necessary requirements, a necessary condition must be present in every occurrence of an outcome (Ragin, 2008). Alternatively, the outcome occurs when that condition occurs, even though the outcome occurs under other conditions (Guedes et al., 2016). The consistency must be larger than 0.9 in order to qualify as a required condition (Schneider & Wagemann, 2012). The overview of antecedent conditions for low-carbon logistics capability is shown in Table 4.. There is no essential condition because, as in this study, none of the conditions have a consistency higher than 0.9.

Table 4. Necessary conditions

Conditions	Consistency	Coverage
ECO	0.820	0.791
~ECO	0.601	0.562
GOV	0.692	0.719
~GOV	0.666	0.582
SOC	0.892	0.844
~SOC	0.523	0.498
DIG	0.873	0.894
~DIG	0.571	0.506
GRE	0.856	0.853
~GRE	0.666	0.582

After the necessary conditions were examined, the sufficient conditions were analyzed. To offer all theoretically feasible configurations of variables in 2k rows (k = number of variables), a truth table should be built, where each row represents a particular configuration. The truth table lists all possible logical combinations as well as cases that satisfy each combination. Logical reminders that contained no cases were not included in the analysis. As stated by Misangyi and Acharya (2014), the raw consistency benchmark of sufficiency analysis should be more than 0.8 and be followed by a standard for PRI (proportional reduction in inconsistency) score of over 0.65 in order to avoid "simultaneous subset" relations of configurations in both the outcome and its absence (Chen et al., 2023; Fiss, 2011). The cut-off consistency and PRI in this study are 0.9 and 0.75, respectively.

Because the intermediate solution meets the theoretical justifications and includes simplifying assumptions, this paper focus on the intermediate solution companied with parsimonious solution. The overall findings for the full panel are shown in Table 5. The recommendation given by Schneider and Wagemann (2006) is that the solution consistency should be greater than 0.75. Additionally, the coverage should range from 0.25 to 0.85 (Ragin, 2008). In terms of this paper, the results met the standards for consistency and coverage, which are 0.973 and 0.780, respectively.

Configuration Paths for Green Business Environment. In Table 5, there are two configurations that illustrated the possible causal relationships that led to high green business environment. The configuration with higher coverage (0.968) and with a very good consistency (0.987) is ECO*GOV*SOC*DIG. This configuration is preferable in terms of government environment, social environment, digital environment, with the economic environment plays supporting roles. The second configuration, SOC*DIG*GRE, has considerable coverage (0.941) and a considerable consistency (0.973). This configuration indicates that provinces with social environment, digital environment, and green environment have better green business environment. Given that both social environment and digital factors are the core conditions in these two configuration paths, the

Table 5. Configuration for high green business environment

	Configuration 1	Configuration 2
ECO	•	
GOV	●	
SOC	●	●
DIG	●	●
GER		●
Consistency	0.987	0.973
Raw coverage	0.968	0.941
Unique coverage	0.499	0.749
Solution consistency	0.973	
Solution coverage	0.780	

Note: ●=Core causal condition (present); •=Peripheral causal condition (present); ⊗=Core causal condition (absent); ⊗=Peripheral causal condition (absent). Blank spaces indicate "don't care."

government environment in the first path and the green environment in the second path are unique. Consequently, the two paths are labeled as *government-led* type and *green innovation* type, respectively.

4.3 Robustness Analysis

This paper employs three different calibration values to examine the results' validity, which are 0.90, 0.50, and 0.10, respectively. The three configurations are returned by the intermediate solution in essentially the same form. The results' validity is thus supported by the robustness analysis.

5 Discussion and Conclusions

Developing a green business environment index and examining its determinants in light of the digital economy's development is especially crucial for achieving the "dual-carbon" goal. This study examines the configuration path to achieve a high green business environment in 30 Chinese provinces using a dynamic fsQCA. The conclusions are provided below. First, individual dimensions of a green business environment do not necessarily lead to high-level green business environment, whereas a combination of multiple conditions can high level of green business. Second, two paths can be found to achieve high level of green business environment: (1) a path driven jointly by government environment, social environment, and digital environment; (2) a path driven collaboratively by social environment, digital environment, and green environment.

The results of this study have several practical implications. First, there are differences in the green business environment amongst provinces. In order to consistently improve the business environment, provinces should coordinate development. Secondly, encouraging the high-quality development of digital economy. The digital environment shows up as a key component in both configuration paths. In order to maximize the business environment, the government should encourage the development of digital infrastructure and raise the bar for digital applications. Finally, choosing the appropriate path to realize high level of green business environment. Provinces can choose their own path according to their resource endowment and social development, so as to achieve a high-level green business environment.

Disclosure of Interests. The authors have no competing interests to declare that are relevant to the content of this article.

References

Beynon, M.J., Jones, P., Pickernell, D.: Country-level entrepreneurial attitudes and activity through the years: a panel data analysis using fsQCA. J. Bus. Res. **115**, 443–455 (2020)

Chen, Y., Hong, J., Tang, M., Zheng, Y., Qiu, M., Ni, D.: Causal complexity of environmental pollution in China: a province-level fuzzy-set qualitative comparative analysis. Environ. Sci. Pollut. Res. **30**(6), 15599–15615 (2023)

Duan, Y.H.: The index of government environment of doing business in the evaluation of local government effectiveness: literature review and index design. Adm. Law **4**, 70–79 (2021)

Economist Intelligence Unit: Business Environment Ranking and Index 2014. T. E. I. Unit (2014)

Fiss, P.C.: Building better causal theories: a fuzzy set approach to typologies in organization research. Acad. Manag. J. **54**, 393–420 (2011)

Garcia-Castro, R., Ariño, M.A.: A general approach to panel data set-theoretic research. J. Adv. Manag. Sci. Inform. Syst. **2**, 63–76 (2016)

Gu, X.Q., Li, Y.D., Yu, H.X.: Policy measures and effect evaluation of business environment in the Yangtze river delta. China Bus. Market **6**, 86–95 (2020)

Guedes, M.J., da Conceição Gonçalves, V., Soares, N., Valente, M.: UK evidence for the determinants of R&D intensity from a panel fsQCA. J. Bus. Res. **69**(11), 5431–5436 (2016)

Gui, H.B., Sun, P., Jiang, M.: An international comparative analysis of competitiveness of innovation culture. Sci. Manag. Res. **39**(4), 159–167 (2021)

Huang, Y., Li, L.: A comprehensive assessment of green development and its spatial-temporal evolution in urban agglomerations of China. Geogr. Res. **36**(7), 1309–1322 (2017)

International Institute for Management Development.: IMD World Competitiveness Yearbook (2020)

Jiang, Q., Wang, Y.: How does administrative monoploy affect the economic growth of Shandong province? Experience analysis based on dual perspective of demand and supply side. Rev. Econ. Manag. **36**(3), 140–151 (2020)

Kent, R., Olsen, W.: Using fsQCA a brief guide and workshop for fuzzy-set qualitative comparative ayalsis (2008)

Liu, Z.Y., Wei, L.L.: A review of business environment construction in China: development track, main achievements and future direction. Contemp. Econ. Manag. **42**(2), 22–27 (2020)

Medina-Molina, C., Pérez-Macías, N., Gismera-Tierno, L.: The multi-level perspective and micromobility services. J. Innov. Knowl. **7**(2), 100183 (2022)

Misangyi, V.F., Acharya, A.G.: Substitutes or complements? A configurational examination of corporate governance mechanisms. Acad. Manag. J. **57**, 1681–1705 (2014)

Pei, R., Hou, G.Y.: The impact of the business environment on the development of the digital economy and paths for enhancement. J. Tech. Econ. Manag. **11**, 23–27 (2023)

People's Daily Press.: Hold High the Great Banner of Socialism with Chinese Characteristics and Strive in Unity to Build a Modern Socialist Country in All Respects: Report to the 20th National Congress of the Communist Party of China. People's Daly Press (2022)

Pineiro-Chousa, J., Lopez-Cabarcos, M.A., Perez-Pico, A.M., Caby, J.: The influence of twitch and sustainability on the stock returns of video game companies: before and after COVID-19. J. Bus. Res. **157**, 113620 (2023)

Ragin, C.C.: Redesigning Social Inquiry: Fuzzy Sets and Beyond. University of Chicago (2008)

Sachs, J.D., Kroll, C., Lafortune, G., Fuller, G., Woelm, F.: Sustainable Development Report 2021 (2021)

Sachs, J.D., Kroll, C., Lafortune, G., Fuller, G., Woelm, F.: Sustainable Development Report 2022 (2022)

Scarlat, N., Dallemand, J.-F., Monforti-Ferrario, F., Nita, V.: The role of biomass and bioenergy in a future bioeconomy: policies and facts. Environ. Dev. **15**, 3–34 (2015)

Schneider, C.Q., Wagemann, C.: Reducing complexity in qualitative comparative analysis (QCA): remote and proximate factors and the consolidation of democracy. Eur J Polit Res **45**(5), 751–786 (2006)

Schneider, C.Q., Wagemann, C.: Standards of good practice in qualitative comparative analysis (QCA) and fuzzy-sets. Comp. Sociol. **9**, 397–418 (2010)

Schneider, C.Q., Wagemann, C.: Set-theoretic Methods for the Social Sciences: A Guide to Qualitative Comparative Analysis. Cambridge University Press (2012)

Shang, D., Li, H.J., Yao, J.: Green economy, green growth and green development: concept connotation and literature review. Foreign Econ. Manag. **42**(12), 134–151 (2020)

Shi, M.J., Fan, X.W., Pang, R., Chen, X.Y.: Perspective of green growth of Chinese cities: an evaluation based on new resource economy index. J. Environ. Econ. **1**(2), 46–59 (2016)

Wagemann, S.: Set-Theoretic Methods for the Social Sciences. Cambridge University Press (2012)

Wanner, T.: The new "passive revolution" of the green economy and growth discourse: Maintaining the "sustainable development" of neoliberal capitalism. New Political Econ. **20**(1), 21–41 (2015)

World Bank Group.: Doing Business 2020: Comparing Business Regulation in 190 Economics. World Bank (2020)

World Economic Forum: The Global Competitiveness Report Special Edition 2020: How Countries are Performing on the Road to Recovery (2020)

Wu, S.H., Li, Y.J.: Research on evaluation of urban competitiveness in western China based on principal component analysis. Econ. Probl. **11**, 115–120 (2021)

Cai, Y.: What are the differences between the various ESG evaluation systems? https://baijiahao.baidu.com/s?id=1725994979055968002&wfr=spider&for=pc. Accessed 28 Feb 2022 (2022)

A Multimodal Analysis of Streaming Subscription

Yi-Cheng Lee[1], Yu-chen Yang[1], Yen-Hsien Lee[2(✉)], and Tsai-Hsin Chu[2]

[1] National Sun Yat-sen University, Kaohsiung 804, Taiwan
[2] National Chiayi University, Chiayi 600, Taiwan
yhlee@mail.ncyu.edu.tw

Abstract. Live streaming, a novel platform enabling users to engage in real-time interactions over the internet, has gained widespread popularity across major social media platforms. This emerging online service involves streamers broadcasting audio and video content to viewers in real-time, making it one of the most sought-after social media experiences in recent years. Live streaming Content creators are usually referred to as "streamers" and viewers participate in the live stream to consume this content. As viewers engage with the streamer during the content consumption, they contribute in diverse ways that actively influence the content creation process. Users engage in multiple interactions while watching live videos, such as subscribing to channels, sending messages, sharing emoticons, and giving virtual gifts to the live streamer and other users. Subscription, donation, and advertising revenue sharing are three main business models for streamers to earn money. The more viewers the streamers can consistently attract, the more revenue they can make. Subscribers to a particular streaming channel will usually be regular viewers, so effectively converting viewers into subscribers will be vital to sustaining the streaming channel. Most prior studies on viewer subscriptions in live streaming focus on the motivation and contributing factors be-hind subscription behavior. They accordingly construct the prediction model for identifying potential subscribers from the viewers. The prediction model is suitable for real-time scenarios, which can assist platforms in timely identifying potential subscribers likely to receive more subscriptions. However, such research has not emphasized live streaming's characteristics as a communication medium. As a result, this study intends to apply foundational communication models to derive critical features related to the sender (streamer), receiver (audience), and message (content) and aims to construct the model for predicting the decrease or increase of subscribers to a particular streaming channel. Specifically, we utilize past empirical findings to identify streamer, audience, and content features from a communication perspective, engineering relevant predictors grounded in communication theory. This multimodal time-series approach combines textual, visual, and audio data to better model the complex interplay of factors driving subscriptions on live streaming platforms. We collect the evaluation dataset from Twitch.tv and conduct an empirical evaluation to evidence the performance of the proposed approach in comparison with the prediction models constructed without considering the characteristics of live streams.

Keywords: Live Stream · Subscription Prediction · Communication Theory · Multimodal Analysis

F. F.-H. Nah and K. L. Siau (Eds.): HCII 2024, LNCS 14721, pp. 200–208, 2024.
https://doi.org/10.1007/978-3-031-61318-0_14

1 Introduction

Live streaming, a novel platform enabling users to engage in real-time interactions over the internet, has been gaining widespread popularity across major social media platforms. This emerging online service involves streamers broad-casting audio and video content to viewers in real time, making it one of the most sought-after social media experiences in recent years.

More recently, becoming a professional gamer or streamer has emerged as an ideal career choice for many young people, often without considering the precariousness of lacking healthcare, unions, or other social safety nets (Guarriel-lo, 2019). This study aims to help streamers look into the features of streamers, subscribers, and streaming content that most attract subscriptions to assist them in building their careers. Specifically, this research focuses on video game streaming, as it represents one of the most popular and lucrative forms of live streaming (Johnson, M. R., and Wood-cock, J., 2019; Sjöblom, M., and Hamari, J., 2019). The success of platforms like Twitch has shown gaming content's ability to attract large audiences and subscriber bases. There exist numerous studies on viewer subscriptions in live streaming, most of which focus on the motivation and contributing factors behind subscription behavior (Lin, Y et al., 2021; Hilvert-Bruce et al., 2018).

Compared to previous work, this study comprehensively considers the influence of multimodal information (text, audio, and visual) on subscription outcomes. The prediction model is suitable for real-time scenarios, which can assist platforms in timely identifying potential streamers likely to receive more subscriptions.

Live streaming represents a new type of social media where audiences interact with broad-casters via chat and re-ward performers, known as social live streaming services (SLSS) (Sjöblom, M., and Hamari, J., 2017). However, cur-rent research on feature selection has not emphasized live streaming's characteristics as a communication medium. Therefore, this study applies foundational communication models (Lasswell, 1948; Shannon & Weaver, 1949) to derive key features related to the sender (streamer), receiver (audience), and message (content). Specifically, we utilize past empirical findings to identify streamer, audience, and content features from a communication perspective, engineering relevant predictors grounded in communication theory. This multimodal time-series approach combines textual, visual, and audio data to better model the complex interplay of factors driving subscriptions on live streaming platforms like Twitch.tv.

In summary, this research aims to predict subscriber growth by considering live streaming as a communication process, selecting predictive features based on communication theory and models. The focus on video game streaming provides a relevant domain for investigating the drivers of subscriptions.

2 Related Work

2.1 Subscription/Donation/Gifting Intention

The live streaming audience can be classified into four distinct categories: viewers, followers, subscribers, and donors. Notably, subscribers and donors can financially support streamers by paying their chosen amount, significantly boosting the streamers' earnings.

Twitch permits users to watch broadcasts and videos without requiring payment. Consequently, for streamers, subscribers and donors stand out as the most valuable consumers, as they make substantial contributions to their income.

Prior studies. Have predominately used questionnaires to examine the intentions behind behaviors such as subscription, donation, and gifting on live streaming platforms, focusing on these actions as forms of social interaction (Wan et al., 2017; Li et al., 2021; Yi Li & Yi Peng, 2021; Sjöblom & Hamari, 2017). While most have utilized questionnaires, some research has developed models to predict subscription, donation, and gifting intentions by extracting features from platform data. For example, features such as totals of tips, comment word count, like counts, and broadcaster and viewer emotions have been used to predict tipping intention (Zhang, Y et al., 2017). However, there remains a lack of research grounded in communication theories for predicting subscription intention on live streaming platforms. This thesis aims to address this gap by selecting features informed by communication theories to predict subscription intention.

In summary, while existing research has focused on social interaction perspectives, there is an opportunity for this thesis to contribute to the literature by developing a subscription intention model grounded in communication theories.

2.2 Ultimodal with Time Series Data

In research on predicting viewers' watching, subscribing, and donating behavior on live streaming platforms, Liu et al. (2022) established a structural equation model to predict the number of gifts after processing viewers' real-time comments through traditional machine learning. Lin et al. (2021) found that streamers' thankful expressions would also inspire viewers to pay for gifts. They chose traditional machine learning methods like logistic regression, random forest, and support vector machines, which are not suitable for multi-modal scenarios (Xi, D et al., 2023). Regarding predicting subscribers, viewers' subscription behavior may be triggered by the content from a previous period of the live streaming rather than cross- sectional time points. Existing methods exploring viewers' subscription behavior cannot fully utilize all the multimodal information available to viewers, such as the live scene, sound, and comments. In this research, we try to address these deficiencies and consider the multimodal time-series information for subscription prediction.

Multimodal fusion refers to processing and integrating information from different modalities so the fused information can be used for downstream prediction tasks like classification and regression (Baltrušaitis et al., 2018). This technology is widely applied in various fields including multimodal emotion recognition (Soleymani et al., 2011). Compared to single modalities, multimodal fusion can often obtain more robust prediction results, as verified in audio-visual speech recognition research (Potamianos et al., 2003).

There are three main architectures for multimodal fusion: early fusion, late fusion, and hybrid fusion (Atrey, P. K et al., 2008). Early and late fusions usually occur at the beginning and end of the learning stage, respectively (Zhang et al., 2021). Their effectiveness depends on the correlation between modalities. Hybrid fusion combines their advantages but increases model complexity. For live streaming subscription prediction, we evaluate these architectures to select the best performing strategy.

3 Proposed Approach

3.1 Data Collection

We have two sources to collect our data. One is the Twitch TV application programming interface (API), which pro-vides developers access to information such as user profiles, stream introductions, and message records in the channel's video on demand (VOD). However, since subscriptions are private to streamers, Twitch only authorizes this data to streamers or their authorized applications. Therefore, we utilize Twitch Ticker to scrape time series subscription data. Additionally, Twitch no longer provides access to live comments after a platform update. Instead, we will use Selenium to scrape live comment data from the VODs. Based on results from the Twitch API, Valorant is the game targeted in our study, and we collect data by weekly viewing from January 3, 2024 to January10, 2023. We build a predictive model to detect subscription numbers. Following Xi, D et al., (2023), who used a ten-minute observation window and outperformed other advanced models by at least 8% in F1 score, we will use five-minute sliding windows to split the data.

In summary, we leverage the Twitch API and web scraping to collect user, stream, and subscription data. We will focus on top streamers and games over a 1 week period, we drew inspiration from a classical model (Baltrušaitis et al., 2018), which delineates the modeling process into 4 distinct steps: Data preprocessing and features extraction, Representation, Fusion, as illustrated in Fig. 1.

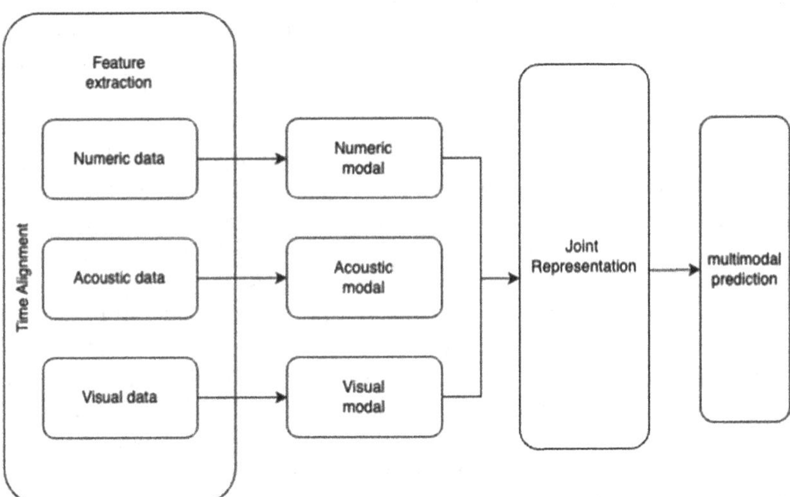

Fig. 1. Overall framework.

3.2 Data Processing and Feature Extraction

According to the results of our data collection, additional variables will need to be incorporated into the model during data preprocessing, as summarized in Table 1.

Table 1. Summary of Features.

Features	Description	Reference
Streamer		
Follower Count	Number of followers for the streamer	Leung et al. (2021)
Content		
Face time.	Amount of time when the streamer's face is visible during the streaming session	Lu et al. (2018)
Volume	Intensity or amplitude of the speech signal	
Speaking rate	Number of words articulated per minute of speech	
Viewer		
Viewer count	Number of viewers who watching the stream	Leung et al. (2021)

We collect numeric data, acoustic data and visual data. First, for the numeric data (follower count), we crawled the changes in viewer count every 10 min from https://twitchtracker.com using Python requests. Second, acoustic information serves as a supplementary channel that facilitates enhanced the performance of model learning. Removing the acoustic information can bring about the most damage to the recall (D. Xi et al., 2023). We use FFmpeg[1] to perform audio conversion and sets the output audio sample rate to 44.1 kHz segments the audio into chunks of 10 s with a 50% overlap (5 s) between consecutive segments, initializes the Wav2Vec2 model and tokenizer from the Hugging Face Transformers library using the Facebook's Wav2Vec2 framework (Baevski et al., 2020). Calculates the total word count for each file based on the transcriptions of individual segments. It also calculates the root mean square (RMS) as a measure of volume. Finally, for visual data, we initialized the Multi-task Cascaded Convolutional Networks (MTCNN) model to extract features related to face time, the model first uses a proposal network to generate bounding box candidates, then refines those candidates with a refinement network, and finally outputs facial landmarks with an output network. As MTCNN may falsely detect game characters as faces, we leveraged a pre-trained MTCNN as a feature extractor and fine-tuned it on 250 real and 250 fake images to aid in human face recognition. This was done by adding a 512-node fully connected layer with ReLU activation and a dropout layer with rate 0.3 after feature extraction to prevent overfitting. This additional classifier enhances vanilla MTCNN's ability to discern real faces in complex gaming footage with non-human characters. The fine-tuned facial recognition model with the added fully connected and dropout layers achieved a validation accuracy of 0.959 (Fig. 2).

3.3 Modal Fusion

In our multimodal fusion strategy, we integrate three modalities: numeric features, audio features, and video features. The goal is to jointly leverage information from these diverse

[1] https://ffmpeg.org/.

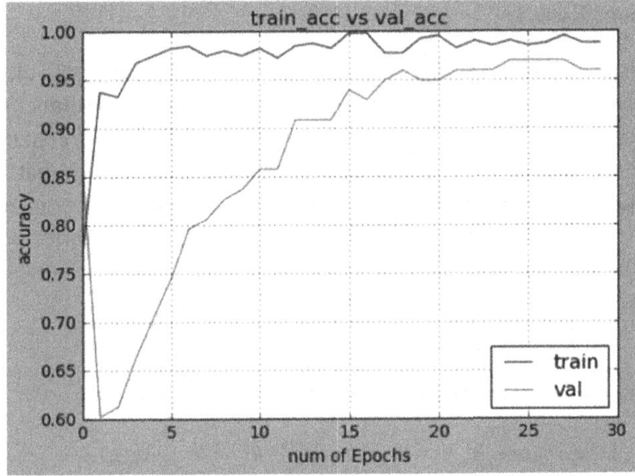

Fig. 2. Validation performance of the fine-tuning model.

modalities to enhance the predictive performance of our regression model. We adopt a late fusion strategy, where the representations from each modality are concatenated before being processed by the neural network.

After aligning signals from different modalities, we define a regression model using a neural network architecture implemented in PyTorch. The model consists of two fully connected layers (Linear layers) with ReLU activation functions. The first layer has 64 neurons, and the second layer (fc2) has 1 neuron, representing the regression output. The model's architecture is structured to handle the concatenated input feature.The model is trained using the Mean Squared Error Loss (nn.MSELoss) and optimized with the Adam optimizer.

4 Discussion

4.1 Data Description

Our dataset is sourced from Twitch, a prominent live streaming video platform widely popular among young audiences. The dataset is constructed by combining information from two primary sources: the live streaming video content and the corresponding follower records. Specifically, we focused on tracking the top 20 live streaming channels based on weekly viewership rankings, with an emphasis on content related to the game "Valorant", a 5v5 character-based tactical first-person shooter (FPS). The data collection period spans from December 27, 2023, to January 3, 2024. During this timeframe, we gathered live streaming videos totaling over 75 h and corresponding to 3306 following records. All collected video clips maintain a frequency of 60 frames per second and a resolution of 1920x1080. This ensures a comprehensive and high-quality representation of the live streaming content for subsequent analysis.

4.2 Experiment Results

In simpler terms, we are assessing the impact of visual and acoustic elements on our model's performance. To do this, we conduct tests by comparing the model's performance with and without these elements. Numeric modalities, which are frequently associated with viewers and videos, serve as control information. We ensure that all modalities are represented in the same vector space through a 2-layer fully-connected network. Analyzing the results, we observed the impact of different modalities on our model's performance. Table 3 presents the R-squared (R2) values and Root Mean Square Error (RMSE) on the test set for various modalities.

Table 2. Results on modalities

Description	R^2	RMSE
Numeric + Acoustic	0.79	396.886
Numeric + Visual	0.875	305.952
Numeric + Acoustic + Visual	0.865	318.144

Table 2 shows that incorporating viewer information along with visual modality yields the highest R2, followed by the late fusion of all three modalities. Interestingly, the combination of visual and acoustic features demonstrates limited complementary effects. Our analysis indicates that, for live streamers, maintaining a visible presence over an extended period is more crucial than optimizing for audio performance.

5 Conclusion and Future Work

Live streaming has rapidly become a dominant form of entertainment, but research on viewer behaviors like subscriptions has been limited due to a reliance on surveys or static analyses. In this work, we aim to model Twitch follower growth through multi-modal time series modeling grounded in communication theory. Much prior work treats live streaming as just another social media, selecting features based on intuition rather than theoretical motivations. However, streaming is highly dynamic and fundamentally involves video, audio, and text interactions between streamers and audiences.

In our task, the collection of modalities from both audio and visual sources plays a pivotal role. These modalities encompass features such as speech rate, volume, frequency of on-screen presence, among others. The gathered data is subsequently transformed into input features for our multimodal model. This model design allows us to achieve 0.865 R2 and 318.144 RMSE.

As this is research in progress, the raw data will be expanded by increasing the time range for data collection. Currently, there are limited features leading to potential over-fitting of the models. To mitigate this, additional features will be incorporated: For the numeric modal, LSTM will be utilized for time series analysis. A textual modal will be added with sentiment analysis of the live stream viewer comments. More features will be extracted from the acoustic modal using Mel-Frequency Cepstral Coefficients (MFCCs) and fine-tuning a Wav2Vec2 model for emotion classification. The visual modal will leverage a pre-trained Vision Transformer (ViT) for facial emotion recognition. With the increase in features, attention mechanisms will be incorporated into the existing models to improve fusion. Fusion strategies previously utilized in relevant research will also be explored, such as Fukui et al.'s (2016) modifications to bilinear transformations to enhance efficiency and Xi et al.'s (2023) orthogonal projection (OP) model to enable cross-modal interaction without additional parameters. Through testing, suitable methods will be selected for integration into the model.

References

Atrey, P.K., Hossain, M.A., El Saddik, A., Kankanhalli, M.S.: Multimodal fusion for multimedia analysis: a survey. Multimedia Syst. **16**, 345–379 (2010)

Baltrušaitis, T., Ahuja, C., Morency, L.P.: Multimodal machine learning: a survey and taxonomy. IEEE Trans. Pattern Anal. Mach. Intell. **41**(2), 423–443 (2018)

Cabeza-Ramírez, L.J., Fuentes-Garcia, F.J., Muñoz-Fernandez, G.A.: Exploring the emerging domain of research on video game live streaming in web of science: state of the art, changes and trends. Int. J. Environ. Res. Public Health **18**(6), 2917 (2021)

Devlin, J., Chang, M.W., Lee, K., Toutanova, K.: BERT: pre-training of deep bidirectional transformers for language understanding. arXiv preprint arXiv:1810.04805 (2019)

Fukui, A., Park, D.H., Yang, D., Rohrbach, A., Darrell, T., Rohrbach, M.: Multimodal compact bilinear pooling for visual question answering and visual grounding. In: Proceedings of the 2016 Conference on Empirical Methods in Natural Language Processing, pp. 457–468 (2016)

Gandolfi, E.: To watch or to play, it is in the game: the game culture on Twitch.tv among performers, plays and audiences. J. Gaming Virtual Worlds **8**(1), 63–82 (2016)

Gros, D., Hackenholt, A., Zawadzki, P., Wanner, B.: Interactions of Twitch users and their usage behavior. In: Social Computing and Social Media. Technologies and Analytics: 10th International Conference, SCSM 2018, Held as Part of HCI International 2018, Las Vegas, NV, USA, July 15-20, 2018, Proceedings, Part II 10, pp. 201–213. Springer, Cham (2018). https://doi.org/10.1007/978-3-319-91485-5_15

Guarriello, N.B.: Never give up, never surrender: Game live streaming, neoliberal work, and personalized media economies. New Media Soc. **21**(8), 1750–1769 (2019)

Hamilton, W.A., Garretson, O., Kerne, A.: Streaming on twitch: fostering participatory communities of play within live mixed media. In: CHI 2014: Proceedings of the SIGCHI Conference on Human Factors in Computing Systems, pp. 1315–1324 (2014)

Hilvert-Bruce, Z., Neill, J.T., Sjöblom, M., Hamari, J.: Social motivations of livestreaming viewer engagement on Twitch. Comput. Hum. Behav. **84**, 58–67 (2018)

Johnson, M.R., Woodcock, J.: 'It's like the gold rush': the lives and careers of professional video game streamers on Twitch.tv. Inform. Commun. Soc. **22**(3), 336–351 (2019)

Leung, F.F., Gu, F.F., Li, Y., Zhang, J.Z., Palmatier, R.W.: Influencer marketing effectiveness. J. Mark. **86**(6), 93–115 (2022)

Li, R., Lu, Y., Ma, J., Wang, W.: Examining gifting behavior on live streaming platforms: an identity-based motivation model. Inform. Manage. **58**(6), 103406 (2021)

Li, Y., Peng, Y.: What drives gift-giving intention in live streaming? The perspectives of emotional attachment and flow experience. Int. J. Hum.-Comput. Interact. (2021). https://doi.org/10.1080/10447318.2021.1885224

Lin, Y., Yao, D., Chen, X.: Happiness begets money: emotion and engagement in live streaming. J. Mark. Res. **58**(3), 417–438 (2021)

Liu, M., Wang, R., Huang, Z., Shan, S., Chen, X.: Self-supervised learning for facial expression recognition using triplet information. In: Proceedings of the 26th ACM international conference on Multimedia, pp. 1083–1091 (2018)

Lu, S., Yao, D., Chen, X., Grewal, R.: Do larger audiences generate greater revenues under pay what you want? Evidence from a live streaming platform. Mark. Sci. **40**(5), 964–984 (2021)

Lu, Z., Xia, H., Heo, S., Wigdor, D.: You watch, you give, and you engage: a study of live streaming practices in China. In: Proceedings of the 2018 CHI Conference on Human Factors in Computing Systems, pp. 1–13. Association for Computing Machinery, New York (2018). https://doi.org/10.1145/3173574.3174040

Pires, K., Simon, G.: YouTube live and Twitch: a tour of user-generated live streaming systems. In: Proceedings of the 6th ACM multimedia systems conference, pp. 225–230 (2015)

Potamianos, G., Neti, C., Luettin, J., Matthews, I.: Audio-visual automatic speech recognition: an overview. Issues Visual Audio-Visual Speech Process. **22**, 23 (2004)

Scheibe, K., Fietkiewicz, K.J., Stock, W.G.: Information behavior on social live streaming services. J. Inform. Sci. Theory Pract. **4**(2), 6–20 (2016)

Shannon, C.E.: A mathematical theory of communication. Bell Syst. Tech. j. **27**(3), 379–423 (1948)

Sjöblom, M., Hamari, J.: Why do people watch others play video games? An empirical study on the motivations of Twitch users. Comput. Hum. Behav. **75**, 985–996 (2017)

Soleymani, M., Pantic, M., Pun, T.: Multimodal emotion recognition in response to videos. IEEE Trans. Affect. Comput. **3**(2), 211–223 (2011)

Sun, C., Qiu, X., Xu, Y., Huang, X.: How to fine-tune BERT for text classification?. In: China National Conference on Chinese Computational Linguistics, pp. 194–206. Springer, Cham (2019).https://doi.org/10.1007/978-3-030-32381-3_16

Twitch: Audience. http://twitchadvertising.tv/audience/. Accessed 27 Jan 2024

Vaswani, A., et al.: Attention is all you need. In: Advances in neural information processing systems, vol. 30 (2017)

Wan, J., Lu, Y., Wang, B., Zhao, L.: How attachment influences users' willingness to donate to content creators in social media: a socio-technical systems perspective. Inform. Manage. **54**(7), 837–850 (2017)

Xi, D., Tang, L., Chen, R., Xu, W.: A multimodal time-series method for gifting prediction in live streaming platforms. Inf. Process. Manage. **60**(3), 103254 (2023)

Zhang, H., Yang, Y., Zhao, J.: Does game-irrelevant chatting stimulate high-value gifting in live streaming? A session-level perspective. Comput. Hum. Behav. **138**, 107467 (2023)

Zhang, Y., Hua, L., Jiao, Y., Zhang, J., Saini, R.: More than watching: an empirical and experimental examination on the impacts of live streaming user-generated video consumption. Inform. Manag. **60**(3), 103771 (2023)

Zhou, J., Zhou, J., Ding, Y., Wang, H.: The magic of danmaku: a social interaction perspective of gift sending on live streaming platforms. Electron. Commer. Res. Appl. **34**, 100815 (2019)

Zhang, H., Xu, H., Tian, X., Jiang, J., Ma, J.: Image fusion meets deep learning: a survey and perspective. Inform. Fusion **76**, 323–336 (2021)

Digital Transformation in Banking–How Do Customers Assess the Quality of Digital Banking Services?

Andrea Müller(✉), Annebeth Demaeght, Larissa Greschuchna, and Joachim Reiter

1Hochschule Offenburg – University of Applied Sciences, Badstrasse 24, 77652 Offenburg, Germany
{andrea.mueller,annebeth.demaeght}@hs-offenburg.de

Abstract. The digital transformation is presenting financial institutions with a number of challenges, not only in terms of technology, but also in terms of delivering new, user-optimized services to customers. In order to provide the best possible touchpoints and access to services, banking institutions need to identify the wishes and needs of their customers. This paper deals with the results of an iterative research with German banks based on the five SERVQUAL model dimensions: tangibles, reliability, responsiveness, assurance and empathy. It presents the results of two quantitative surveys of bank customers in Southern Germany and provides recommendations for financial institutions for their future digital activities.

Keywords: Digital Transformation · Quantitative Marketing Research · SERVQUAL Model · Financial Services · Banking Institutions

1 Digital Transformation–Specific Challenges for Banking Institutions

In the ever-evolving landscape of the financial sector, the advent of digital transformation has become a driving force reshaping the banking industry. Digital transformation refers to the integration of digital technologies into all aspects of business operations, fundamentally altering the way organizations operate and deliver value to their customers. In the context of banking, this transformation has particularly impacted traditional banking service processes and customer experiences. Four specific aspects are highly relevant:

Aspect 1: User Experience. One of the most significant outcomes of digital transformation in banking is the need for an enhanced customer experience: Traditional banking methods have given way to seamless and personalized digital interactions, enabling customers to access a new dimension of services with just a few clicks and wherever they are. Mobile banking apps, online account management, and 24/7 customer support have become standard offerings, providing convenience to users.

Aspect 2: Data-based Service Personalization. Digital transformation has empowered banks to use and automize the potential of big data analytics. The vast amount of data

F. F.-H. Nah and K. L. Siau (Eds.): HCII 2024, LNCS 14721, pp. 209–220, 2024.
https://doi.org/10.1007/978-3-031-61318-0_15

generated from customer interactions is now being used to gain valuable insights into user behaviour, preferences, and various financial trends. This data-driven approach enables banks to improve decision-making for new services and product strategies, as well as identify specific risks in banking interactions.

Aspect 3: Acceleration of Service Performance. Digital transformation has led to the automation of numerous banking processes, reducing face-to-face interaction, and accelerating operational efficiency. From customer onboarding to fraud detection, automation technologies streamline workflows, minimize errors, and free up human resources to focus on tasks, that cannot (yet) be fulfilled by technology.

Aspect 4: Safe Banking Processes. As banking operations become more digital, the need for robust cybersecurity measures becomes increasingly important. The increased connectivity and reliance on digital platforms make banks prime targets for cyberattacks. As a result, banks are investing high amounts in advanced security protocols, encryption technologies, and continuous monitoring to safeguard customer data and maintain trust in the digital platforms.

Digital transformation has forced the banking industry to focus on on-going innovation and customer-centricity. The evolution of technology will continue to shape the future of banking, creating growth opportunities for financial institutions that develop capabilities in areas such as artificial intelligence. As banks continue to adapt to these changes, those that successfully embrace digital transformation will thrive in this dynamic and competitive landscape.

To gain insight into the specific relevance of these issues from the customer's perspective, the evaluation of satisfaction with current service quality is a major component for future success of banking institutions. The usage of the quantitative data-based SERVQUAL-model is one of the essential approaches to achieve this task.

2 Quantitative Research with the Five SERVQUAL Model Dimensions in Banking Institutions

In order to find out more about the current situation in the banking sector, a study concept was developed by Offenburg University and a banking institution also based in Offenburg. As part of the lectures "Service- and B2B-Marketing", "Quantitative Methods" and "Marketing Research", students of the bachelor's program in business administration at Offenburg University conducted two sequential quantitative online survey during two teaching periods in May and November 2023.

2.1 Emerging Relevance of Quantitative Research in Banking Business Management

In the dynamic landscape of banking business management, the role of quantitative research is becoming increasingly important. Quantitative research involves the systematic collection and analysis of numerical data, providing valuable insights into various aspects of banking business operations. It has become an essential tool for decision-makers, providing a structured approach to understanding complex phenomena and informing strategic choices. Five reasons can be summarized, why quantitative research is becoming increasingly relevant to the banking sector:

Reason 1: Making Data-Driven Decisions. Quantitative research allows organizations to move beyond intuition by relying on hard data. Decision-makers in banking institutions can base their strategies on empirical evidence, reducing the inherent risks associated with subjective decision-making in areas such as market trends, customer behaviour or operational efficiency.

Reason 2: Improving Accuracy. Quantitative research provides precise and accurate data. Through statistical analysis, banks can measure variables with a high degree of precision, enabling a new and sometimes surprising understanding of relationships and patterns. This precision is critical for identifying trends, forecasting outcomes, and developing strategies.

Reason 3: Measuring Performance and Benchmarking. Banking business management requires continuous evaluation of performance metrics to evaluate success and identify areas for improvement. Quantitative research enables banks to establish specific key performance indicators (KPIs) and benchmarks. This data-driven approach facilitates the identification of strengths, weaknesses, opportunities, and threats, fostering a culture of continuous improvement within an organization.

Reason 4: Gaining Customer Insights. Understanding customer preferences, behaviours, and market dynamics is essential for sustainable business growth. Quantitative research methods, such as surveys, experiments, and data analysis, provide a systematic means to collect customer data. This information helps financial institutions tailor their banking products and services to meet customer needs and stay competitive in the dynamic financial market.

Reason 5: Integrating Technological Innovations. Advancements in technology, particularly in data analytics and artificial intelligence, have propelled the relevance of quantitative research in banking business management. Powerful tools and algorithms enable banks to process large datasets efficiently, extract meaningful patterns, and derive valuable insights in real-time. This technological integration further enhances the agility and responsiveness of banks to changing market dynamics.

In conclusion, the emerging relevance of quantitative research in banking business management reflects a broader shift toward evidence-based decision-making and strategic planning. As especially financial institutions increasingly recognize the value of quantitative insights, the integration of data-driven methodologies will likely continue to shape the future of effective and efficient banking business management practices in the financial sector. Quantitative research with the SERVQUAL model offers a specific insight into the assessment of service quality by customers. All future business activities should be focused on the needs of the customers. This method offers a valuable insight into these customer wishes and requirements through an empirical study and data analysis based on quantitative approaches.

2.2 Introduction of the SERVQUAL Model in Banking Service Quality Assessments

The SERVQUAL Model is one of the most widespread and best-known models for measuring service quality in service companies. It was developed in the 1980s by the researchers Parasuraman/Zeithaml/Berry. The model is based on the assumption that service quality consists of five dimensions with a total of 22 items [1, 2].

- Tangibles - Appearance of the physical environment (premises, staff,
- advertising materials)
- Reliability - ability to adequately fulfil the advertised service
- Responsiveness - willingness to help the customer and provide quick solutions
- solutions
- Assurance - knowledge, courtesy and trustworthiness of the staff
- Empathy - individuality of the service for customers

To measure service quality, customers are surveyed and their perceptions of the company's actual performance are compared with their expectations. In this way, the company can identify areas for improvement and take targeted measures to improve service quality. One of the basic ideas behind the model is that it can be modified and thus applied in different industries.

Numerous studies have already shown that the SERVQUAL approach can be used very effectively in the banking sector, especially as service quality plays a decisive role in the financial services industry. Overall, the SERVQUAL approach can help banks to improve service quality, better understand customer needs and build long-term customer relationships. The following are examples from selected studies of how the adapted SERVQUAL approach can be used in the banking sector [3–8]:

- Measuring customer satisfaction: by applying the SERVQUAL approach, banks can measure customer satisfaction by comparing perceived service quality with customer expectations. This enables banks to identify weaknesses, and take targeted measures to improve service quality.
- Identification of areas for improvement: The SERVQUAL approach helps banks to identify specific areas, where perceived service quality falls short of customer expectations. This can help to plan targeted training for employees, process optimization or investments in technology and infrastructure.
- Competitive advantage through service quality: High service quality can lead to a competitive advantage for banks, as customers are more willing to stay with the bank, utilize additional services and generate positive word of mouth if the service quality is high.
- Customer loyalty and trust: By improving service quality, banks can increase customer trust, boost satisfaction and increase customer loyalty. Satisfied customers are more likely to build long-term relationships with their bank and use additional services.

These and other recent studies show that the SERVQUAL approach is still a relevant and important tool for measuring and improving service quality in the banking sector. However, the model has not remained without criticism. For example, it has been questioned whether the five dimensions are independent enough of each other to fulfil the validity criteria [9]. However, studies show that respondents differentiate between the factors. The influence that a measurement of expectations can have on respondents is criticized because a corresponding "claim inflation" is possible here, as expectations are often stated too high in the comparison situation [10, 11]. Many of the critics criticize the survey design and the formation of differences from the answers to the expectation and perception questions, although there is insufficient empirical evidence that the separation of the two variables is necessary [12]. The comparison between target and actual values in the sense of the confirmation/disconfirmation paradigm is closest to the concept of assessing service quality or customer satisfaction [11].

Despite all the criticism, the numerous studies have shown that the SERVQUAL approach is still a relevant and important tool for measuring and improving service quality in the banking sector. For this reason, we have chosen the SERVQUAL approach for our study.

2.3 Research Process for Marketing Problem Solving

The market research project took place in cooperation with a bank that was in the process of analyzing its service quality and wanted to use market research techniques to gain a better understanding of customer satisfaction factors.

The project allowed the students to gain practical insight into the systematic phases of a marketing research process by using quantitative methods (Fig. 1) [13].

Fig. 1. Phases of the marketing research process based on Meffert et al. [13]

Phase 1: Briefing and Problem Definition. A clear formulation of the marketing problem is an important prerequisite for determining the data collection requirements [13]. Therefore, the project started with a briefing by the decision makers of the cooperating bank on the objectives and requirements of the study.

The marketing problem was defined in relation to the new challenges faced by financial institutions in providing services through additional digital channels, such as apps or websites. The bank's main objective was to determine consumer assessments of service quality across traditional and new communication channels which included the dimensions.

- tangibles
- reliability
- responsiveness
- assurance and
- empathy.

The key questions were: What digital and traditional services do customers use? How satisfied are they with service performance? Which SERVQUAL dimensions are the most important for bank customers? What are the expectations and wishes of bank customers? How do customers assess their bank with regard to its state in the digital transformation process?

With the defined research questions at hand, the following stages of planning and designing of the quantitative data collection started.

Phase 2: Data Collection. The second phase of a marketing research process is to collect data to answer to the central research questions. The researchers determine which secondary or primary research tools are appropriate, conceptualize the research design, and collect data.

Step 1 Research Method. In order to answer the central research questions, a survey was conducted among the target group (primary research). The students decided to use an online survey instead of a face-to-face interview. The advantages of this method are that it can be implemented at low cost, respondents can be contacted quickly, a high reach can be achieved, and automated data collection is possible [13]. Another advantage is that these surveys can be conducted quickly and can include people in several locations. However, this means that the response situation cannot be strictly controlled.

After reviewing a wide range of online survey tools, it was decided to use the software LimeSurvey [14] to conduct the survey. LimeSurvey is a free online survey application that allows researchers to develop and publish online surveys without programming knowledge. The tool offers a comprehensive package of functions that also meet scientific requirements. A particular advantage is that data privacy can be guaranteed since the software and the surveys are hosted on your own server.

Step 2 Research Design. During the survey design the students dealt with different types of scales and question formulations. The SERVQUAL model survey framework comprises a total of 22 questions. These were used in the first survey, in the second one TANGIBLES item was split into two items (digital and traditional channels) to ensure a clearer differentiation of the assessment.

Step 3 Research Scope. In the first study, the cooperating bank provided an e-mail address database with approximately 15,000 e-mail addresses of people belonging to the target group. The mailing of the invitation to participate was handled by the bank. Students provided the subject line, the text and the link to the survey. To achieve the highest possible response rate, a prize was raffled among the respondents. A total of 1,116 persons took part in the survey. The second study based on an open online-survey, which was shared by the two study organizer teams via social media platforms.

Phase 3: Data Analysis. The results of the surveys were both times analyzed using SPSS and EXCEL as statistical and analysis software. For this purpose, the data was first exported from LimeSurvey and imported into SPSS and EXCEL.

Phase 4: Communication. In the final phase, the study organizer teams prepared the results and presented them to the bank's decision makers. The students interpreted the figures and used them to draw up valuable recommendations for upcoming decisions.

This four-phased process was applied two times by two different student semester groups with the same banking institution: first time with customers of the bank (Study 1: 1,116 respondents), second time with persons contacted via social media channels by the students (study 2: 106 respondents). In this paper we focus in the following chapters on the empirical results of study 2.

3 Results of the Quantitative Research

Our survey among bank customers brought a return of n = 106 completely filled questionnaires for the SERVQUAL-items. The research was focused on the evaluation of the SERVQUAL-dimensions as well as on the evaluation of the overall service quality in order to examine the influence of the SERVQUAL-dimensions on the overall service quality.

In our sample 98.1% of the respondents were active online banking users with at least one online contact per month, 89.2% of the respondents had a higher online services usage than local branch service usage, and 85.2% of the respondents were not older than 40 years. So, this suits to our research topic to analyse the younger people's satisfaction with services in digital channels of the bank sector.

Each SERVQUAL-dimension was measured by different survey items. Cronbach-Alpha values show, that these items reliably measured the same corresponding construct (see Table 1).

In a first approach, we investigated the assessed quality by mean values. Given a scale from 1 to 7, we see that all SERVQUAL-dimensions are rated rather high, with *assurance* and *reliability* at the top. There is no significant difference between *assurance* and *reliability* but between each of them and all other dimensions. Although on a rather high level (normalised mean value 0.71), banks could step up efforts to improve the perceived *empathy*. The following table shows an overview of the key figures.

Table 1. SERVQUAL-dimension's key figures based on the evaluation of the survey items assigned in advance.

SERVQUAL-dimension	No. Items	Cronbach α	Mean value	Normalised mean value
Assurance	3	0.916	5.859	0.81
Reliability	5	0.929	5.759	0.79
Tangibles	5	0.877	5.505	0.75
Responsiveness	4	0.801	5.414	0.74
Empathy	4	0.874	5.287	0.71

Two *tangibles*-items were used to differentiate the assessment of digital and traditional channels. It concerns the visual designs of the local branch bank and the bank's website. Here again, we have rather high mean values (local: 5.16 – website: 5.46), with the rating of the website being even higher than that of the local branch. Additionally, the mean values' difference is statistically significant. Banks seem to successfully attach great importance to their websites and digitalisation in general.

In a second step, we tried to find out, which SERVQUAL dimension determines overall service quality at most, and as a consequence, which survey item determines the corresponding SERVQUAL dimension at most, in order to derive specific recommendations to improve overall service quality.

As expected, we observed significant correlations between the survey items, so that we carried out an exploratory factor analysis. As also already known from literature [2], the SERVQUAL-dimensions, that are identified in this manner, might show some overlaps. In our case there were high correlations between the *tangibles*- and the *reliability*-survey items, so that in a first factor analysis with all survey items as input, these items were represented by the same factor.

Thus, we focused in the first instance on the relationship between *tangibles*- and *reliability*-survey items. To quantify these SERVQUAL-dimensions, we sequentially carried out two factor analyses, each with the corresponding survey-items. This resulted in a one factor solution respectively, so that the SERVQUAL-dimensions could be measured by the corresponding factor scores. On this base, we tested the hypothesis that the *tangible*-dimension determines the *reliability*-dimension by means of a regression analysis. Obtaining a highly significant ($p < 0.001$) r^2-value of 0.726, that shows a high explanatory power, we conclude our hypothesis to be confirmed[1].

Without the *tangibles*-survey items we obtain four factors, where each factor has a unique assignment to the survey items belonging to a SERVQUAL-dimension (see Table 2). The dimensionality reduction's quality is rather high as the factors explain 67% of the total variance with only 5% non-redundant residuals between observed and reproduced correlations.

Altogether, this leads to our adapted causal model for SERVQUAL-dimensions and overall service quality as illustrated subsequently in Fig. 2.

To investigate the factors as influence variables on the overall service quality, we used the four factors, named by the corresponding SERVQUAL-dimension, as independent variables in a regression model. Thereby, the factors were represented by their factor scores, calculated in the factor analysis. Overall Service Quality as the dependent variable was measured directly by a single item.

As main result, the SERVQUAL-dimensions show a high explanatory power, measured by an adjusted r^2-value of 0.715 that is also highly significant ($p < 0.001$). All regression coefficients are positive, confirming the general hypothesis, that the higher the SERVQUAL-dimensions are evaluated, the better overall service quality is rated. Furthermore, each SERVQUAL-dimension factor individually shows a significant ($p < 0.001$) influence on overall service quality, except of *responsiveness*, that we had to eliminate from the model therefore.

---

[1] Basically, we obtained the same result ($r^2 = 0.71$) using the data of the first study half a year ago with a much higher sample size of $n = 782$.

Table 2. Factor assignment to the survey items

Item ID	Item	Extracted factor[2]	Factor loading
Rel1	My bank fulfils the specified time frame for responding to my enquiries	F1	0.669
Rel2	The reactions of the bank staff are appropriate to the situation I find myself in	F1	0.662
Rel3	Services are carried out within a reasonable time frame	F1	0.861
Rel4	The bank's services are offered reliably during the specified service hours	F1	0.793
Rel5	Sensitive customer data is stored securely	F1	0.645
Res1	I expect my bank to tell me exactly when it will process my customer enquiries	F3	0.499
Res2	I expect an immediate response from the service staff when I ask questions	F3	0.748
Res3	Customer service staff must always be willing to help me	F3	0.754
Res4	I expect the service staff to deal with my request immediately	F3	0.728
A1	Employees of the bank appear trustworthy	F4	0.625
A2	I feel safe when dealing with the bank's customer service	F4	0.587
A3	The bank's customer service staff is polite	F4	0.638
E1	The bank's employees pay personalised attention to me	F2	0.684
E2	The service staff at my bank fully understand my needs	F2	0.853
E3	The employees in customer service act in my best interests	F2	0.701
E4	The bank offers its services at times that are convenient for customers	F2	0.596

The relative importance of a factor concerning its explanatory power is measured by BETA-values. In this case, we detected *empathy* as the most important influence factor on overall service quality with a BETA value of 0.562 followed by *reliability* (0.395) and *assurance* (0.380).

[2] Notation according to the order of factor extraction. Factor with highest factor loading to the corresponding item.

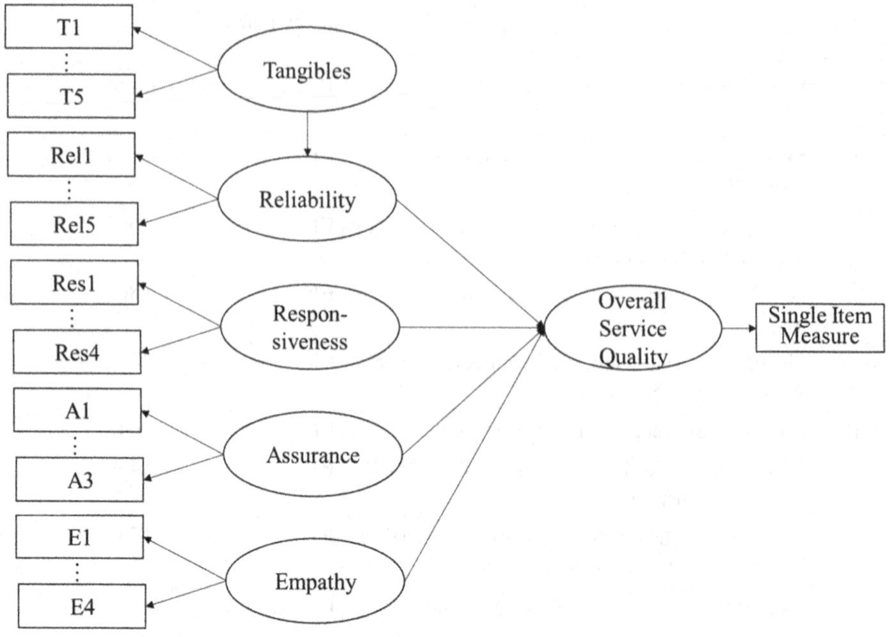

Fig. 2. Adapted causal model for overall service quality

That means, that banks should focus on the improvement of the *empathy*-dimension in order to improve the overall service quality. Of course, the two other factors aren't irrelevant. Usually, we need to improve more than one factor in order to reach a substantial improvement of the overall service quality. However, as *empathy* was evaluated with the lowest mean value (see Table 1), this seems to be a suitable starting point for future actions.

At last, we must focus on the survey items again, in order to derive recommendations, how to influence the SERVQUAL-dimension in a positive and substantial manner. Here we have to refer to the coefficients of the factor scores, that show how much the original item value determines the factor score.

For each dimension we refer to the top item, because there is a clear winner respectively. At first, banks should focus on the service staff's full understanding of the client's needs (see E2) in order to improve the evaluation of the *empathy*-dimension. Most important to improve the *reliability* is, that services are carried out within a reasonable time frame (see Rel3). Related to *assurance* banks should concentrate on the employees' trustworthiness (see A1).

4 Final Evaluation and Digital Service Performance Optimization Recommendations for Banks

In the digital age, where banking services are increasingly offered through online platforms and mobile applications, optimizing digital service performance is crucial for ensuring a seamless and satisfying customer experience. Banks face the challenge of meeting customer expectations for the reviewed SERVQUAL-dimensions reliability, responsiveness, assurance, and empathy - corresponding with the tangible elements - in their digital interactions.

Based on our research and accompanied by the findings of our empirical study here are seven key recommendations for banks to optimize their digital service performance:

Recommendation 1: Robust Infrastructure for the Customers Access Needs. Investing in a robust and scalable IT infrastructure is fundamental to digital service performance. By leveraging e. g. cloud services, banks can ensure high availability, faster response times, and seamless scalability to meet evolving customer demands, which was addressed by the participants in our recent empirical studies with the banking institution.

Recommendation 2. Performance Monitoring for the Customers Service Needs Implementing comprehensive performance monitoring tools is essential for identifying issues and enhancing the acceptance of digital services by the customers. Real-time analytics provide insights into user behaviour, system performance, and potential problems. By continuously monitoring KPIs, banks can enhance overall service quality.

Recommendation 3: Mobile App Optimization for Customer Experience Needs. Given the increasing reliance on mobile banking apps, optimizing the performance of these applications is critical: Banks should focus on minimizing app load times, ensuring smooth navigation, and providing a user-friendly interface. The study stated the growing usage intensity of digital banking channels in all age groups. It is likely that older people will need to be supported by the system in their interaction process. Regular updates, bug fixes, and usability testing are crucial to maintaining a high-performance mobile banking experience.

Recommendation 4: Secure Technology for Customer Trust Needs. As digital banking services expand, so do the potential cybersecurity threats. Ensuring robust cybersecurity measures is essential not only to protect customer data and maintain the overall performance of digital services, but also to maintain confidence in the bank's trustworthiness. Banks should regularly update security protocols, conduct penetration tests, and invest in advanced threat detection technologies to safeguard their digital infrastructure.

Recommendation 5: User Centred Design for Customer Acceptance Needs. A seamless user experience is a key factor in optimizing digital services. Banks should invest in intuitive UX design to simplify navigation, enhance usability, and streamline customer interactions. Even though several customers in our study were quite satisfied with the ease and convenience of using digital banking applications, there are still several issues to be addressed. In particular, cross-channel processes, such as submitting digitally delivered forms printed with original handwritten signatures, need to be completely redesigned and user-optimized. Conducting user testings can help identify areas for improvement, ensuring that the digital services align with customer expectations and preferences.

Recommendation 6: Technology Integration for Customer Personalization Needs. Implementing personalized features and leveraging artificial intelligence can significantly enhance digital service performance. Digital services like chatbots, recommendation engines, and predictive analytics can streamline customer interactions, providing automized personalized and efficient services. This not only improves customer satisfaction but also optimizes the overall performance of digital platforms, which was also a finding of our empirical study.

Recommendation 7: Scalable Service Quality for the Customer Satisfaction Needs. Optimizing digital services is an ongoing process that requires continuous testing and quality assurance. Banks should implement continuous testing procedures, including performance testing to identify and address potential issues before they impact customers. This proactive approach helps maintain consistent service quality.

In conclusion, as banking services continue to evolve in the digital realm, optimizing digital service performance becomes a strategic imperative for banks. By implementing these seven recommendations, financial institutions can not only meet but exceed customer expectations, fostering loyalty and maintaining a competitive edge in the rapidly changing landscape of digital banking.

References

1. Parasuraman, A., Zeithaml, V.A., Berry, L.L.: A Conceptual model of service quality and its implications for future research. J. Mark. **49**(4), 41–50 (1985)
2. Parasuraman, A., Berry, L.L., Zeithaml, V.A.: Refinement and reassessment of the SERVQUAL scale. J. Retail. **67**(4), 420–450 (1991)
3. Kumar, M.: Mohapatra, A: assessing the service quality of banks by a modified SERVQUAL model. Bimaquest **23**(3), 36–54 (2023)
4. Mawi, H.L., Blanco, E.M., Buenafe, W.R.J., Nababan, S.D.R., Amoguis, J.A.: The Influence of Service Quality (SERVQUAL) Dimensions in Electronic Banking on Customer Satisfaction Among International Students (2023). https://doi.org/10.13140/RG.2.2.35426.71369
5. Mishra, S.: Impact of SERVQUAL dimensions on consumer satisfaction from mobile banking in mid-size banks of india shweta mishra. Asian Resonance **8**(2), 118–123 (2019)
6. Roy Dutta, M., Das, A.: Service quality analysis of digital banking services in India: a SERVQUAL approach. Indian J. Econ. Develop. **6**(1), 89–95 (2018)
7. Yesmin, M.N., et al.: SERVQUAL to determine relationship quality and behavioral intentions: an SEM approach in retail banking service. Sustainability **15**(8), 6536 (2023)
8. Parekh, P., Marvadi, C.: Evaluating effect of demonetization on quality of E-banking services using SERVQUAL model. IJRAR – Inter. J. Res. Analyt. Rev. **6**(1), 10–15 (2019)
9. Hoffman, K.D., Bateson, J.E.G.: Services marketing. In: Concepts, Strategies, & Cases, 6th edn. South-Western Cengage Learning, Mason, Ohio (2023)
10. Kaiser, M.O.: Erfolgsfaktor Kundenzufriedenheit. Dimensionen und Messmöglichkeiten. 2nd edn. Erich Schmidt, Berlin (2005)
11. Bauer, H. H., Falk, T., Hammerschmidt, M.: Messung und Konsequenzen von Servicequalität im E-Commerce: Eine empirische Analyse am Beispiel des Internet -Banking. Marketing ZFP 26(Sonderheft Dienstleistungsmarketing), pp. 45–57 (2004)
12. Nerdinger, F.W., Neumann, Ch., Curth, S.: Kundenzufriedenheit und Kundenbindung. In: Moser, K. (ed.): Wirtschaftspsychologie, 2nd edn, pp. 119–138, Springer, Heidelberg (2015)
13. Meffert, H., Burmann, C., Kirchgeorg, M.: Marketing. Grundlagen marktorientierter Unternehmensführung; Konzepte - Instrumente - Praxis-beispiele. 10nd edn. Springer Gabler, Wiesbaden (2008)
14. LimeSurvey: An Open Source survey tool. LimeSurvey GmbH, Hamburg, Germany. http://www.limesurvey.org. Accessed 30 Jan 2024

The Innovation Generated by Blockchain in Accounting Systems: A Bibliometric Study

Javier Alfonso Ramírez[1(✉)], Evaristo Navarro[2], Joaquín Sierra[3], Johny García-Tirado[1], Rosmery Suarez Ramirez[4], and Carlos Barros[5]

[1] Corporación Universitaria Taller Cinco, 58 North Highway Kilometer 19, Chia, Colombia
Javier.ramirez@taller5.edu.co
[2] Universidad de la Costa, 58 Street #55-66, Barranquilla, Colombia
[3] Universidad Libre de Colombia, Seccional Cartagena, 177 Street #30-20, Bolívar, Colombia
[4] Universidad de La Guajira, Km 5, Maicao - Riohacha, Manaure, Colombia
[5] Institución Universitaria de Barranquilla, 45th Avenue #48-31, Barranquilla, Colombia

Abstract. This study aims to identify new scientific trends related to the impact of Blockchain technology in the field of accounting by providing an innovative system that can transform the way financial and business records are maintained in the Scopus database. To conduct this research, we propose a descriptive and quantitative methodology based on bibliometric tools that effectively measure the scientific production of the variables related to accounting and Blockchain within the Scopus database. This process necessitates the formulation of a search equation covering the essential elements of the study. This research found a total of 777 documents published between 2019 and 2023. In summary, the findings conclude that in the exercise of scientific observation, Blockchain technology is revolutionizing accounting by offering a more secure, efficient, and transparent method of managing financial data. Its impact on the accounting industry continues to expand, providing opportunities to enhance accuracy, reliability, and efficiency in financial record management.

Keywords: Blockchain · Accounting Systems · Bibliometric

1 Introduction

New technologies have been a turning point for all industries worldwide, because they allow people and organizations to be interconnected, processes automated, and large data analytics to be obtained in a matter of seconds [1, 2]. In this way, within the corporate world, these technologies have a positive impact on global competitiveness, not only by allowing any entity to enter markets they never thought of, but also it requires existing competitors to strengthen themselves in order to achieve that competitive differential that allows them to stand out [3, 4].

Accounting is undergoing a revolution, thanks to Blockchain technology, which is introducing secure and innovative solutions in the field of financial information management [5]. This decentralized and distributed technology offers the ability to record transactions transparently and securely through linked blocks, ensuring the integrity and immutability of data [6, 7].

F. F.-H. Nah and K. L. Siau (Eds.): HCII 2024, LNCS 14721, pp. 221–232, 2024.
https://doi.org/10.1007/978-3-031-61318-0_16

It is undeniable to recognize that the work of accounting and financial control has undergone major changes in recent years, creating new opportunities for improvement within the sector and requiring companies and people in the sector to be aligned with these trends [8]. Within the accounting field, Blockchain emerges as an innovative tool that revolutionizes transaction verification and auditing, eliminating the need for intermediaries and significantly reducing the risks associated with errors and fraud. This technological advancement has raised important challenges and questions regarding the regulation and protection of financial data, as well as its compatibility with conventional accounting systems [9].

With this premise, we embarked on a scientific observational exercise to gain a precise understanding of the current state of this important topic. Various studies such as those by Bonsón & Bednárová titled "Blockchain and its implications for accounting and auditing" and Yu, Lin & Tang "Blockchain: The introduction and its application in financial accounting" lay the foundations for this process. Understanding how the blockchain through its tracking and authentication system becomes an important tool for achieving better transparency in accounting and auditing processes, presenting in turn an important path to follow in its optimization and improvement of security. For its widespread use as a tool in organizations [10, 11].

As a result, this study aims to identify new scientific trends related to the impact of Blockchain technology in the field of accounting by providing an innovative system that can transform the way financial and business records are maintained in the Scopus database; this will become a study of utmost relevance as it will reveal the process of evolution of the trends of scientific knowledge related to the blockchain applied to accounting processes, thus serving as a technological surveillance tool of this cutting-edge technology that is impacting the world business.

2 Material and Method

To conduct this research, a descriptive and quantitative methodology it is proposed, which is based on bibliometric tools that effectively measure the scientific production of the variables related to accounting and Blockchain within the Scopus database. The development of this biometric research allows the development of an effective mapping of scientific trends in various areas of knowledge. This type of documentary approach, supported by quantitative tools, is gaining great validity and recognition worldwide thanks to the high-impact publications that support it [12, 13]. In this sense, for this research process, an approach from the Scopus database is proposed as mentioned, which was selected for its rigor and high academic impact. Subsequently, an identification of the keywords related to the search variables to be considered in the search equation is made. These keywords are shown below (Table 1):

Table 1. Keywords

Variables	Keywords
Blockchain	"Blockchain", "smart contract"
Accounting	"Accounting", "integrated systems"

From this process, the following search equation was formulated: (TITLE-ABS-KEY ("accounting") OR TITLE-ABS-KEY ("integrated systems") AND TITLE-ABS-KEY ("blockchain") OR TITLE-ABS-KEY ("smart contract")) AND PUBYEAR > 2018 AND PUBYEAR < 2023. It should be noted that for the review process, only publications in the time frame from 2019 to 2023 are considered, in order to obtain an overview of the most recent scientific contributions in this area of knowledge. The data is processed via the bibliometrix package of the R software, and the Vos Viewer software.

3 Results

Below is a general report on the search results of the equation posed in the Scopus database (Table 2):

Table 2. Results overview

Description	Results
MAIN INFORMATION ABOUT DATA	
Sources (Journals, Books, etc.)	453
Documents	777
Document Average Age	2,66
Average citations per doc	8,27
References	26084
DOCUMENT CONTENTS	
Keywords Plus (ID)	2725
Author's Keywords (DE)	1785
AUTHORS	
Authors	1941
Authors of single-authored docs	115
AUTHORS COLLABORATION	
Single-authored docs	122
Co-Authors per Doc	2,94
International co-authorships %	15,83
DOCUMENT TYPES	
Article	347
Book	5

(continued)

From the table above, it is possible to identify the main elements of the review process carried out in the Scopus database on the use of blockchain technologies within the accounting process. Among these findings, a total of 777 investigations located in

Table 2. (*continued*)

Description	Results
Book Chapter	88
Conference Paper	265
Conference Review	35
Erratum	2
Note	5
Retracted	2
Review	28

453 different sources that are part of the Scopus database are observed, with an average citation rate of 8.27 and 2725 keywords. The most common typologies within the area of knowledge addressed are articles with 347, conference paper with 265 and book chapter with 88; highlighting the predominance of Articles within the type of sources.

On the other hand, when reviewing the annual production index contrasted by the total average citations per article each year, it can be observed that the number of publications has increased significantly, but the average number of citations has been decreasing. This phenomenon could be explained by the life span of the documents in the scope of the average impact within the area of knowledge. This can be seen in the figure below (Fig. 1):

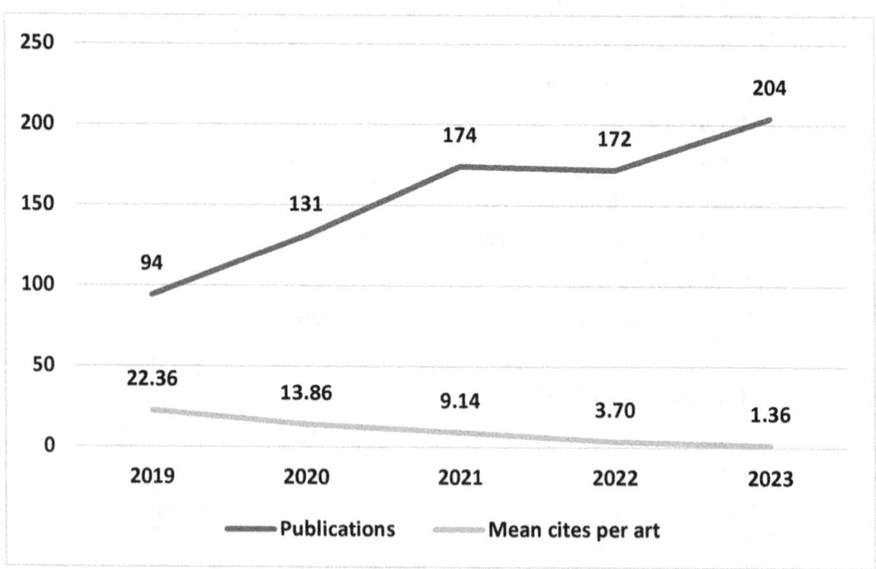

Fig. 1. Track and Wheelbase of a formula vehicle upper view.

In this sense, the most relevant sources by frequency of publication can be seen in Fig. 2 as the most outstanding is ACM INTERNATIONAL CONFERENCE PROCEEDING SERIES with 33 publications, followed by LECTURE NOTES IN NETWORKS AND SYSTEMS with 23, LECTURE NOTES IN COMPUTER SCIENCE with 20; the three sources of categories are: conferences, paper or book chapter; while the JOURNAL OF EMERGING TECHNOLOGIES IN ACCOUNTING with 18 investigations and ACCOUNTING, AUDITING AND ACCOUNTABILITY JOURNAL with 12; both of them are in the Journal category:

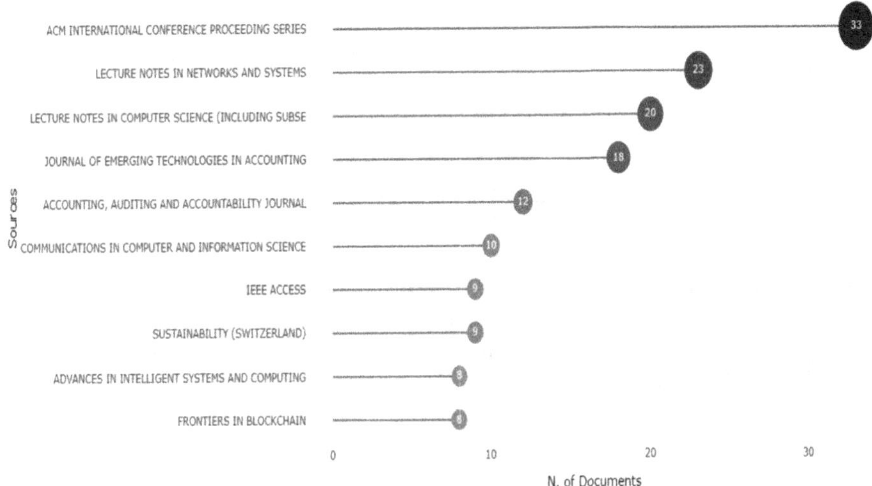

Fig. 2. Most relevant sources

On the other hand, when reviewing these outstanding sources, the impact of these within the area of knowledge is highlighted. To corroborate this, Bradford's law of dispersion is applied; which shows how the main sources are the ones that concentrate the largest number of publications associated with blockchain technologies within accounting processes (Figs. 3 to 5) (Table 3):

Table 3. Bradford's Law

Zone	Total sources	Freq papers	%
1	29	258	33,20%
2	168	263	33,85%
3	256	256	32,95%

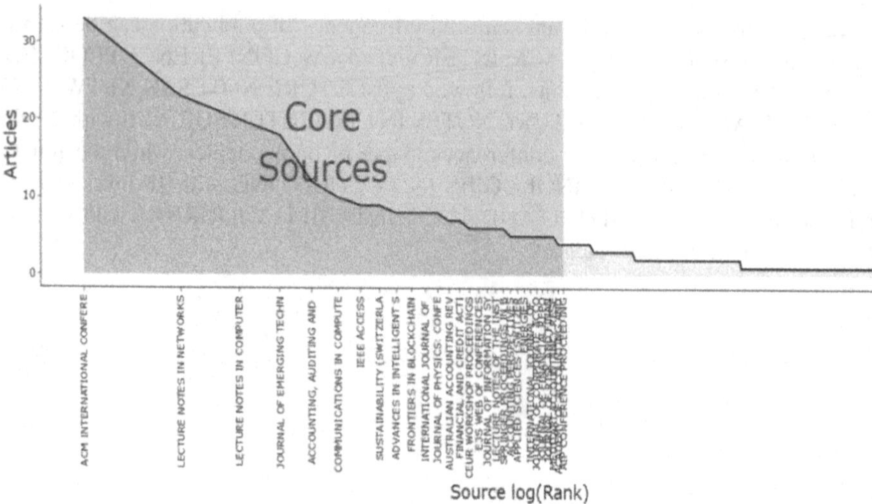

Fig. 3. Bradford's Law.

In this sense, when reviewing the most relevant authors within the area of knowledge under study based on the number of publications, it is observed that the most outstanding author is ZHANG Y with 9 contributions, followed by ZHANG J and ZHANG W with 8 documents published each one:

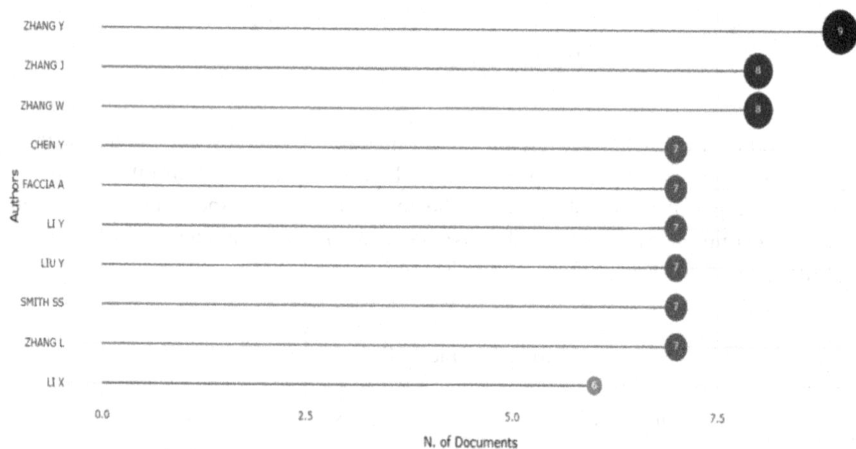

Fig. 4. Most relevant authors.

Delving into the understanding of the trends of authors' contributions within studies related to the application of blockchain technology to accounting processes today, Lotka's Law Curve it is observed; which stipulates the decrease in the number of authors the greater the number of contributions per author. In this case, 1721 authors have participated in at least one document, decreasing to 163 with 2 contributions, 32 with 3, 8 with 6 until reaching a single author with nine reported contributions:

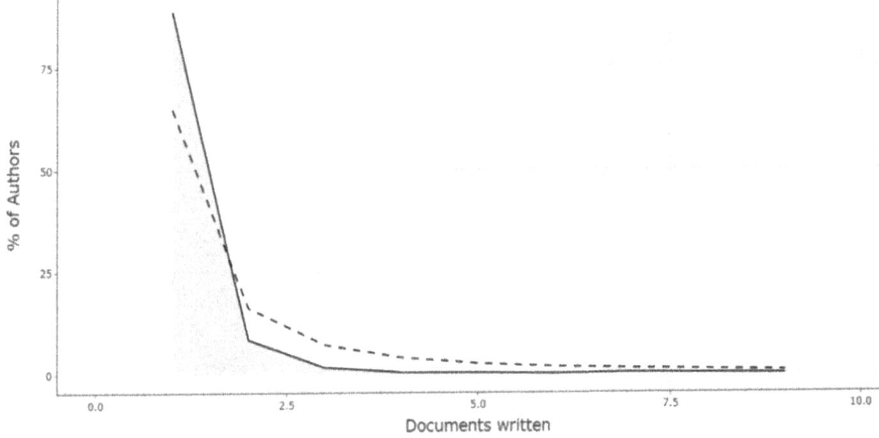

Fig. 5. Lotka's Law.

In this sense, below are the studies considered to have the greatest impact based on the number of citations reported (Table 4):

Table 4. Top cited documents

Document	DOI/LINK	Total Cites		
ESPOSITO C, 2021, INF PROCESS MANAGE	https://doi.org/10.1016/j.ipm.2020.102468	262	65.50	28.65
MOLL J, 2019, BR ACCOUNT REV	https://doi.org/10.1016/j.bar.2019.04.002	203	33.83	9.08
SCHMITZ J, 2019, AUST ACCOUNT REV	https://doi.org/10.1111/auar.12286	171	28.50	7.65
DEMIRKAN S, 2020, J MANAG ANAL	https://doi.org/10.1080/23270012.2020.1731721	124	24.80	8.94
BONSÓN E, 2019, MEDITARI ACCOUNT RES	https://doi.org/10.1108/MEDAR-11-2018-0406	116	19.33	5.19
KWILINSKI A, 2019, ACAD ACCOUNT FINANC STUD J	https://www.proquest.com/openview/524fee4b6a17c197550a834524260124/1?pq-origsite=gscholar&cbl=29414	112	18.67	5.01

(*continued*)

Table 4. (*continued*)

Document	DOI/LINK	Total Cites		
KSHETRI N, 2021, INT J INF MANAGE	https://doi.org/10.1016/j.ijinfomgt.2021.102376	110	27.50	12.03
GALLERSDÖRFER U, 2020, JOULE	https://doi.org/10.1016/j.joule.2020.07.013	100	20.00	7.21
FU Y, 2019, IEEE ACCESS	https://doi.org/10.1109/ACCESS.2019.2895327	96	16.00	4.29
MOSTEANU NR, 2020, QUAL ACCESS SUCCESS	https://pureportal.coventry.ac.uk/en/publications/sustainability-integration-in-supply-chain-management-through-sys	92	18.40	6.64

Among the documents with the highest number of citations within the area of knowledge is "Blockchain-based authentication and authorization for smart city applications", which reports a technological proposal for applications based on blockchain technology that allow optimizing the implementation of the so-called smart cities [14]. At the same time, the study "The role of the internet-related technologies in shaping the work of accountants: New directions for accounting research" conducts a review of the literature related to new technologies within the accounting profession worldwide, not only addressing blockchain but also mentioning other technologies such as AI, big data, among others [15].

On the other hand, the study "Accounting and Auditing at the Time of Blockchain Technology: A Research Agenda" it focuses prominently on the use of this technology as a tool in the process of accounting and financial audits, where the benefits of the traceability capacity of the processes are taken into account [16]. Meanwhile, the investigation "Blockchain technology in the future of business cyber security and accounting" it highlights the application of these technologies in accounting transparency processes based on the implementation of Smart Contracts as a tool for monitoring the safety of information and the support of blockchain as a database with higher levels of security than other conventional means [17]. This information is also supported by the study "Blockchain and its implications for accounting and auditing", also focused on modern authentication in the face of new technologies [10].

Similarly, authors such as Kwilinski, author of "Implementation of blockchain technology in accounting sphere" it also recognizes the great potential of these technologies in terms of information security [18]. From another perspective, the research "Blockchain and sustainable supply chain management in developing countries", addresses how these types of technologies also transcend in the field of inventory processes in supply chains, a key element of accounting control, where they are alluded to as a catalyst for the achievement of sustainable ecosystems within the industry [19].

However, the study "Energy Consumption of Cryptocurrencies Beyond Bitcoin" gives an important revision of the high electrical consumption that involve the start-up and existence of crypto-based platforms and their standards [20]. These authors recommend having a clear emphasis between the differentiation of energy-efficient blockchain systems and those that involve dangerous energy consumption compared to the needs of today's world.

In this sense, the research "Big Production Enterprise Supply Chain Endogenous Risk Management Based on Blockchain" like other studies already mentioned, it establishes how blockchain technologies can improve the transparency of supply chains, increasing the rigor of control processes to prevent fraud and/or information errors [21]. The latter, also supported by the study "Sustainability integration in supply chain management through systematic literature review" through a systematized literature review [22].

In this way, once the main contributions of the most outstanding research in the implementation of blockchain technology within accounting processes have been understood; The mapping of the existing networks between the keywords of these investigations is carried out (Fig. 6):

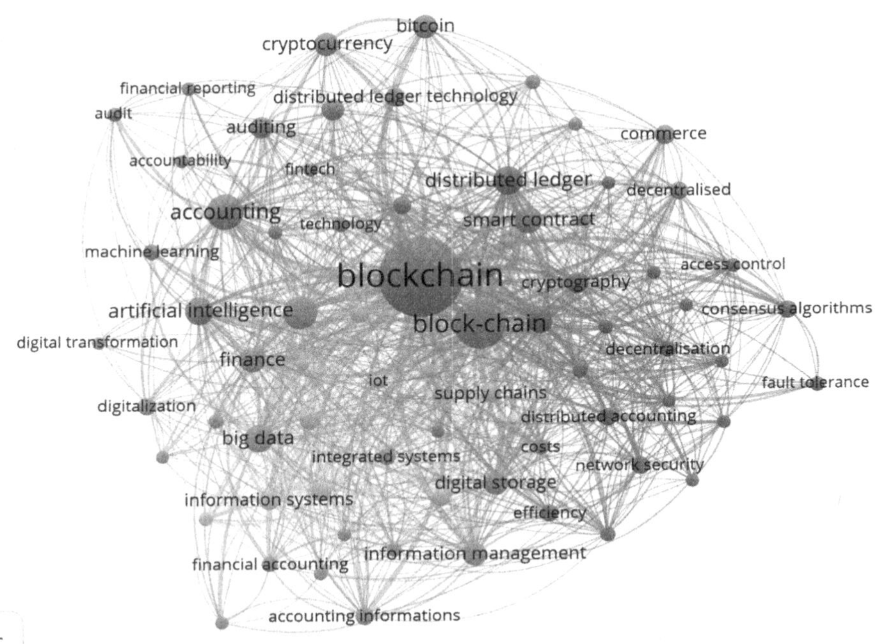

Fig. 6. Keyword co-occurrence.

Within the analysis of the co-occurrence of the keywords, it is observed how the term "blockchain" appears centrally and with a high number of links with other nodes; showing its high impact and importance. Next, other terms such as "artificial intelligence", "digital transformation", "big data" or "Digitalization" evidence the high relevance of the technological component identified within the research in the area of knowledge. On the other hand, keywords such as "Fintech", "accounting", "accounting information" or "Audit", allow reference to the accounting component. However, as has been observed, among the most cited research, an allusion is made to decentralized and transparent supply chains with keywords such as "supply chain" and "decentralization" (Fig. 7).

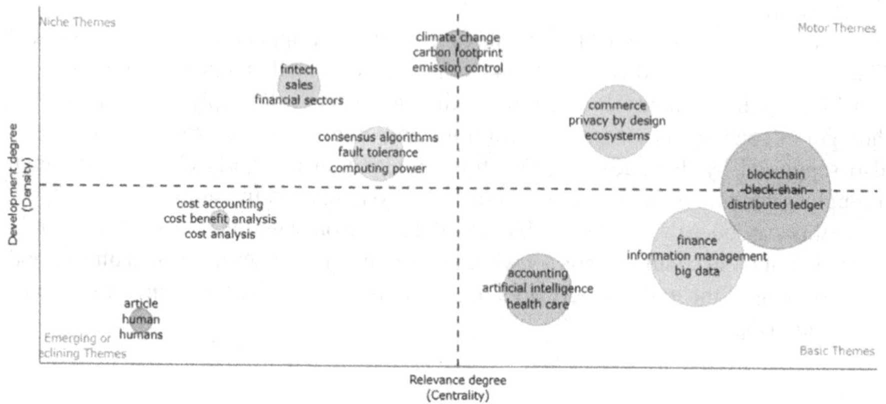

Fig. 7. Thematic map.

Within the studies analyzed, it can be observed that the topics with the greatest relevance and density are identified from the line of "blockchain, block-chain, distributed ledger", "commerce, privacy by design, ecosystems" and "climate change, carbon footprint, emission control"; while topics related to "cost accounting, cost benefit analysis, cost analysis" show less impact within the study area.

4 Conclusions and Discussions

This bibliometric study offers a comprehensive overview of the impact of Blockchain technology on accounting systems, highlighting its relevance in the transformation of financial and business processes. Through the analysis of 777 studies, a growing interest in this area of knowledge has been evidenced, reaching its peak of scientific production in 2023 with 204 published research.

Research reveals that Blockchain is revolutionizing accounting by providing a more secure, efficient, and transparent method of managing financial data [23]. Its ability to transparently and securely record transactions through linked blocks ensures the integrity and immutability of data, eliminating the need for intermediaries and significantly reducing the risks associated with errors and fraud [24].

The analysis of the results obtained through the bibliometric search shows a significant increase in the number of publications on Blockchain and accounting, this reflects the growing interest and recognition of the importance of this technology in the accounting field. However, there is a decrease in the average number of citations per document, this could be attributed to the lifetime of the documents in relation to their average impact on the field of knowledge.

One of the highlights of this study is the mention and identification of the relevance of supply chains as a focal point in the implementation of blockchain within the industry. Several previous studies corroborate the important role of new technologies in the optimization, control and transparency of business logistics processes [25, 26].

In conclusion, this study provides a clear overview of the trends offered by this new technology, from its challenges and strengths; It also points out the most important elements facing the horizon of this area of scientific knowledge worldwide.

References

1. Ding, S., Tukker, A., Ward, H.: Opportunities and risks of internet of things (IoT) technologies for circular business models. J. Environ. Manage. **336**, 117662 (2023)
2. Arevalo, O., Parra, M., García-Tirado, J., Ramírez, J., Ballestas, M., Rondón Rodríguez, C.: Dynamic adaptive capacity of metalworking companies in barranquilla-colombia: analysis from the export process in the 4.0 Era. In: International Conference on Computer Information Systems and Industrial Management, pp. 127–138. Springer Nature Switzerland, Cham (2023). https://doi.org/10.1007/978-3-031-42823-4_10
3. Sullivan, Y., Fosso Wamba, S., Dunaway, M.: Internet of things and competitive advantage: a dynamic capabilities perspective. J. Assoc. Inf. Syst. **24**(3), 745–781 (2023)
4. Falkenreck, C., Leszczyński, G., Zieliński, M.: What drives the successful launch of IoT-related business models? J. Bus. Indus. Market. **38**(13), 180–194 (2023)
5. Jiang, L.: The use of blockchain technology in enterprise financial accounting information sharing. PloS One **19**(2) (2024)
6. Duran, J.A.R., Ramírez, J.I., Núñez, L.N.: Tokens no fungibles: una revisión sistematizada sobre el panorama desde la evidencia científica. Revista Colombiana de Tecnologias de Avanzada (RCTA) **1**(41) (2023)
7. Sreenivasan, A., Suresh, M.: Start-up sustainability: does blockchain adoption drives sustainability in start-ups? A systematic literature reviews. Manag. Res. Rev. **47**(3), 390–405 (2024)
8. Barth, M.E., Li, K., McClure, C.G.: Evolution in value relevance of accounting information. Account. Rev. **98**(1), 1–28 (2023)
9. Surya, B.T.M.: Revolutionizing accounting through digital transformation: the impact of technology. Eng. Sci. Lett. **3**(01), 6–10 (2024)
10. Bonsón, E., Bednárová, M.: Blockchain and its implications for accounting and auditing. Meditari Accountancy Res. **27**(5), 725–740 (2019)
11. Yu, T., Lin, Z., Tang, Q.: Blockchain: the introduction and its application in financial accounting. J. Corporate Account. Finance **29**(4), 37–47 (2018)
12. Niebles-Nunez, W., Niebles-Nunez, L., Babilonia, L.H.: Energy financing in Colombia: a bibliometric review. Inter. J. Energy Econ. Policy **12**(2), 459–466 (2022)
13. Goodell, J. W., Kumar, S., Lahmar, O., Pandey, N.: A bibliometric analysis of cultural finance. Intern. Rev. Financial Anal. **85** (2023)
14. Esposito, C., Ficco, M., Gupta, B. B.: Blockchain-based authentication and authorization for smart city applications. Inform. Process. Manag. **58**(2) (2021)
15. Moll, J., Yigitbasioglu, O.: The role of internet-related technologies in shaping the work of accountants: New directions for accounting research. British Accounting Rev. **51**(6) (2019)
16. Schmitz, J., Leoni, G.: Accounting and auditing at the time of blockchain technology: a research agenda. Aust. Account. Rev. **29**(2), 331–342 (2019)
17. Demirkan, S., Demirkan, I., McKee, A.: Blockchain technology in the future of business cyber security and accounting. J. Manag. Analyt. **7**(2), 189–208 (2020)

18. Kwilinski, A.: Implementation of blockchain technology in accounting sphere. Acad. Account. Financ. Stud. J. **23**, 1–6 (2019)
19. Kshetri, N.: Blockchain and sustainable supply chain management in developing countries. Inter. J. Inform. Manag. **60** (2021)
20. Gallersdörfer, U., Klaaßen, L., Stoll, C.: Energy consumption of cryptocurrencies beyond bitcoin. Joule **4**(9), 1843–1846 (2020)
21. Fu, Y., Zhu, J.: Big production enterprise supply chain endogenous risk management based on blockchain. IEEE Access **7**, 15310–15319 (2019)
22. Mosteanu, N.R., Faccia, A., Ansari, A., Shamout, M.D., Capitanio, F.: Sustainability integration in supply chain management through systematic literature review. Quality-Access Success **21**(176), 117–123 (2020)
23. Groenewald, E., Kilag, O. K.: Automating finances: balancing efficiency and job dynamics in accounting and auditing. Inter. Multidisciplinary J. Res. Innovat. Sustainabil. Excellence (RISE) **1**(2), 14–20 (2024)
24. Sakib, S.N.: Blockchain Technology for Smart Contracts: Enhancing Trust, Transparency, and Efficiency in Supply Chain Management. In: Editor, F. (ed.) Achieving Secure and Transparent Supply Chains With Blockchain Technology, pp. 246–266. IGI Global (2024)
25. Ramírez, J., Gallego, G., Niebles, W., Tirado, J.G.: Blockchain technology for sustainable supply chains: a bibliometric study. J. Distribut. Sci. **21**(6), 119–129 (2023)
26. Purwaningsih, E., Muslikh, M., Suhaeri, S., Basrowi, B.: Utilizing blockchain technology in enhancing supply chain efficiency and export performance, and its implications on the financial performance of SMEs. Uncertain Supply Chain Manag. **12**(1), 449–460 (2024)

EV Explorer 2.0: An Online Vehicle Cost Calculator for Gig Drivers Considering Going Electric

Angela Sanguinetti(✉), Kate Hirschfelt, Debapriya Chakraborty, Matthew Favetti, Nathaniel Kong, Eli Alston-Stepnitz, and Angelika Cimene

Consumer Energy Interfaces Lab, University of California, Davis, Davis, CA 95616, USA
asanguinetti@ucdavis.edu

Abstract. This paper describes the design and development of EV Explorer 2.0, an online vehicle cost calculator (VCC). EV Explorer 2.0 was designed to support electric vehicle adoption among both the general population of vehicle buyers and gig drivers, such as those providing ridehailing or food delivery services. It includes several features important for the gig driver use case that are not available in other VCCs: (1) estimation of total cost of ownership (TCO) for used vehicles (other tools focus on new cars); (2) inclusion of gig-driving income estimates for comparison cars; and (3) a more precise accounting of the costs of public charging. EV Explorer 2.0 also includes important features found in other VCCs, such as electric vehicle incentive estimation, and innovative features, such as animations conveying the social and environmental impacts of electric versus gas vehicles. The tool was informed by user testing with gig drivers who had expressed interest in electric vehicles.

Keywords: Vehicle Cost Calculator · Electric Vehicles · Consumer Adoption · Gig Drivers · Transportation Network Companies

1 Introduction

Vehicle electrification is an important strategy in moving toward a more sustainable transportation future (Knobloch et al., 2020). Regulations are emerging that promote electrification of ridehailing services (e.g., California Senate Bill 1014 "clean miles standard"). This is challenging because ridehail drivers are independent contractors for transportation network companies (TNCs) who are responsible for supplying their own vehicle.

Even with available incentives, upfront costs for electric vehicles (EVs) are typically higher than those of comparable internal combustion engine vehicles (ICEVs) and EV adopters tend to be higher income (Erdem et al., 2010, Hjorthol, 2013; Saarenpää et al., 2013), whereas ridehail drivers predominantly self-identify as low income (Taiebat et al., 2022). However, EVs are cheaper to fuel and maintain, which can make total costs of ownership (TCO) lower than for comparable gas cars in the long run. TNC and other gig drivers who use their vehicles intensively may reap greater benefits from the operational

F. F.-H. Nah and K. L. Siau (Eds.): HCII 2024, LNCS 14721, pp. 233–252, 2024.
https://doi.org/10.1007/978-3-031-61318-0_17

cost savings of EVs relative to gas cars, thus EVs can ultimately yield much more favorable TCO and higher earnings, and this will increasingly be the case as batteries become more affordable and EVs reach cost parity with ICEVs (Chakraborty et al., 2021; Hamza et al., 2021; Pavlenko et al., 2019).

EVs can also meet the range requirements of most gig drivers. Taiebat et al. (2022) recently estimated that a fully charged battery electric vehicle (BEV) with a 250-mile range (BEV250) could meet the daily driving requirements for more than 86% of Lyft drivers on at least 95% of days, and that an incentive total of $5,700 for a new BEV250 would result in a lower TCO relative to a comparable gas car for those drivers. Higher-mileage drivers could achieve a lower TCO with a new BEV250 without any subsidy, and all drivers could achieve a lower TCO with a used BEV250, which had the lowest TCO of all vehicle types per their models.

However, TCO calculations are complex (Egbue & Long, 2012; Hagman et al., 2016) and there have been calls for efforts to make this information more accessible to car-buyers, including gig drivers (Taiebat et al., 2022). Vehicle cost calculators (VCCs) have been recommended as a promising informational strategy (Eppstein et al., 2011; Wu et al., 2015). There are at more than a dozen online VCCs available to consumers (Sanguinetti et al., 2020), but until recently none focused on the gig driver use case. This paper describes the design and development of a VCC to aid gig drivers considering EV purchase or lease.

2 Background and Methodology Overview

Our VCC development built on the foundation of EV Explorer (gis.its.ucdavis.edu/evexplorer/#!/), a vehicle *energy* cost calculator (providing estimated fuel costs for up to four vehicles of any drivetrain) developed by a member of the research team. The new tool, EV Explorer 2.0, builds upon the original by designing for the considerations of gig drivers as well the general population and factoring in vehicle acquisition costs, incentives, depreciation, and maintenance costs to provide the full picture of TCO. The TCO model used is adapted from past work of another member of the research team (Chakraborty et al., 2021) to better account for factors that can be important to gig drivers, i.e., leasing, used vehicle purchase, maintenance, and public charging. Sections 4 and 5 provide further detail about web development and the TCO model.

The research team has conducted the only published research testing the effectiveness of VCCs in promoting EV adoption (Sanguinetti et al., 2016), and in another project articulated best practices for VCC user experience design based on a systematic analysis of VCCs and extensive usability research (Sanguinetti et al., 2020). Our VCC design drew on these works as well as the team's understanding of the needs of ridehail drivers from a focus group and large-scale survey of EV drivers on the Uber platform (Sanguinetti & Kurani, 2021).

This paper also reports on a small sample VCC usability study ($N = 12$) with current and former gig drivers to inform the design of EV Explorer 2.0, troubleshoot for programming bugs and errors, and test the effectiveness of the final design in terms of user experience, including interest, ease of use, enjoyment, and comprehension. We recruited drivers we knew or through word-of-mouth as well as a contact list Uber Technologies, Inc. Provided with email addresses for 45 of their drivers who had rented an EV and snowball method with respondents. Participants ranged from 19 to 61 years old ($M = 41$) and ranged in gig-driving experience, from 1–3 months to 5–6 years ($M = 1$–2 years), and intensity–driving between 5 and 60 h per week ($M = 20$).

Using an unmoderated usability testing approach in UserZoom software platform, participants visited EV Explorer 2.0 and Uber's VCC (EV Cost Calculator Guide), where they completed programmed tasks and answered questions on their own while their voices and screens were recorded. Uber's VCC was included to compare users' response to elements in each tool that are geared toward gig drivers. A series of questions repeated in pre- and post-session questionnaires assessed participants' intention to buy or lease a vehicle of each drivetrain for their next car (gas, hybrid, PHEV or BEV), like our past research with EV Explorer (Sanguinett et al., 2017).

3 User Interface Design

The following sections describe the EV Explorer 2.0 user interface design. Overall goals were to minimize demand on users but maximize the potential for customization; tell a clear and compelling story about potential TCO savings with an EV compared to a gas vehicle, including for the specific case of gig drivers, while also prioritizing accuracy and transparency; and highlight the social and environmental benefits of EVs. The initial design relied heavily on the best practices articulated in our prior research (Sanguinetti et al., 2020). Other sources of inspiration are mentioned where applicable. We also weave discussion of insights from usability testing into the description as rationale for design choices, while the usability testing methodology and some broader analyses are presented in a subsequent section.

3.1 Minimize Demand

The landing page is a simple introduction to the tool that quickly orients the user (Fig. 1). We considered adding facts about the non-cost related benefits of EVs (e.g., fun to drive) that could potentially increase users' knowledge, interest, and intention to adopt. However, VCC users seek these tools out because they are already interested in EVs (confirmed in our prior research: Sanguinetti et al., 2016, 2020); VCCs might be considered a "pull" marketing strategy as opposed to a "push". Programs that promote the use of VCCs among TNC drivers who would not seek them out independently should leverage the opportunity to increase users' awareness of non-cost related EV benefits. One strategy along these lines that we did use, based on feedback from usability testing, was to change our vehicle graphic on the landing page from a small hatchback (similar to a Chevy Bolt) to a more appealing Tesla-esque sedan.

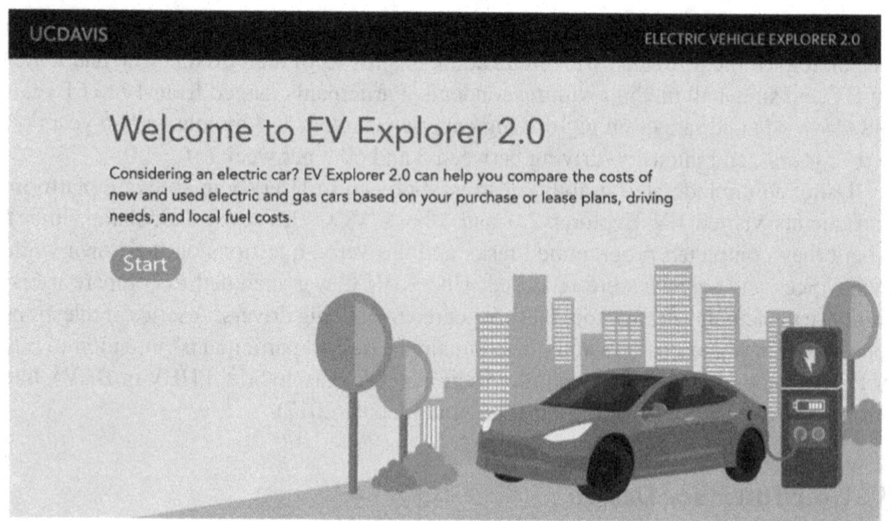

Fig. 1. EV Explorer 2.0 landing page

Beyond the simple landing page that allows the user to move quickly into the main purpose of the VCC tool (i.e., cost estimates and comparisons), a short series of modal screens collects user inputs designed to be minimal and high leverage in terms of their ability to personalize outputs. It is at this point that paths diverge for the gig-driving (Fig. 2) versus non-gig-driving user (Fig. 3). In both pathways, the user must choose a vehicle model or body style. Our past research showed users prefer to choose a vehicle upfront rather than after an initial cost output. The flexibility of allowing users to choose a specific model or a body style category lowers demand because it provides options suited to two different mental models: one where the user is interested in a specific vehicle (whether electric or gas) and one where the user may be unaware of specific EV models and interested in exploring the options available for a preferred body style. A notification appears beneath this modal screen for the gig driver pathway if the user chooses a two-door vehicle model or body style or a vehicle older than 2007 that states the vehicle or style may not be eligible to provide rides on Uber or Lyft platforms.

The modal also collects some input on how much the user drives, a variable with a big impact on TCO and a particularly important factor when comparing ICEVs and EVs. Gig drivers are also asked to report the companies for which they drive, their income, and (if they drive for Uber) the number of trips they make. The latter is used to estimate potential additional income from EV trip bonuses that only Uber currently provides ($1 per trip). For non-gig drivers, the driving inputs are flexible, allowing users to report mileage information (over a user-selected time interval: week, month, or year) or specify a most frequent trip. Someone familiar with their mileage can easily use that input, whereas someone unsure of their mileage can enter their trip information to have it calculated (this works best for a commuter, as others may not have a single trip that dominates their driving patterns).

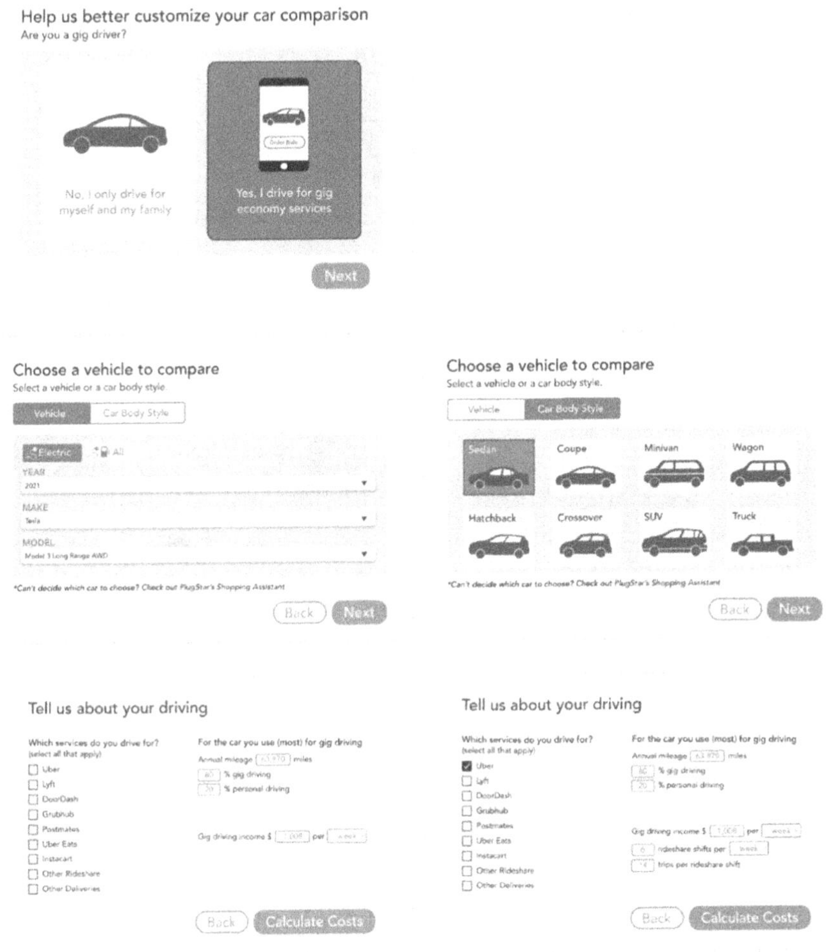

Fig. 2. Modal screens with required user inputs, gig driver pathway

After selecting a vehicle model or body style and inputting a minimal amount of information about their driving in the modal, the user reaches the output screen. From this point forward, every output and input window is available via one click on the visible screen. This is meant to facilitate easy access and use of features and to minimize the potential for a user to get lost. Further user interactions follow a non-linear process. Users can edit information in the customization panel along the bottom of the screen, including all their initial inputs, to change the cost outputs.

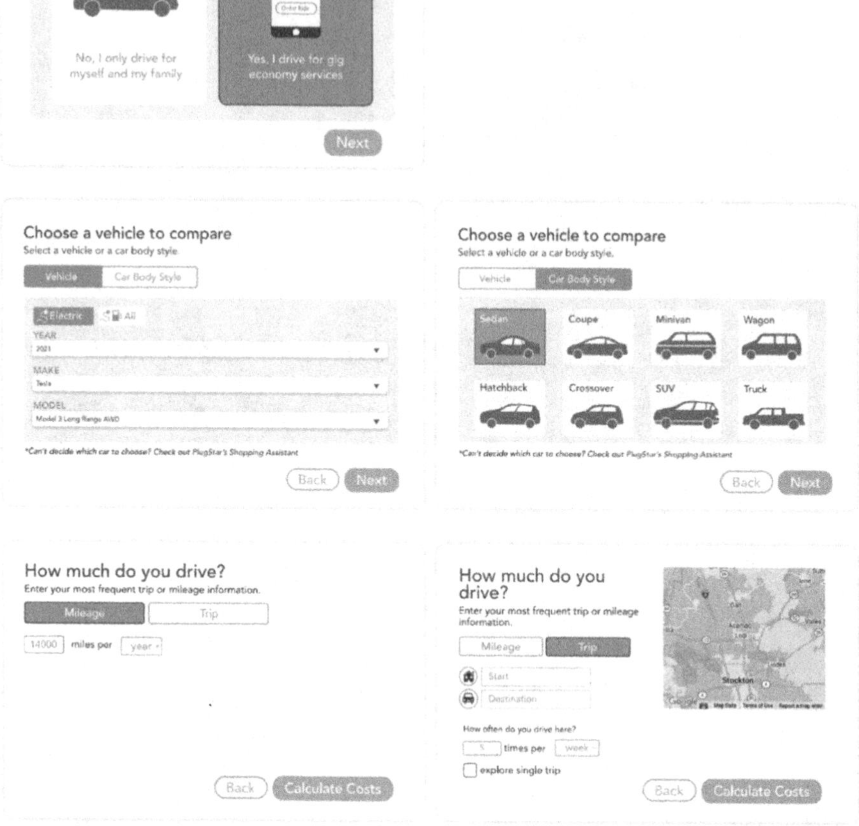

Fig. 3. Modal screens with required user inputs, non-gig driver pathway

3.2 Tell a Clear and Compelling Story

The output provides cost estimates in a two-car comparison format. The two cars must be one electric—either plug-in hybrid (PHEV) or BEV, and one gas-powered (either ICEV or hybrid). Our past usability research showed that users preferred to be able to compare more than two cars at once, but this functionality was sacrificed in the interest of simplicity and telling a clear story; future goals include building in this capability in an elegant way. The main goal of conveying the potential TCO savings with an EV compared to a gas car is accomplished in a two-car comparison, and users can change the cars in the comparison to obtain the information for as many cars as they wish.

The comparative cost estimates are divided into four categories, each with its own data visualization, navigable via tabs at the top of the screen: Vehicle Costs, Driving

Income & Costs (just Driving Costs for non-gig drivers), Total Costs, and Cumulative Costs. Each has a narrative component to the right of the data visualization that interprets the chart to help less numerate users more easily understand the information. Sacrificing the ability to compare multiple cars at once was related to this priority, since the narrative gets more complex with more than two cars.

The first cost output displayed is Driving Income & Costs for gig drivers (Fig. 4) and Driving Costs for non-gig drivers (Fig. 5). The strategy is to lead with the benefits of EVs for each user group. Operating costs (termed driving costs in the tool to be more colloquial), including fuel and maintenance, are where EVs save users the most money compared to gas cars. For gig drivers, situating operating costs within the context of driving income may be particularly compelling as it highlights greater earning potential.

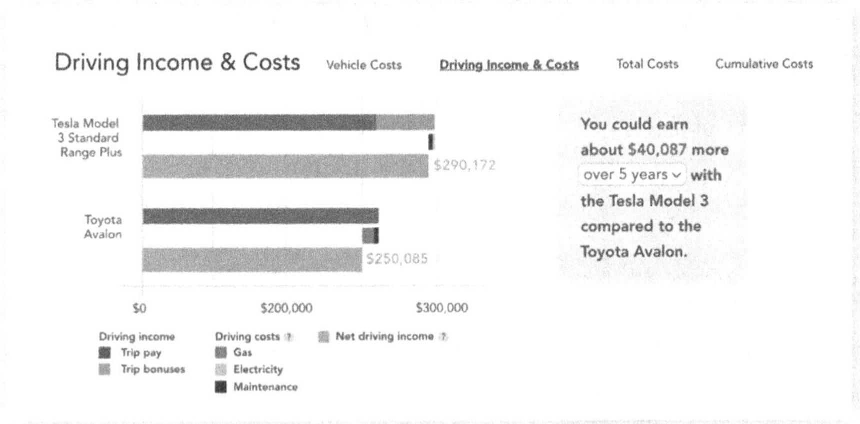

Fig. 4. Driving income and costs output for gig drivers

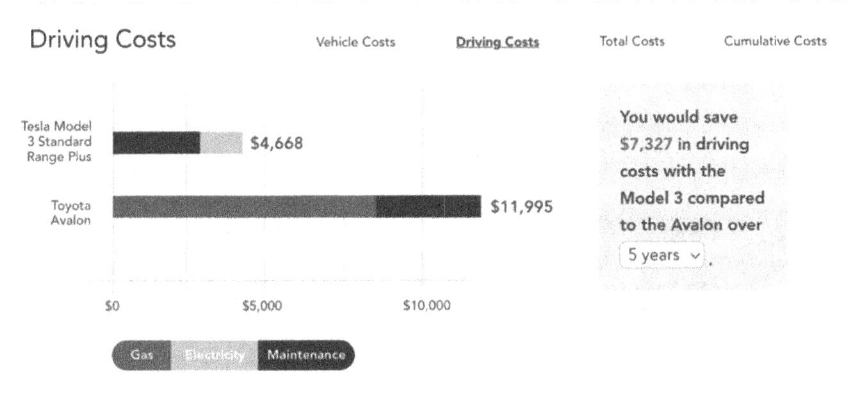

Fig. 5. Driving costs output for non-gig drivers

The other intention behind this strategy of splitting up types of costs is to match users' mental models. For the general population of car buyers, operational costs, if considered at all, might be more likely than part of weekly or monthly budgeting, separate from a large purchase decision (Allcott, 2011; Turrentine & Kurani, 2007). For the gig driver, operational costs factor into their earning potential, and thus something they are highly likely to consider and compare to their income (Sanguinetti & Kurani, 2021); the Driving Income and Costs output matches this mental model.

The Vehicle Costs output (Figs. 6, 7 and 8) speaks to the mental model that focuses on the upfront costs of buying or leasing a vehicle, e.g., if a consumer has a price range in mind and they consider this independently from TCO. Incentives are relevant here as they can be considered a deduction from upfront costs, even though the delay in receiving some incentives may not address all upfront cost barriers. Again leading with EV benefits, the first output displayed under Vehicle Costs is Incentives (Fig. 6). The user can then navigate to Purchase or Lease Cost (Fig. 7) and Net Cost (Fig. 8).

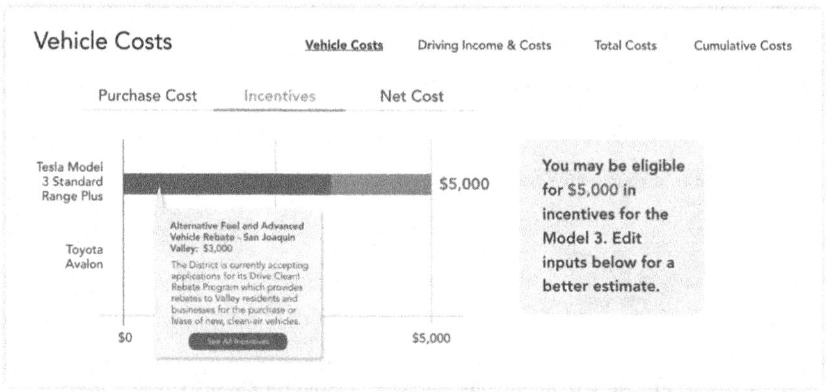

Fig. 6. Vehicle costs output: Incentives view

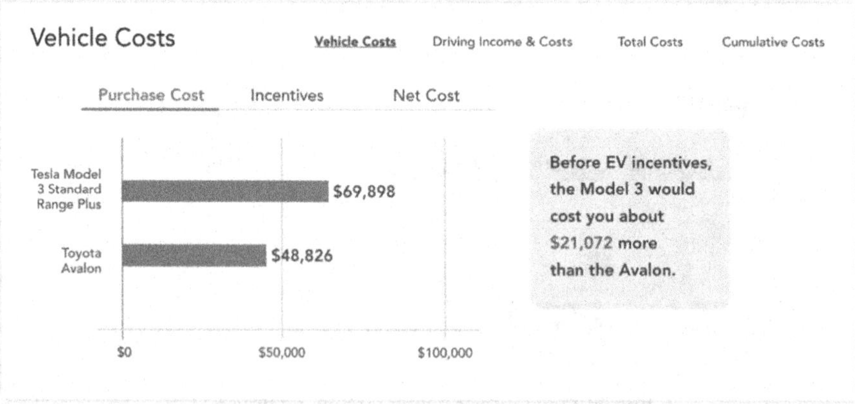

Fig. 7. Vehicle costs output: Purchase or lease costs view

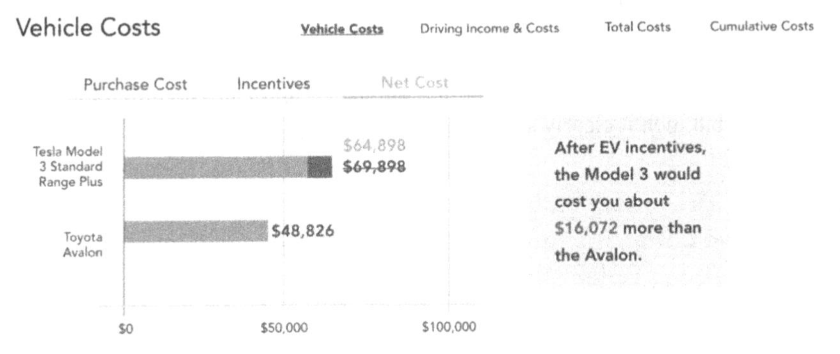

Fig. 8. Vehicle costs output: Net cost view

The last two cost outputs, Total Costs and Cumulative Costs, reflect TCO. The Total Costs output (Fig. 9) was inspired by Zappy Ride's (https://www.zappyride.com/) tools that provide a concise summary of TCO in a stacked bar chart. Our usability testing in this and prior projects indicated users find this output visually appealing, though they have difficulty comprehending all the information at once. In particular, the contribution of vehicle ownership to TCO (purchase cost minus incentives and resale value at the end of a specified timeframe) is complex. Our Total Cost output attempts to make this concept more comprehensible by using simpler language and a toggle for the user to subtract resale value from the vehicle cost factor. The idea is that a user can first orient to the familiar concept of vehicle costs and then consider resale value.

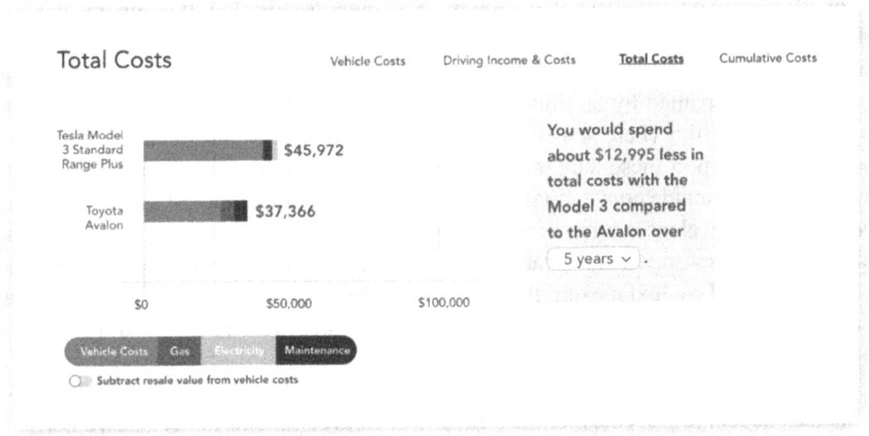

Fig. 9. Total costs output

The Cumulative Costs output (Fig. 10) was inspired by WattPlan (wattplan.com). In the (common) case that an EV costs more to purchase or lease than a comparable gas vehicle (even after incentives), this output highlights the breakeven point. The benefits of being able to see costs over time in a straightforward manner were noted by drivers during the usability tests. One user noted, "…Break even in 5.9 years, interesting, that is really meaningful information… The Tesla does seem like a significant savings, obviously not for six years, but then it clearly skyrockets."

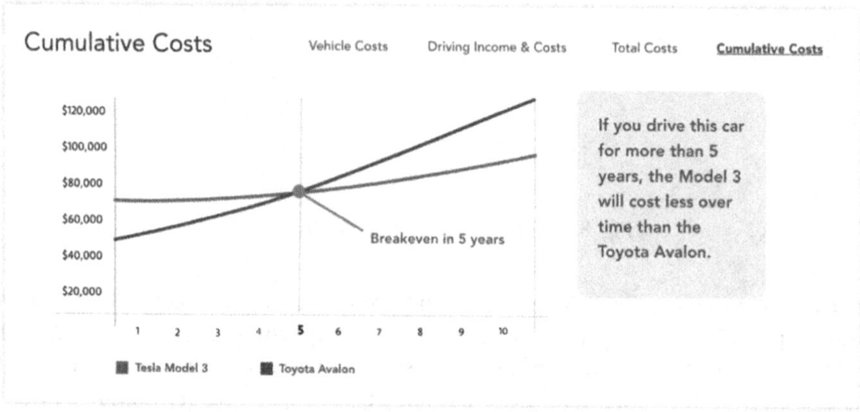

Fig. 10. Cumulative costs output

3.3 Highlight Social and Environmental Impacts

EV Explorer 2.0 includes salient information about the environmental and social benefits of EVs relative to gas cars alongside the cost outputs (Fig. 11). It compares the two displayed vehicles in terms of three types of impacts: carbon emissions from vehicle production and operation (and equivalent ice melt), tailpipe local air pollutants, and gasoline use, each accompanied by an illustrative animation ("empathetic gauge") to make the data more compelling (Petersen et al., 2016). This approach contrasts sharply with other tools that only report these data in numeric form and scientific data visualizations, and even those that include conversions and small graphics like number of trees required to sequester an equivalent amount of emissions with a tree icon. We made this information salient because environmental awareness has been a key driver of EV adoption (Rezvani et al., 2015), and by juxtaposing these benefits with vehicle costs, we hope to support users in framing any additional financial costs associated with an EV as an investment in these benefits.

Usability testing revealed generally positive reactions to these outputs. Remarks about the imagery (e.g., "cute pictures, "nice vibe") and comments implying an emotional impact ("makes me want to drive an electric car", "makes you stop and think") validated our choice to use a more artistic data visualization style for this information. One user remarked, "I really like the social and environmental benefits. I would never take a lot of time to look up what the benefits would be, but seeing that, that's extremely interesting and really eye opening actually." Different users were drawn to different

Fig. 11. Social and environmental impacts outputs

things, confirming the benefit of showing diverse social and environmental impacts and equivalencies. The glacial ice melt CO2 equivalency was particularly compelling to one user, while another was more interested in seeing the pounds of CO2. The air pollution information resonated with a user whose grandson has asthma.

Some comments suggested room for improvement in terms of clarifying the information and providing resources to learn more about CO2 emissions and air pollution, which led us to add more salient information hovers with more details and links to other resources in those outputs. Two users expressed skepticism about the emissions impacts which are difficult to tackle within the site: One indicated the impact of their personal actions pales in comparison with those of large corporations (suggesting it may not matter), and the other was not convinced the lifecycle impact (beyond emissions) of EVs is lighter on the environment when considering production and waste issues.

3.4 Maximize Customization

Although there are (strategically) few inputs required before a user reaches the initial output, a customization panel located along the bottom of the screen allows further customization. We aimed to include inputs for all factors that would significantly impact costs, particularly anything that differs substantially for EVs vs ICEVs. This was to ensure reasonable accuracy overall and in terms of the relative costs of EVs and ICEVs. The latter is crucial to prevent misleading people, e.g., over-flattering EVs.

Wherever possible, we included meaningful default values for optional inputs so that users can assess and consider whether they want to change them. "Meaningful" refers to both providing the source for default values so users can gauge whether it is credible, as well as tailoring the information when possible (e.g., default fuel prices based on user location). Another way we add meaning to the inputs, prompted by insights from our usability testing, is to describe how optional inputs impact costs so that the user might be more motivated to change them or at least understand why they are there.

The initial view of the customization panel upon landing on the output page displays a prompt in large, bold font to notify users of the ability to customize the output (Fig. 12). There is a vertical menu on the left side of the panel with five icons depicting different categories of inputs: Car Manager, Financing, Driving, Fueling, and Driving Income (displayed for gig drivers only). Car Manager inputs (Fig. 13) allow the user to change the electric and/or gas vehicle compared in the outputs, and to edit basic information about vehicle fuel efficiency. Financing inputs (Fig. 14) include information related to EV incentive eligibility to provide users more tailored estimates for incentives. There is also a section in the panel for how the user plans to acquire their vehicle (purchase with cash or loan, or lease).

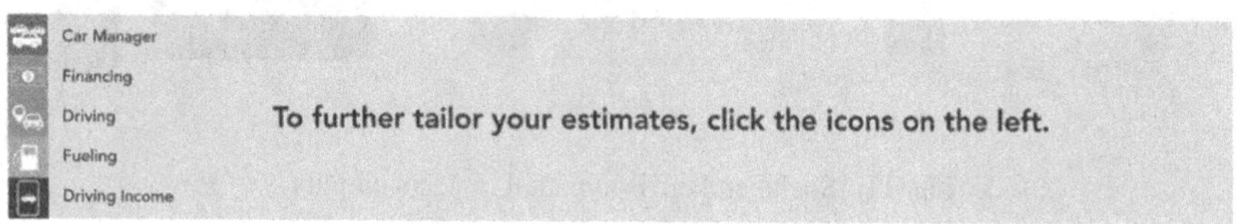

Fig. 12. Landing view of customization panel at bottom of output page

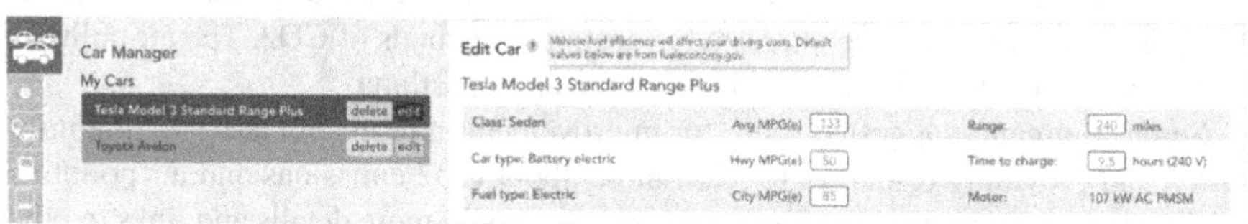

Fig. 13. Car manager customization panel: editable inputs for vehicle efficiency and range

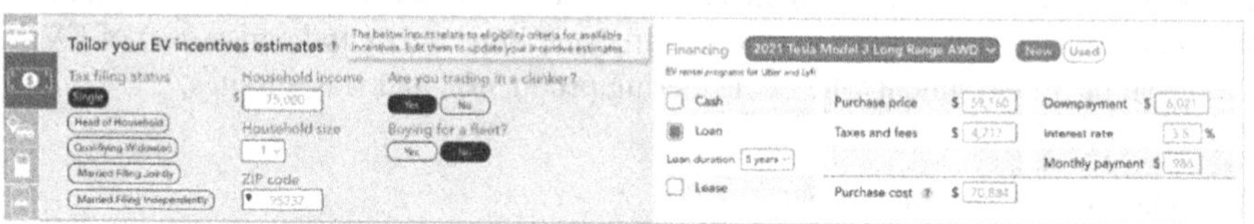

Fig. 14. Financing and incentives customization panel

Driving inputs (Fig. 15) include the ability to specify mileage, traffic conditions, and frequency for two daily profiles (heavier and lighter driving days) as well as long trips. Fueling inputs (Fig. 16) allow users to specify local fuel prices (the default is the average prices in the State where the user is located) and charging information. Charging information can be input in a basic or advanced window. The basic charging inputs specify what percent of charging users intend to do at home versus at workplace or public charging stations, whereas the advanced inputs allow users to specify percent of charging at multiple public stations and enter charging membership, session fees, and a one-time home charger installation cost. For gig drivers only, links are provided in this panel to TNC charging programs, in the basic view.

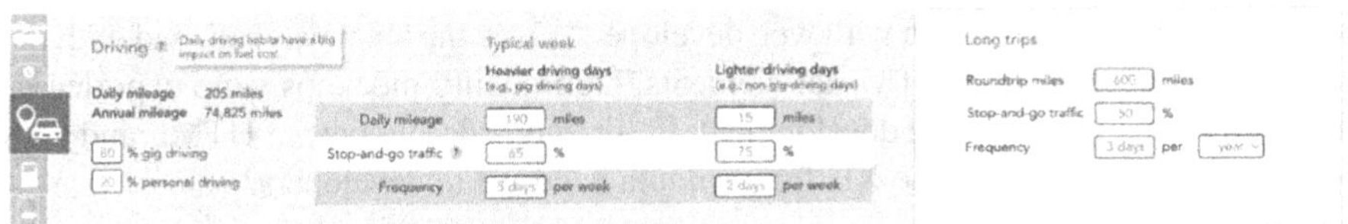

Fig. 15. Driving information customization panel for more precise driving cost estimates

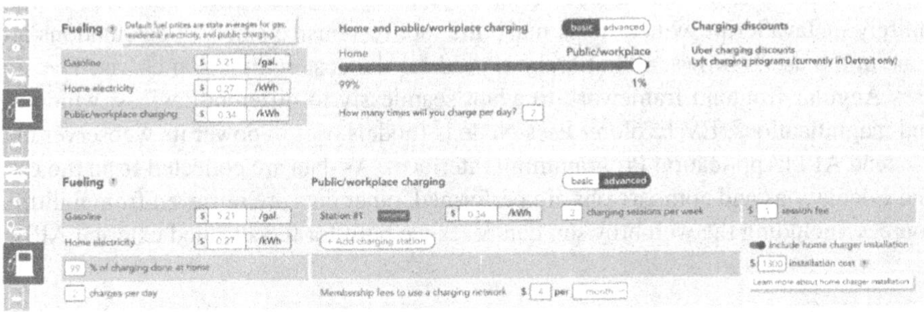

Fig. 16. Fueling information customization panel for more precise driving cost estimates.

For gig drivers only, Driving Income inputs (Fig. 17) allow users to edit the information they entered in the initial modal regarding their income. They can also edit the amount of rideshare driving and distribution of ridesharing driving across companies to see the impact of EV trip bonuses, currently offered only by Uber. An additional notification appears in this panel if the EV in the comparison is eligible for upgraded services on Uber or Lyft that offer higher pay rates (e.g., Uber Comfort, Lyft Lux).

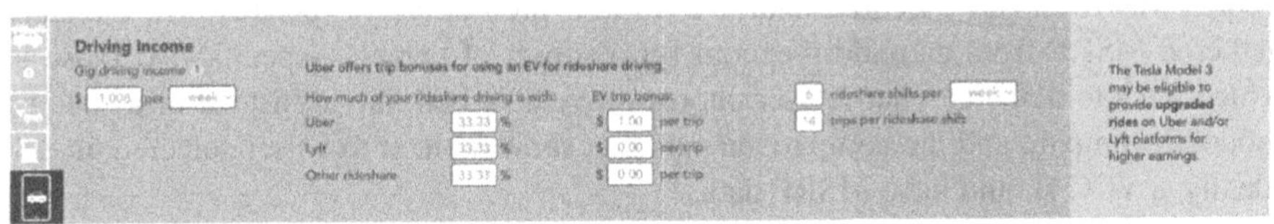

Fig. 17. Driving income customization panel for more precise driving income estimates

4 Website Development

The following sections detail the three key components of the development process: mockups, programming, and the TCO model.

4.1 Website Mockups and Clickable Prototype

Website mockups and a prototype for EV Explorer 2.0 were created in Adobe XD to enable mockups and a hi-fidelity clickable prototype for team discussion, design iteration, and communication with web developers. Once the team was satisfied with the prototype's functionality and visual elements, high-quality mockups were exported as image references for website development. Front-end code snippets in HTML and CSS were also generated in Adobe XD for a smooth hand-off to developers.

4.2 Programming Overview

EV Explorer 2.0 is hosted on a server at the authors' research institution. It is written entirely in JavaScript, which helps make the tool responsive since all calculations are done in the user's browser. D3 library is used for the cost comparison charts. The site uses Angular frontend framework to adapt seamlessly to different devices, windows, and magnifications. EV Explorer uses Node.js (nodejs.org) to power its webserver and backend API (Application Programming Interface). As data are collected from the user and calculations and comparisons are performed, other data are retrieved from multiple sources, including the web browser, databases we built for the site, and external APIs.

4.3 Data Sources

EV Explorer 2.0 makes use of several freely available public APIs. Google Maps API is used to display user-input trip origin, destination, and route on a map as an aid to estimate mileage (in the non-gig driver pathway only). An API from fueleconomy.gov, a site co-sponsored by the US DOE and EPA, is used for vehicle fuel economy and electric range information. Default (but modifiable) fuel prices in the tool are state averages taken from the US Energy Information Administration (EIA). The user's state is determined based on their IP address using ipstack (ipstack.com).

Building out EV Explorer 2.0 to account for TCO required identifying additional data sources for vehicle acquisition and maintenance costs, as well as resale value. Vehicle MSRP and EV incentive data sources were prohibitively expensive. We manually gathered MSRP from caranddriver.com for a subset of vehicles–specifically, those to which the site defaults when a user chooses a vehicle body style instead of a specific model to compare and the comparison cars that show up next to a user-selected model. Ideally, a VCC should have MSRP data.

It was a high priority to include comprehensive incentive information for all EVs since users may not know how to find it otherwise, whereas vehicle purchase and lease prices are readily available via familiar channels (e.g., manufacturer websites and local car dealerships). We created our own comprehensive (at the time) database for vehicle incentives, manually gathering the data on federal, state, and local incentives. Federal incentive amounts were gathered from fueleconomy.gov. For state and local credits, we gathered the incentive names, descriptions, links to program information, amount levels, and eligibility criteria from the DOE's Alternative Fuels Data Center (AFDC; afdc.energy.gov); the appropriate incentive(s) and amount(s) populate for the user based on the following inputs related to the eligibility criteria: tax filing status, household income, household size, zip code, if a clunker could be traded in, and fleet status.

We relied heavily on the 2020 AFLEET tool by the Argonne National Laboratory. This is where we gathered data for calculating vehicle maintenance costs and resale value as well as social and environmental impacts (GHG emissions, local air pollution, and gasoline use). The formulas we use for maintenance and resale value are detailed in the next section on the TCO model. Our calculations for vehicle lifetime GHG emissions are not as precise as is possible with the AFLEET tool. We simplified by using the average emissions for a model year 2020 vehicle of each drivetrain, so the resulting comparisons reflect user mileage and grid energy sources based on user location but not vehicle age.

EV Explorer 2.0 also provides users with links to pertinent sources, starting in the first modal screen when the user is tasked with selecting a vehicle model or body style to explore. There is a prompt, “Not sure which car to choose?” with a link to Plug-in America’s PlugStar Shopping Assistant tool (by Zappy Ride) to help users looking to identify an EV suited to their needs. The incentives output provides links to each incentive program for which the user may be eligible (from AFDC). As previously mentioned, news article links are provided to learn more about vehicle lifetime GHG emissions, local air pollution, and oil dribble. In the Fueling Customization Panel, we link to a news article about home charger installation costs and incentives.

For the gig driver pathway, and ridehail drivers specifically, the site includes links to various Uber and Lyft webpages for information about: vehicle eligibility requirements if the user selects a vehicle or body style that may be ineligible for driving on those platforms, EV charging programs in the Fueling Customization Panel, upgraded ride services when the EV in the comparison may be eligible to provide these services, and TNC rental programs next to the inputs for vehicle purchase or lease. For the latter, we created a database of EVs eligible to provide upgraded ride services by referencing Uber and Lyft websites.

5 Total Costs of Ownership Model

TCO has three main components: the capital cost associated with purchasing a vehicle, operating cost of the vehicle, and a resale value. Operating costs include fuel, insurance, vehicle registration, and maintenance. The resale value of the vehicle depends on the depreciation rate that may vary by fuel type and vehicle segment.

The TCO model developed for EV Explorer 2.0 builds from the framework in Chakraborty et al. (2021), which was developed for the vehicle purchase scenario (vehicle purchased new with zero downpayment and financed for a 5 year loan period). Inputs for the capital cost component came from a teardown analysis and other public sources of data; operating cost estimates were based on fuel cost data from the EIA database and home v. public charging prevalence from a survey of general population EV drivers; and resale value estimates from the 2020 AFLEET tool. The aim was to adapt this TCO framework to the gig driving use case so that EV Explorer 2.0 can provide accurate cost estimates for gig drivers that fully account for the potential savings with an EV compared to a gas car.

The first major goal of our adaptation was to increase the precision of operating (fuel and maintenance) cost estimates. Fuel cost is a major differentiating factor between ICEVs and EVs and particularly important when miles traveled by the vehicle owner is high as in the case of gig drivers. Fuel costs for EVs are largely affected by charging location; e.g., gig drivers generally rely much more on public fast charging which is typically much more expensive than charging at home. EVs are much cheaper to maintain than gas cars and, like fuel costs, this difference is amplified for high-mileage drivers. Our past VCC usability research indicated maintenance cost estimates for EVs in existing VCCs were higher than EV users reportedly experienced, diminishing estimated savings with an EV compared to a gas car, so a new method for calculating maintenance costs was a priority.

The other major adaptation was to build in consideration of alternative vehicle acquisition systems. Rather than solely focus on the vehicle purchase model, we needed to include used vehicle purchase and vehicle lease acquisition methods, which required modifying the upfront capital cost component in the model. The calculations were done in Excel and used in the VCC as inputs for the appropriate output tab. The main steps in the estimation process, the data used, and the assumptions for the three scenarios are described next.

5.1 New Vehicle Purchase

Capital Costs. The capital cost component of the TCO calculation for the Vehicle Cost output only accounts for the cost associated with vehicle purchase. Assuming a loan term of 5 years (by default, but modifiable by the user), the loan amount is estimated to be user-inputted purchase price - downpayment - incentive rebates (if eligible). An annual interest rate of 5% is assumed for the loan financing calculation, though in practice the rate depends on the loan term and credit score of the borrower (thus, this is also a user-editable input).

Operating Costs. For ICEVs, fuel efficiency depends on the vehicle model year and fuel price depends on the PADD region of the driver. In case of BEVs, range and MPGe of a vehicle depends on the model year of the vehicle and electricity price vary by utilities. As mentioned in the previous section on Data Sources, the data for vehicle efficiency is obtained from fueleconomy.gov while electricity price data is from EIA. Fuel cost for PHEV, depend on utility factor (share of electric miles driven) that in practice depends on the vehicle range and on the charging behavior of the driver. For estimation purposes, we assume the utility factor based on the driving information a user inputs in the model (i.e., trip or daily mileage or annual mileage divided by 365) until and unless the user specifies more nuanced daily driving patterns in the Driving Customization Panel, at which point the utility factor is changed to reflect those daily profiles and provide more tailored estimates.

Other VCCs use an average per-mile cost for maintenance, which differs by drivetrain and may increase with vehicle age. However, as previously mentioned, we found that VCC users with EV experience were shocked by what they reported as unrealistically high estimates for EV maintenance costs. To try to correct for this, we developed a different method for estimating maintenance costs, based upon user-inputted mileage and cost of specific maintenance measures using data from AFLEET on the recommended number of miles for the following measures/replacements: oil change, oil filter, tire rotation, fuel filter, air filter, cooling system, tires, brake pads, headlights, transmission service, timing belt, spark plugs, brake rotors, battery replacement for conventional engines, and battery replacement for EVs. Total cost is calculated based on total user-inputted mileage.

We do not include other operational cost components, i.e., insurance and annual vehicle registration costs. These are mainly dependent on driving behavior and characteristics of the user and the vehicle registration rate of the area where the vehicle is registered respectively. The difference by fuel/powertrain type is marginal and does not contribute a significant difference in total or cumulative costs between EVs and ICEVs. We allow the user to add a one-time home charger installation cost to the Driving Cost output via a toggle and editable input in the Fueling Customization Panel, advanced window (this is often considered a capital cost).

Resale Value. For resale value estimation, we assume a depreciation rate of 23% in the first year and 15% in the subsequent years (Burnham, 2021). As EVs (short-range BEVs in particular) can have a higher depreciation rate than ICEVs, the depreciated value of a BEV with less than a 100-mile range is multiplied by an adjustment factor of 13% (Rush et al., 2022).

5.2 Used Vehicle Purchase

Capital Costs. For the capital cost of used vehicles, we would reduce the purchase cost based on depreciation rate described above given the vehicle model year if we had a data source for vehicle MSRP. Since we do not currently have an MSRP data source, the Vehicle Cost output relies on the user inputting a purchase price for a used car, just as they must for most new cars.

Operation Costs. Regarding operating costs, both fuel efficiency and battery efficiency deteriorates with vehicle age. Therefore, ideally these two metrics can be adjusted based on scaling factors available in the literature. For now, EV Explorer uses the statistics from fueleconomy.gov which do not vary based on vehicle age. Maintenance is calculated in the same manner as for new vehicles.

Resale Value. The resale value of a used vehicle is estimated assuming a 15% year-in-year depreciation rate with the 13% adjustment factor for short-range BEVs.

5.3 Vehicle Lease

Capital Costs. The vehicle acquisition cost in case of lease contract includes the downpayment and the monthly lease payment, which is calculated in the tool from the downpayment and interest rate (both editable user inputs). In case of EVs, the monthly lease payment also accounts for any purchase incentive(s) for which it is eligible. Upfront costs include the downpayment and monthly lease payment for the first month as well as an initial drive-off cost that comprises of an acquisition fee, documentation fee, and vehicle registration fee. We consider average values for these cost components for a default total of $645, which the user can edit.

The monthly lease payment also accounts for the residual value of the vehicle estimated by the dealer/financer. To estimate residual value we assume the same depreciation rate as in case of new or used vehicle purchase and a lease term of 36 months. The accuracy of the lease payment estimate depends on the accuracy of vehicle price reported by the user since in practice the residual value is estimated on the basis of MSRP. Therefore, again, an MSRP data source is crucial to both maximize user-friendliness and value of a VCC as well as accuracy. Another thing to note, residual value calculation in lease contracts also depends on the mileage limit set for the vehicle, but it was difficult to find a consistent formula in the literature that dealers may use for the estimation process, so we do not account for it.

Operational Costs. Operational costs for a leased vehicles are estimated like the other two cases (purchase new or used vehicles). A lessee can pay for a maintenance package that includes all the required maintenance costs, but we do not include a way to account for this at this time; it is assumed that the lessee behaves like a regular vehicle owner in terms of maintenance.

Resale Value. In case of lease, there is no resale value, but the lessee needs to pay a disposition fee. In addition, he or she may have to pay a mileage limit fee if the mileage limit set in the lease contract is exceeded and a fee for excess wear and tear of the vehicle when applicable. We account for these end of lease costs in the TCO estimation assuming average cost estimates of a $350 disposition fee and $0.175/mile fee for exceeding mileage limits. These are both accounted for as "end-of lease fees" factoring into the Total Costs and Cumulative Costs outputs.

6 Conclusion

EV Explorer 2.0 is the first online vehicle cost calculator (VCC) developed for gig drivers broadly and the second intended to meet the special needs of ridehail drivers. It was designed to embody best practices as well as innovative features not seen in other tools. It supports users interested in purchasing used vehicles and estimates the impact of different vehicles on gig driver income. Future goals include updating the incentive database, acquiring a data source for vehicle MSRP, and further user testing. We are also sharing our work with VCC developers in government and industry in hopes they can use the strategies and insights reported herein to develop or improve their own VCCs. Ultimately, we hope this tool and research will support and empower gig drivers by informing them about the potential TCO savings and social and environmental benefits of EVs and support gig driving companies to continue developing goals and programs to transition to zero-emissions services.

Acknowledgments. This study was funded by a grant from the National Center for Sustainable Transportation (NCST), supported by the U.S. Department of Transportation (USDOT) through the University Transportation Centers program. The authors would like to thank the NCST and the USDOT for their support of university-based research in transportation, and especially for the funding provided in support of this project. Additional support for website development work was provided by the EV Research Center's STEPS+ consortium. The authors would also like to thank

Uber Technologies, Inc. For assisting with recruitment of drivers to participate in the usability testing and Zappy Ride for their feedback and interest in this research.

Disclosure of Interests. The authors have no competing interests.

References

Allcott, H.: Consumers' perceptions and misperceptions of energy costs. Am. Econ. Rev. **101**(3), 98–104 (2011)

Burnham, A.: AFLEET: Assess the impacts of conventional and alternative fuel vehicles. (DOE-ANL-CH11357). Argonne National Lab.(ANL), Argonne, IL (United States) (2021)

Chakraborty, D., Buch, K., Tal, G.: Cost of plug-in electric vehicle ownership: the cost of transitioning to five million plug-in vehicles in California (2021)

Egbue, O., Long, S.: Barriers to widespread adoption of electric vehicles: an analysis of consumer attitudes and perceptions. Energy Policy **48**, 717–729 (2012)

Eppstein, M.J., Grover, D.K., Marshall, J.S., Rizzo, D.M.: An agent-based model to study market penetration of plug-in hybrid electric vehicles. Energy Policy **39**(6), 3789–3802 (2011)

Erdem, C., Şentürk, İ, Şimşek, T.: Identifying the factors affecting the willingness to pay for fuel-efficient vehicles in Turkey: a case of hybrids. Energy Policy **38**(6), 3038–3043 (2010)

Hagman, J., Ritzén, S., Stier, J.J., Susilo, Y.: Total cost of ownership and its potential implications for battery electric vehicle diffusion. Res. Transp. Bus. Manag. **18**, 11–17 (2016)

Hamza, K., Laberteaux, K.P., Chu, K.C.: On modeling the cost of ownership of plug-in vehicles. World Electr. Veh. J. **12**(1), 39 (2021)

Hjorthol, R.: Attitudes, ownership and use of Electric Vehicles–a review of literature. TØI report **1261**(2013), 1–38 (2013)

Knobloch, F., et al.: Net emission reductions from electric cars and heat pumps in 59 world regions over time. Nat. Sustain. **3**(6), 437–447 (2020)

Pavlenko, N., Slowik, P., Lutsey, N.: When does electrifying shared mobility make economic sense. Int. Council Clean Transp. (2019)

Petersen, J.E., Frantz, C., Shammin, R.: Using sociotechnical feedback to engage, educate, motivate and empower environmental thought and action. Solutions **5**(1), 79–87 (2014)

Rezvani, Z., Jansson, J., Bodin, J.: Advances in consumer electric vehicle adoption research: a review and research agenda. Transp. Res. Part D: Transp. Environ. **34**, 122–136 (2015)

Rush, L., Zhou, Y., Gohlke, D.: Vehicle residual value analysis by powertrain type and impacts on total cost of ownership (No. ANL/ESD-22/2). Argonne National Lab. (ANL), Argonne, IL (United States) (2022)

Saarenpää, J., Kolehmainen, M., Niska, H.: Geodemographic analysis and estimation of early plug-in hybrid electric vehicle adoption. Appl. Energy **107**, 456–464 (2013)

Sanguinetti, A., Kurani, K.: Characteristics and experiences of ride-hailing drivers with electric vehicles. World Electr. Veh. J. **12**(2), 79 (2021)

Sanguinetti, A., Alston-Stepnitz, E., Cimene, A.: Facilitating electric vehicle adoption with vehicle cost calculators (No. NCST-UCD-RR-20–13). National Center for Sustainable Transportation (NCST)(UTC) (2020)

Sanguinetti, A., Nicholas, M.A., Tal, G., Favetti, M.: EV Explorer: evaluating a Vehicle Informational Tool (2016)

Taiebat, M., Stolper, S., Xu, M.: Widespread range suitability and cost competitiveness of electric vehicles for ride-hailing drivers. Appl. Energy **319**, 119246 (2022)

Turrentine, T.S., Kurani, K.S.: Car buyers and fuel economy? Energy Policy **35**(2), 1213–1223 (2007)

Wu, G., Inderbitzin, A., Bening, C.: Total cost of ownership of electric vehicles compared to conventional vehicles: a probabilistic analysis and projection across market segments. Energy Policy **80**, 196–214 (2015)

Blockchain for Food Traceability - Consumer Requirements in Austria

Robert Zimmermann[1](✉), Magdalena Richter[2], and Patrick Brandtner[1]

[1] University of Applied Sciences Upper Austria, 4400 Steyr, Austria
robert.zimmermann@fh-steyr.at

[2] Wacker Neuson, Flughafenstraße 7, 4063 Hörsching, Austria

Abstract. Global food supply chains are becoming increasingly complex and the demand for transparency and traceability in food is growing. As a result, there is increasing pressure on the food industry to track food from production to sale. However, current traceability systems are inadequate, with blockchain technology offering a promising approach to food traceability. However, little is known about consumers' interests and requirements for traceability information. As such, this paper examines the requirements of Austrian consumers for food traceability and the advantages and disadvantages of satisfying the identified requirements with blockchain technology. In this regard, the requirements of consumers mainly relate to the welfare of animals and the precise indication of the origin of animal products. As such, the use of blockchain technology exhibits significant potential for enhancing transparency and traceability within the supply chain, however, requiring a nuanced strategy to maximize its advantages and overcome inherent limitations such as efficiency, cost-effectiveness, and scalability.

Keywords: Blockchain · Food Traceability · Consumer Requirements · Austria

1 Introduction

The networks of the food supply chain (FSC) are becoming increasingly complex and fragmented. This complexity makes the identification and tracking of products and processes along the supply chain very difficult [1]. Moreover, the current transport routes of food from producer to consumer have become longer due to the globalization of food trade. This raises more concerns about origin and quality as well as food safety on the part of consumers [2]. Apart from the increasing interest of consumers in the origin of their food, there are also growing problems within the current FSC [3]. These problems include delayed response times, inadequate diagnoses of outbreaks of foodborne diseases, food adulteration, such as the addition of undeclared ingredients, and fraud with seafood [4].

In addition, the global food supply chains are under increasing pressure due to rising consumer demands [5]. Consumers demand safe and healthy products of high and consistent quality. In addition, they require a certain availability of products in a wide range of assortments throughout the year at competitive prices [6]. Simultaneously,

F. F.-H. Nah and K. L. Siau (Eds.): HCII 2024, LNCS 14721, pp. 253–275, 2024.
https://doi.org/10.1007/978-3-031-61318-0_18

consumers are very demanding and place more value on quality and the health benefits of products. As such, food quality has become a central issue and the last decades have shown that the concern for a healthier lifestyle and the protection of the environment are the driving forces behind food shopping [7].

Therefore, much depends on the perception of food quality by consumers. The environmental impacts of the entire FSC, from the producer to the trash can, are influenced by the decisions of consumers [7]. And they, among other things, want to be informed about the origin, the production methods, the safety level, the application of pesticides, and the environmental aspects [8].

The use of blockchain technology along the supply chain is versatile. For example, it can be used in the production sector and enable intelligent automation [9]. Similarly, the technology has great potential to establish sustainable processes in the supply chain [10]. Also, with the help of blockchain technology, the traceability of products, such as food, is also possible. As such, the use of blockchain technology has the potential to create transparency along the entire FSC, to reduce inefficiencies and to improves food safety [11].

Conducting a first feasibility study with a limited sample size, this paper reflects on the mentioned issues and possibilities of using blockchain technology for food traceability. As such, this paper aims to provide first insights into the traceability requirements that are most relevant for Austrian consumers and how well blockchain technology is suited to satisfy these demands. We chose the region of Austria as to the best of our knowledge no study exists investigating the region of Austrian regarding food traceability using blockchain. In addition, by focusing on Austria, the research aims to offer insights that may have broader applicability and relevance to regions (Germany, Switzerland) facing similar challenges and exhibiting comparable consumer behaviors. As such, this paper investigates the following research questions:

RQ1: What are the consumer requirements for food traceability information in Austria?

RQ2: What are the advantages and disadvantages of satisfying the identified consumer requirements for food traceability in Austria via blockchain?

Consequently, this paper makes a significant contribution to the discourse on human-computer interaction in the context of digital marketing and commerce by showcasing how blockchain technology can play a pivotal role in enhancing transparency, reducing inefficiencies, and improving food safety within the omnichannel setting of the food supply chain.

The paper is structured in the following way. To ensure a general understanding of the blockchain technology Sect. 2 provides a brief overview of the blockchain technology as well as its advantages and disadvantages. A structured literature review about how the blockchain technology can be used to generate traceability along the FSC is presented in Sect. 3. Section 4 introduces the methodology used to uncover the consumer requirements for food traceability information in Austria. The results of this analysis are presented in Sect. 5. They are discussed in combination with the results gathered from the structured literature review in Sect. 6, which also presents possible limitations and avenues for future research. Lastly, Sect. 7 provides a conclusion of the paper.

2 Blockchain Technology - Advantages and Disadvantages

To ensure a general understanding of blockchain technology this chapter gives a brief overview of the blockchain technology as well as its advantages and disadvantages.

Blockchain is a distributed ledger that records transactions validated and maintained by a network of computers worldwide. Instead of relying on a single central authority (e.g., a bank), all records are monitored by the network participants (peer-to-peer network). Information cannot be manipulated as in conventional central databases, as the transactions have to be confirmed by the participants [12].

Blockchain technology is categorized into three types: Blockchain 1.0, focused on cryptocurrencies like Bitcoin; Blockchain 2.0, is mostly applied in financial services with the introduction of smart contracts; and Blockchain 3.0, offering enhanced security and scalability, applied across various industries [13]. Additional types such as public, private, semi-private blockchains, and sidechains are discussed in more detail by Sarmah [12] but exceed the scope of this paper.

The main advantage of blockchain technology is decentralization. The system operates without a third organization or a central administrator. Instead, all participants of the blockchain make the decisions. The conventional process of securing a database against third parties or hackers can be time-consuming and costly. By employing blockchain technology, this can be avoided, as the transactions of the blockchain have their own validity and authorization proof. This implies that the transactions can be independently verified and processed [14].

Another advantage of this technology is the recording of every action and its data, which are available to every participant but cannot be altered or deleted. The outcomes of this recording confer the blockchain transparency, immutability, and trustworthiness [15].

Moreover, the blockchain is designed to identify and correct all problems if necessary, as all transactions are interconnected. This advantage makes the blockchain technology traceable and allows the verification of the authenticity of assets and products and thus prevent fraud. Since all parties involved have to agree on transactions before they are recorded, security is also a major advantage of the blockchain [16].

The biggest disadvantage of blockchain technology is the high energy consumption required for the implementation of a real-time ledger. A considerable amount of computing power is required to validate transactions. Additionally, each transaction must be signed with an encrypted scheme. This signature verification is a challenge of the blockchain and requires a large computing power as well as time for the calculation process of the signature [17].

Another major disadvantage of the blockchain are the high costs. The costs for an average transaction range between 75 and 160 dollars. The largest part of this is caused by the energy consumption [18].

Although the processing time at banks is often longer, the pure transaction process is much faster than with the common blockchain networks. The Bitcoin system currently allows only seven transactions per second. In contrast, Mastercard or Visa payment infrastructures allow several thousand transactions per second [19].

Summarizing, the blockchain technology brings many advantages such as transparency and immutability. However, the disadvantages, such as the enormous energy demand, must not be overlooked.

3 Literature Review

To get an inside how blockchain technology can enable traceability along the FSC a structured literature review was conducted from April to Mai 2023. In accordance with Vom Brocke et al. [20] and Cooper [21], the focus of the literature review was to uncover research outcomes, theories, and applications. Our goal was thus to find central issues and we organized or findings in a conceptual way (see also Table 2). We took a neutral perspective as we collect rather than criticize available literature. The literature review aims at general scholars who want to get an overview of potential blockchain technologies, theoretical concepts, and practical applications. We give an exhaustive coverage of the literature, as we want to summarize the current status of blockchain technology used along the food supply chain.

For this review the keyword "food traceability" was chosen to achieve the most accurate search results possible on the topic of food traceability. The terms "blockchain" and "application" were also used in order to obtain research articles that examine the technology and applications. As the literature search focuses on the possible applications along the FSC, the search term "supply chain" was also used. As such, the following search string was defined for the structured literature review:

```
"food traceability" AND blockchain AND application AND "supply
chain"
```

The search string was applied in the following data bases: EBESCO Business Source Elite, Emerald Collections, IEEE/IEL, ScienceDirect, Springer Link, Taylor and Francis, Wiley Online Library, WISO. In total, 750 papers could be identified. These were screened if they contain information about how the blockchain can be used for traceability along the FSC. As such, 18 papers were selected containing such information. An overview of the data bases reviewed, the number of papers found, and the number of papers selected can be found in Table 1.

Based on the 18 identified papers, the current state of research on the topic of how blockchain technology can enable traceability along the food supply chain was analyzed. This analyzes revealed three main concepts ("Blockchain Technology and the Food Supply Chain in General", "Theoretical Blockchain Concepts in the Food Supply Chain", "Practical Blockchain Concepts in the Food Supply Chain"), containing twelve subconcepts (see also Table 2). It became evident that, most of the identified literature focuses on the use of blockchain technology along the food supply chain in general, focusing on specific technologies used for generating transparency, chances and challenges, and the impact on consumers. In contrast, practical and theoretical blockchain concepts received very little interest in the current body of research, mostly depicting specific applications (tracing eggs/Halal meat) or presenting a specific model or case study.

In the following Subsects. (3.1, 3.2, 3.3), the content of the papers found in the literature review is described in detail.

Table 1. Data bases reviewed, papers reviewed and selected

Data base	Papers found	Papers selected	Source
EBESCO Business Source Elite	8	0	N/A
Emerald Collections	147	2	[22, 23]
IEEE / IEL	12	2	[24, 25]
ScienceDirect	211	7	[1, 26–31]
Springer Link	303	1	[11]
Taylor and Francis	33	4	[32–35]
Wiley Online Library	32	2	[36, 37]
WISO	4	0	N/A

Table 2. Concepts found in the literature

Concept	No. of papers	Sub-Concept	No. of papers	Source
Blockchain Technology and the Food Supply Chain in General	12	Technology	11	[1, 11, 23, 27, 29–31, 33, 35, 36]
		Consumers	4	[23, 28, 30, 36]
		Companies	2	[28, 30]
		Chances/ Challenges	5	[23, 29, 33, 35, 37]
Theoretical Blockchain Concepts in the Food Supply Chain	4	Model	1	[1]
		Case Study	1	[32]
		Ethereum based	2	[24, 25]
Practical Blockchain Concepts in the Food Supply Chain	5	Eggs	1	[26]
		Agriculture/ food industry	1	[27]
		Halal Meat	1	[34]
		Aquaculture	1	[31]
		Staple food	1	[22]

3.1 Blockchain Technology and the Food Supply Chain in General

Butt & Fahim [11] emphasize the blockchain's role in addressing food traceability challenges, including quality, safety, and transparency. Using five foodborne outbreak examples, they illustrate how blockchain overcomes issues like information asymmetry, fraud, adulteration, and contamination in the FSC. Their findings suggest that blockchain enhances FSC transparency, accountability, efficiency, food safety, and consumer trust.

Kamilaris et al. [29] investigate blockchain's impact on agriculture and FSC, showcasing ongoing projects. While endorsing blockchain as a solution for a transparent and sustainable FSC, they acknowledge hurdles like technical, educational, political, and legal barriers impeding its widespread adoption.

Roy et al. [33] address challenges in the current food supply chain (FSC), including food waste and fraud, proposing blockchain technology as a solution to enhance transparency and traceability. Their case study, based in India, showcases a blockchain-based FSC platform, highlighting its potential to revolutionize the FSC, fostering sustainability, efficiency, and trust.

Similarly, Tayal et al. [37] and Xu et al. [35] explore the potentials of blockchain technology in the FSC, emphasizing its benefits. Despite recognizing its promise, both studies highlight significant challenges, including costs, privacy, standardization, and complexity.

Cozzio et al. [28] comprehensively address blockchain integration in the FSC for both consumers and suppliers. They emphasize the perceived advantages, including enhanced trust. Proposing blockchain for food safety and traceability, the authors acknowledge challenges like regulatory barriers and data-sharing concerns. They categorize key issues into trust, efficiency, sustainability, and financial/technological constraints.

Patel et al. [30] address blockchain technology's potentials and challenges. They discuss its role in enhancing supply chain security by ensuring goods' authenticity and traceability. Despite limited research, several companies, such as AgriDigital, Demeter, and Ripe, have initiated blockchain solutions in agriculture to address specific challenges. For example, AgriDigital offers a commodity management solution using smart contracts, Demeter facilitates micro-field management globally, and Ripe creates a transparent digital supply chain for quality food data mapping.

Rogerson & Parry [23] explore the potentials and challenges of blockchain technology adoption in the FSC, emphasizing the crucial role of consumers. The study reveals that consumer adoption depends on perceived benefits, trust, and affordability. Consumers are more inclined to pay for blockchain if it enhances food safety and prevents fraud. The authors suggest that companies must educate consumers on blockchain benefits, addressing concerns about data protection and security to boost their willingness to pay.

Shew et al. [36] assess blockchain's potential in the U.S. beef sector, emphasizing consumer preference for USDA certification [38] (organic certification in the USA) over blockchain traceability. The study recommends prioritizing the value of product data rather than focusing solely on the technology managing it.

3.2 Theoretical Blockchain Concepts in the Food Supply Chain

Casino et al. [1] introduce a blockchain-based food traceability model leveraging IoT technology. The model enhances information collection, processing, and exchange among FSC stakeholders. Using decentralized storage, participants store product details in a table of content locally. Traceability involves accessing this table to obtain necessary information. Performance analysis using indicators reveals substantial benefits for all supply chain participants.

Granillo-Macías et al. [32] investigate the grain supply chain, analyzing its actors and processes from cultivation to customer delivery. Their case study suggests that implementing a blockchain solution can decrease food waste by up to 20%, enhancing inventory management and reducing spoilage. Utilizing smart contracts and a public geographic information system, the proposed solution facilitates interaction and information exchange among supply chain parties, ensuring seamless traceability. Despite these benefits, the study identifies challenges, including technical, regulatory, and cultural barriers to blockchain adoption in the food industry.

Sai & Rekha [25] present an Ethereum-based smart contract system for soybean traceability, enhancing trust, efficiency, security, and reliability while eliminating intermediaries. Each supply chain entity is assigned a unique identifier, monitored through high-resolution cameras from production to consumption. Collected data are stored in a personal blockchain network, with hashes and links stored publicly. Customers can trace the entire supply chain by scanning the product identifier, showcasing blockchain's potential to revolutionize the agricultural sector.

Kim et al. [24] present the Harvest Network, an end-to-end application for food traceability integrating blockchain, smart contract tokens, and IoT to overcome information asymmetry and standardization issues in the current FSC. The authors highlight the benefits of these technologies in ensuring data standardization and regulatory compliance globally. However, successful implementation necessitates collaboration among all FSC stakeholders, addressing challenges related to data protection, security, and scalability.

3.3 Practical Blockchain Concepts in the Food Supply Chain

Bumblauskas et al. [26] investigate a Midwest company's use of blockchain to enhance its egg delivery system. In a retail-level test, egg cartons were traceable back to the supplier and pickup date through scanning and a web application. The study revealed higher-than-expected consumer engagement, stakeholder interest, and scan rates (21.2%) with an average usage time of 2 min and 48 s. The findings suggest opportunities for enhanced traceability and supply chain analysis, emphasizing blockchain's potential to improve food distribution efficiency, transparency, and security. The authors stress the need for further research to fully comprehend blockchain's impact on the industry.

Compagnucci et al. [27] explore blockchain integration into agriculture and the food supply chain, presenting "Trusty," a blockchain-based traceability platform enhancing data collection. The study assesses Trusty's adoption in two small Italian food companies, revealing positive impacts on data management, stakeholder/customer interaction, and market access. Despite these benefits, technical and economic challenges were identified, emphasizing the considerable potential of blockchain technology for food traceability.

Gupta & Shankar [22] study the potential of blockchain in addressing food insecurity in India. Employing blockchain for a sustainable traceability system, they find it effective in closing supply chain gaps, ensuring end-to-end traceability from raw product procurement to consumers.

Tolentino-Zondervan et al. [31] discuss blockchain's application in fisheries and aquaculture, highlighting its use for traceability and payments. They refer to the "Provenance" project as an example, employing blockchain to track captured fish, prevent certificate double spending, and establish an open traceability system.

Tan et al. [34] explore blockchain's use in tracing Halal food within Malaysian supply chains. Their decentralized system ensures security, transparency, and compliance with Halal standards using smart contracts. This approach aims to prevent fraud, minimize waste, enhance efficiency, and build consumer trust.

4 Method

To examine the consumer requirements for food traceability information in Austria an Online survey was conducted between April and Mai 2023. In the following it is explained how the survey was designed, conducted, and what the demographics of the study participants are.

4.1 Design of the Survey

To identify relevant questions for examining consumer requirements for food traceability information in Austria, we extracted traceability elements from literature, primarily relying on Román et al.'s meta-study [39], which encompassed 72 studies from 32 countries. Additional sources included Petrescu et al. [7], the European Food Safety Authority [40], Meyerding et al. [41], and Xu et al. [42], resulting in a total of 157 traceability attributes. After accounting for duplicates and synonymous attributes, the final count reduced to 39.

The 39 attributes were categorized into five top-level groups: origin, sustainability, freshness, cultivation and production methods, and transportation. From these categories, a 16-question questionnaire (see also Table 3) was developed. The first section, "Shopping Behavior," aims to understand respondents' general shopping habits, and a screening question ensures the inclusion of Austrian customers. The subsequent section, "Interest in Traceability Information," addresses the 39 traceability elements, each represented by one question within its corresponding category. Every question queried multiple attributes corresponding to the respective category. This section provides insights into food traceability interests. The final part collects sociodemographic data for analysis purposes.

The questionnaire was generated using the service "SurveyMonkey" [43]. For the questions regarding "Interest in traceability information" a random sorting of the possible answers was used in order not to influence the voting behavior by the order of the attributes. The questionnaire was distributed on personal and institutional social media platforms (Facebook, Instagram, LinkedIn) by sharing the survey link. On LinkedIn and Facebook, specific groups in which surveys can be shared were used. In addition, the website "Surveycircle" [44] was also used to reach additional survey participants. After 2 weeks, another reminder was sent out or shared.

4.2 Survey Demographics

A total of 186 entries were recorded. From these, 60 entries were not completed in full or were aborted. A further 2 entries were disqualified in the filter question as to whether

Table 3. Survey Outline

Question	Choices	Type
Shopping Behavior		
Current country for main purchase	2	Closed
Purchasing manager	4	Closed
Number of people for whom purchases are made	6	Closed
Household description	6	Closed
Frequency of shopping	5	Closed
Places of purchase	8	multiple selection; closed
Interest in traceability information		
Information relating to the origin of the food	5	Likert scale; closed
Information relating to the freshness of the food		
Information relating to production methods and cultivation of food		
Information relating to sustainability		
Information relating to the transportation of food		
Demographics		
Gender	3	Closed
Age	Free text	Open
Highest educational qualification	7	Closed
Occupational situation	6	Closed
Net household income	12	Closed
All questions translated from German		

the main purchase had been made in Austria in the last six months. Leaving 124 entries which were used for further analysis.

The evaluation was carried out using Microsoft Excel and JASP [45].

Regarding participants demographics, the age distribution revealed that 31% were less than 25 years old, 52% were between 25 and 35 years, 9% were aged 36–50, and 9% were above 50. Regarding education, 58% held a university degree, 27% had a baccalaureate, 10% completed an apprenticeship, 3% had a high school diploma, and 2% had other educational backgrounds. The gender distribution showed 29% male and 71% female participants. In terms of occupation, 55% were employed, 36% were students, 6% were in retirement, and 2% were unemployed, with 1% being pupils. All participants were of Austrian nationality. Regarding household composition, 37% were couples without children, 23% were single households, 14% were couples with children, 9% were single individuals living with parents, 2% were single individuals with children, and 15% identified as other household types. The net monthly household income varied, with 43% having incomes below €2,400, 23% between €2,400 and €3,600, 13% between €3,600 and €4,800, and 21% above €4,800.

In terms of shopping behavior among the participants, 58% shared grocery shopping responsibilities with household members, while 34% shopped alone, and 8% involved other household members. The preferred locations for grocery shopping varied, with 33% favoring supermarkets, 25% choosing discount stores, 11% opting for specialty stores, 10% selecting drugstores, and the remaining percentages spread across farmer/producer markets, farmers' markets, organic food stores, and other options. Regarding the frequency of grocery shopping per week, 65% shopped 1–2 times, 25% shopped 3–4 times, 3% shopped 5–6 times, and 1% shopped daily. Additionally, 7% reported shopping more than once a day.

It must be noted that the sample includes a substantial bias towards female respondents (71%) and a slight bias towards people with at least higher education. Other biases could not be detected.

5 Results

Resulting from the analysis of the questionnaire, the individual traceability elements were evaluated for each category and then sorted according to their mean value. After that, the categories themselves were also ranked by importance according to their mean values (see also Table 5).

In addition, the categories were evaluated with a Friedman Test to highlight which categories significantly differ from each other. The Friedman Test showed that there was a significant difference between the rating of the categories with a small to moderate effect size (X2 (4) = (95.345), $p < 0.001$, Kendall's W = 0.192). A Conover's Post Hoc Comparisons (see also Table 4) showed that all categories significantly differ from each other with the exception of the category pairs "Origin – Freshness", and "Origin – Transport".

Table 4. Conover's post hoc comparisons

Category	Comparison	T-Stat	df	p	p_{bonf}	p_{holm}
Origin	Freshness	1.103	492	0.271	1.000	0.271
	Cultivation and Production	4.024	492	<.001	<.001	<.001
	Sustainability	6.905	492	<.001	<.001	<.001
	Transport	1.818	492	0.070	0.697	0.139
Freshness	Cultivation and Production	2.921	492	0.004	0.036	0.018
	Sustainability	5.802	492	<.001	<.001	<.001
	Transport	2.921	492	0.004	0.036	0.018
Cultivation and Production	Sustainability	2.880	492	0.004	0.041	0.018
	Transport	5.842	492	<.001	<.001	<.001
Sustainability	Transport	8.723	492	<.001	<.001	<.001

The category (1) sustainability has the highest mean value of 4.12, followed by (2) cultivation and production methods with a mean value of 3.91. (3) origin follows with a mean value of 3.66 and (4) freshness with 3.65. The last category is (5) transportation. As such it can be argued that there is interest in traceability information and that this interest differs significantly between the categories. Only the category origin seams not to differ significantly from the categories "Transport" and "Freshness" indicating a similar perception of importance according to the survey participants.

Table 5. Interest in traceability information

Category	Traceability Attribute	Mean	ER
Sustainability (M = 4.12)	Species-appropriate animal husbandry	4,59	2
	Environmentally friendly production & environmental impact	4,29	5
	Recycling	4,19	8
	Workers' rights & working conditions, child labor, etc	4,08	11
	CO2 footprint of the product	4,06	12
	Seal (e.g., organic, fair trade)	3,98	15
	Amount of packaging, waste generated	3,90	18
	Consumed resources	3,87	19
Cultivation And Production Methods (M = 3.91)	Type of animal husbandry (free-range/stable/etc.)	4,60	1
	Organic/conventional farming	4,24	6
	Use of pesticides for fruit and vegetables	4,21	7
	Type of farming method for fish	3,94	17
	Type of fishing method for fish	3,59	26
	Cultivation method	3,44	28
	Processing steps of the product	3,34	33
Origin (M = 3.66)	Country of origin for meat	4,55	3
	Country of origin for fruit & vegetables	4,52	4
	Exact place of origin for fish	4,14	9
	Country of aquaculture/fishing area for fish	4,13	10
	Additional information on the animal species/fish species	4,00	13
	Exact place of origin for fruit & vegetables	3,97	16
	Inspection protocol of the food	3,73	23
	Name of the producer(s)	3,35	32
	Address of the producer(s)	3,11	35
	Name of the supplier(s)	2,91	37
	Address of the supplier(s)	2,79	38
	Additional information about the company	2,67	39

(*continued*)

Table 5. (*continued*)

Category	Traceability Attribute	Mean	ER
Freshness (M = 3.65)	Best before date	4,00	13
	Meat slaughter date	3,77	21
	Catch date fish	3,76	22
	Packing date of the food	3,54	27
	Laying date for eggs	3,42	29
	Harvest/picking date for fruit & vegetables	3,40	30
Transport (M = 3.48)	CO2 consumption due to transportation	3,81	20
	Tracking compliance with the cold chain	3,66	24
	Distance from the field to the shelf in km	3,65	25
	Time from field to shelf	3,40	30
	Main means of transport used	3,27	34
	All means of transportation used	3,08	36

ER = Element rank
M = Mean interest value

The literature review reveals that blockchain technology can generally trace all identified traceability attributes. However, each attribute has its own set of advantages and disadvantages when implemented using blockchain (see also Table 6).

Table 6. Advantage and Disadvantage of using Blockchain for Traceability Element

Category	Traceability Attribute	Blockchain Advantage	Blockchain Disadvantage
Sustainability	Species-appropriate animal husbandry	Provides transparency and verification of the animal welfare standards along the supply chain	Not the most efficient or cost-effective solution compared to other certification schemes
	Environmentally friendly production & environmental impact	Enables the measurement and reporting of the environmental impact of the production processes	High environmental cost due to its energy consumption and carbon footprint

(*continued*)

Table 6. *(continued)*

Category	Traceability Attribute	Blockchain Advantage	Blockchain Disadvantage
	Recycling	Facilitates the tracking and verification of the recycling process	Recycling may not need the high level of security and immutability that blockchain provides
	Workers' rights & working conditions	Enhances traceability and accountability of the labor practices along the supply chain	Blockchain alone cannot guarantee the accuracy and validity of the data, and it may face ethical and legal issues regarding the privacy and consent of the workers
	CO2 footprint of the product	Enables the calculation and verification of the carbon footprint of the product, based on the data from the production, transportation, and consumption stages	Significant carbon footprint, and it may not be the most efficient or scalable solution for this purpose
	Seal	Provides transparency and verification how seal got granted	May not add much value to the existing certification schemes
	Amount of packaging, waste generated	Facilitates the tracking and reduction of the packaging and waste generated along the supply chain, by providing incentives and feedback mechanisms	Requires the complex integration of physical and digital assets. Packaging and waste may not need the high level of security and immutability that blockchain provides
	Consumed resources	Enables the measurement and optimization of the resources consumed along the supply chain	High resource consumption, and thus may not be the most efficient or scalable solution for this purpose

(continued)

Table 6. *(continued)*

Category	Traceability Attribute	Blockchain Advantage	Blockchain Disadvantage
Cultivation and Production Methods	Type of animal husbandry	Provides transparency and accountability for animal welfare and environmental impact	May not be the most efficient or cost-effective solution
	Organic/conventional farming	Verifies the authenticity and quality of organic products and prevent fraud and mislabeling	May not add much value to the existing certification schemes
	Use of pesticides for fruit and vegetables	Tracks the origin and amount of pesticides used and ensure food safety and consumer trust	May not be the most efficient or scalable solution, depends on the accuracy of the data input
	Type of farming method for fish	Improves the traceability and sustainability of aquaculture and reduce the risk of disease and contamination	May not add much value to the existing certification schemes
	Type of fishing method for fish	Enhance the transparency and efficiency of the seafood supply chain and combat illegal, unreported, and unregulated fishing	May face social and cultural barriers in engaging and educating the stakeholders
	Cultivation method	Enables the traceability and verification of different cultivation methods	Require a high level of standardization and coordination among the participants and the regulators
	Processing steps of the product	Provides a complete and immutable record of the processing steps of the product, and facilitate quality control and recall management	May increase the complexity and overhead of the processing operations and the data management

(continued)

Table 6. (*continued*)

Category	Traceability Attribute	Blockchain Advantage	Blockchain Disadvantage
Origin	Country of origin for meat	Ensures food safety, quality, and authenticity, and prevents fraud and mislabeling	May not add much value to the existing certification schemes
	Country of origin for fruit & vegetables	Ensures food safety, quality, and authenticity, and prevents fraud and mislabeling	May not add much value to the existing certification schemes
	Exact place of origin for fish	Ensures food safety, quality, and authenticity, and combat illegal, unreported, and unregulated fishing	May not add much value to the existing certification schemes
	Country of aquaculture/fishing area for fish	Ensures food safety, quality, and authenticity, and combat illegal, unreported, and unregulated fishing	May not add much value to the existing certification schemes
	Additional information on the animal species/fish species	Ensures food safety, quality, and authenticity, and combat illegal, unreported, and unregulated fishing	May not be the most efficient or cost-effective solution
	Exact place of origin for fruit & vegetables	Ensures food safety, quality, and authenticity, and prevents fraud and mislabeling	May not add much value to the existing certification schemes
	Inspection protocol of the food	Ensures food safety, quality, and authenticity, and prevents fraud and mislabeling	May face regulatory and legal challenges for certification and compliance

(*continued*)

Table 6. (*continued*)

Category	Traceability Attribute	Blockchain Advantage	Blockchain Disadvantage
	Name of the producer(s)	Ensures food safety, quality, and authenticity, and prevents fraud and mislabeling	May not be the most efficient or cost-effective solution
	Address of the producer(s)	Ensures food safety, quality, and authenticity, and prevents fraud and mislabeling	May not be the most efficient or cost-effective solution
	Name of the supplier(s)	Ensures food safety, quality, and authenticity, and prevents fraud and mislabeling	May not be the most efficient or cost-effective solution
	Address of the supplier(s)	Ensures food safety, quality, and authenticity, and prevents fraud and mislabeling	May not be the most efficient or cost-effective solution
	Additional information about the company	Ensures food safety, quality, and authenticity, and prevents fraud and mislabeling	May not be the most efficient or cost-effective solution
Freshness	Best before date	Ensures accuracy and validity and prevents fraud and mislabeling	May not add much value to the existing certification schemes
	Meat slaughter date	Ensures food safety, quality, and authenticity, and prevent fraud and mislabeling	May not add much value to the existing certification schemes
	Catch date fish	Ensures food safety, quality, and authenticity, and combat illegal, unreported, and unregulated fishing	May not add much value to the existing certification schemes

(*continued*)

Table 6. *(continued)*

Category	Traceability Attribute	Blockchain Advantage	Blockchain Disadvantage
	Packing date of the food	Ensures accuracy and validity and prevents fraud and mislabeling	May not add much value to the existing certification schemes
	Laying date for eggs	Ensures accuracy and validity and prevents fraud and mislabeling	May not add much value to the existing certification schemes
	Harvest/picking date for fruit & vegetables	Ensures accuracy and validity and prevents fraud and mislabeling	May not add much value to the existing certification schemes
Transport	CO2 consumption due to transportation	Enables the calculation and verification of the carbon footprint of the product	Significant carbon footprint, and it may not be the most efficient or scalable solution for this purpose
	Tracking compliance with the cold chain	Ensure the integrity and transparency of the temperature and humidity data along the cold chain, which can improve food safety and quality	May face challenges in integrating with the IoT devices and sensors that monitor the cold chain, thus making it not cost effective
	Distance from the field to the shelf in km	Provides accurate real-time information on the origin and location of the products, which can enhance consumer trust	May not be the most efficient or cost-effective solution
	Time from field to shelf	Streamlines the processes and transactions involved in the supply chain, such as documentation, verification, and payment, thus reducing time to shelf	May increase the time from field to shelf in some situations, such as when the network is congested, the consensus is slow, or the data is large

(continued)

Table 6. *(continued)*

Category	Traceability Attribute	Blockchain Advantage	Blockchain Disadvantage
	Main means of transport used	Can help optimize the logistics efficiency and reduce the costs	May not be the most efficient or cost-effective solution
	All means of transportation used	Traces and audits all the means of transportation used for each product, which can help prevent fraud, theft, and loss	May require a high level of coordination and collaboration among all the means of transportation used increasing complexity and diversity along the supply chain

6 Discussion

Reflecting on the obtained results, it became clear that consumers have different preferences and priorities for traceability information in the food industry. The results show that consumers rank the categories of traceability information in the following order of importance: (1) sustainability, (2) cultivation and production methods, (3) origin, (4) freshness, and (5) transportation. When the individual traceability elements were analyzed, regardless of their category, consumers expressed the highest interest in the type and quality of animal husbandry, followed by the country of origin of meat, fruit, and vegetables. Moreover, consumers indicated a preference for detailed information on the products regarding sustainability and resource consumption. These findings suggest that consumers are more concerned about the ethical and environmental aspects of the food products, and that they value the authenticity and quality of the products. However, these findings also indicate that consumers may have different expectations and needs for different products, and that they may not be aware of or interested in some of the attributes that blockchain can trace, such as "Consumed resources", the "Processing steps of the product", or, more general, the packaging, and the transportation.

Regarding the advantages and disadvantages to make the identified traceability attributes traceable via blockchain a multifaceted picture emerged.

In the pursuit of **sustainability** within supply chain traceability, blockchain technology emerges as a promising tool. The transparency and verification capabilities of blockchain offer substantial benefits in ensuring species-appropriate animal husbandry and enhancing accountability in labor practices, providing stakeholders with unprecedented visibility. However, its application faces challenges, notably in terms of efficiency and cost-effectiveness compared to established certification schemes. Additionally, while blockchain enables the measurement of environmental impact, recycling processes, and resource consumption, its own energy-intensive nature and significant carbon footprint may pose counterproductive challenges. The technology's role in certifying seals and reducing packaging waste raises questions about its added value compared to existing systems. Moreover, the intricate integration requirements of physical and digital assets underscore potential limitations in applying blockchain to certain aspects of traceability, where the high level of security and immutability it provides may exceed practical necessities. Thus, while blockchain holds promise in advancing sustainability, careful consideration is required to optimize its benefits and navigate inherent limitations in the quest for a more environmentally responsible and ethically sound supply chain.

In examining traceability attributes related to **cultivation and production methods**, blockchain technology demonstrates considerable promise alongside notable challenges. The transparency and accountability it offers for animal husbandry practices contribute to enhanced animal welfare and reduced environmental impact, although concerns arise regarding its efficiency and cost-effectiveness compared to alternative certification schemes. For organic and conventional farming, blockchain's potential to verify product authenticity and prevent fraud may be limited, raising questions about its added value within existing certification frameworks. Tracking the origin and usage of pesticides for fruits and vegetables through blockchain can ensure food safety and consumer trust, contingent upon accurate data input, but scalability challenges may arise. In the context of fish farming and fishing methods, blockchain's role in improving traceability and sustainability faces skepticism about its value addition to current certification schemes and potential cultural barriers. The traceability of cultivation methods requires a high level of standardization and coordination among participants and regulators, potentially limiting its applicability. While the provision of immutable records for processing steps via the blockchain aids quality control and recall management, concerns about increased complexity and operational overhead warrant careful consideration. Thus, while blockchain offers notable advantages in cultivating transparency, traceability, and accountability, its application in these contexts necessitates careful evaluation of efficiency, scalability, and integration challenges to realize its full potential in advancing responsible and sustainable practices within the supply chain.

In evaluating traceability attributes related to the **origin** of products, blockchain technology emerges as a potential solution to ensure food safety, quality, and authenticity, while mitigating fraud and mislabeling risks. The application of blockchain to specify the country of origin for meat, fruit, and vegetables, as well as the exact place of origin for fish, and aquaculture/fishing areas, holds the promise of enhancing traceability and combating illegal activities in the supply chain. However, it must be noted that the technology's value addition to existing certification schemes might be low, especially regarding its efficiency and cost-effectiveness. Additional information on animal and fish species, as well as details about producers, suppliers, and companies, can be safeguarded through blockchain, ensuring transparency and authenticity. Nonetheless, the technology again may face challenges in terms of efficiency and cost-effectiveness, raising questions about its optimal utilization. The use of blockchain in documenting inspection protocols for food aims at ensuring safety, quality, and authenticity, yet regulatory and legal challenges may impede seamless certification and compliance. In summary, while blockchain technology exhibits considerable potential in enhancing traceability and securing the authenticity of product origins, careful consideration is required to balance its advantages against concerns related to efficiency, cost-effectiveness, and integration within existing certification frameworks.

In assessing traceability attributes related to **freshness**, blockchain technology emerges as a potential tool to ensure accuracy, validity, and prevent fraud and mislabeling in various product categories. The incorporation of blockchain for specifying the best-before date, meat slaughter date, catch date for fish, packing date of food, laying date for eggs, and harvest/picking date for fruits and vegetables holds the promise of enhancing traceability and safeguarding against fraudulent activities in the supply chain. However, one should be concerned regarding the technology's value addition to existing certification schemes, especially about its efficiency and cost-effectiveness. While blockchain ensures accuracy and validity in recording freshness attributes, its optimal utilization is questionable in terms of providing substantial advantages beyond established certification frameworks. The technology's potential may not be fully realized if it does not significantly enhance existing practices. In summary, while blockchain technology showcases potential benefits in improving traceability and preventing mislabeling, careful consideration is required to balance its advantages against concerns related to efficiency, cost-effectiveness, and integration within existing certification frameworks.

In examining traceability attributes related to **transportation**, its capability to calculate and verify a product's carbon footprint based on transportation is advantageous, but concerns arise about blockchain's own significant carbon footprint, potentially affecting efficiency. Tracking cold chain compliance with blockchain has the potential to improve food safety, though integrating with IoT devices and sensors poses challenges. Providing real-time information on the distance from field to shelf can enhance consumer trust, but concerns about blockchain's efficiency and cost-effectiveness for this purpose exist. While streamlining processes can optimize logistics, challenges like network congestion may arise. Blockchain's potential in tracing and auditing transportation for fraud prevention introduces complexity, requiring coordination. In summary, while blockchain enhances transparency, considerations are needed for its efficiency, cost-effectiveness, and integration challenges within existing supply chain practices.

As every research this research comes with certain limitations. While an extensive literature review was conducted, it cannot guarantee the identification of all relevant literature, especially regarding blockchain applications for food supply chain traceability and traceability element collection. The sample size of the online survey ($n = 124$) is limited, with most participants being female, under 35 years old, and highly educated, affecting the general applicability of the data. The use of convenience sampling may further limit result generalizability.

Future research might thus conduct survey with a greater sample size to investigate if different consumer groups have different requirements for traceability items. In addition, research from the perspective of retail companies on the topic of traceability and the use of blockchain technology is required to investigate the advantages and disadvantages for retailers when using blockchain technology. Furthermore, the scalability of blockchain technology is a challenge, especially when it comes to processing large volumes of transactions in real time. Further research is thus needed to develop efficient and scalable blockchain applications for food traceability that meet the requirements of the food industry. From a methodological perspective, future research might extent our search string to include more synonyms and thus find an even broader spectrum of papers describing possible applications of blockchain technology in the food supply chain.

7 Conclusion

The aim of this paper was to give insights which traceability information are most relevant for Austrian consumers (RQ1) and to investigate what the advantages and disadvantages of satisfying the identified consumer requirements for food traceability in Austria via blockchain are (RQ2).

In regard to RQ1, an online survey was conducted. The survey revealed that interest in (1) sustainability information was greatest, followed by (2) cultivation and production methods. This is followed by information on (3) origin, (4) freshness and finally (5) transportation. If the individual traceability elements are considered, regardless of which category they belong to, consumers are most interested in information on the type of animal husbandry and species-appropriate animal husbandry, followed by information on the country of origin of meat, fruit, and vegetables. However, respondents also want detailed information on products regarding sustainability and resource consumption.

In regard to RQ2, the integration of blockchain technology for tracing the identified traceability attributes shows promise but requires careful consideration of its advantages and challenges. Blockchain's transparency and verification capabilities hold significant benefits for ensuring animal welfare standards, reducing environmental impact, and enhancing accountability in labor practices. However, challenges such as efficiency and cost-effectiveness compared to existing certification schemes, and the technology's own environmental footprint, need thoughtful navigation. The application of blockchain to trace cultivation and production methods, product origins, freshness attributes, and transportation details reveals potential benefits but raises concerns about scalability, integration, and added value to existing systems. While blockchain demonstrates considerable potential in advancing transparency and traceability within the supply chain, a nuanced approach is essential to optimize its benefits while addressing inherent limitations.

Acknowledgments. This research is sponsored by the Government of Upper Austria as part of the excellence network for logistics Logistikum.Retail.

References

1. Casino, F., Kanakaris, V., Dasaklis, T.K., et al.: Modeling food supply chain traceability based on blockchain technology. IFAC-PapersOnLine **52**(13), 2728–2733 (2019)
2. Aung, M.M., Chang, Y.S.: Traceability in a food supply chain: Safety and quality perspectives. Food Control **39**, 172–184 (2014)
3. Astill, J., Dara, R.A., Campbell, M., et al.: Transparency in food supply chains: a review of enabling technology solutions. Trends Food Sci. Technol. **91**, 240–247 (2019)
4. K. Hoelzer, A.I., Switt, M., Wiedmann, M., et al.: Emerging needs and opportunities in foodborne disease detection and prevention: from tools to people. Food Microbiol. **75**, 65–71 (2018)
5. Qian, J., Ruiz-Garcia, L., Fan, B., et al.: Food traceability system from governmental, corporate, and consumer perspectives in the European Union and China: a comparative review. Trends Food Sci. Technol. **99**, 402–412 (2020)
6. Trienekens, J.H., Wognum, P.M., Beulens, A., et al.: Transparency in complex dynamic food supply chains. Adv. Eng. Inform. **26**(1), 55–65 (2012)

7. Petrescu, D.C., Vermeir, I., Petrescu-Mag, R.M.: Consumer understanding of food quality, healthiness, and environmental impact: a cross-national perspective. Int. J. Environ. Res. Public Health **17**(1), 169 (2019)
8. Trienekens, J.H. (ed.): European pork chains: Diversity and quality challenges in consumer-oriented production and distribution. Wageningen Academic Publishers, Wageningen (2009)
9. Gunasekaran, A., Yusuf, Y.Y., Adeleye, E.O., et al.: Agile manufacturing: an evolutionary review of practices. Int. J. Prod. Res. **57**(15–16), 5154–5174 (2019)
10. Dutta, P., Choi, T.-M., Somani, S., et al.: Blockchain technology in supply chain operations: applications, challenges and research opportunities. Transp. Res. Part E, Logistics Transp. Rev. **142**, 102067 (2020)
11. Butt, R.M., Fahim, S.M.: Tractability, the mantra of block chain technology in the food supply chain. In: Blockchain Driven Supply Chain Management: A multi-dimensional. Mubarik, M.S., Shahbaz, M., (Eds.), pp. 101–118, Springer, [S.l.] (2023). https://doi.org/10.1007/978-981-99-0699-4_7
12. Sarmah, S.S.: Understanding blockchain technology. Comput. Sci. Eng. **2018**, 23–29 (2018)
13. Swan, M.: Blockchain: Blueprint for a new economy, O'Reilly, Sebastopol, CA (2015)
14. Golosova, J., Romanovs, A.: The advantages and disadvantages of the blockchain technology. In: 2018 IEEE 6th Workshop on Advances in Information, Electronic and Electrical Engineering (AIEEE), pp. 1–6. IEEE (2018)
15. Niranjanamurthy, M., Nithya, B.N., Jagannatha, S.: Analysis of blockchain technology: pros, cons and SWOT. Clust. Comput. **22**(S6), 14743–14757 (2019)
16. Hülsbömer, S., Genovese, B.: Was ist Blockchain? COMPUTERWOCHE **2016**(02), 05 (2016)
17. Song, W., Shi, S., Xu, V., et al.: Advantages & Disadvantages of Blockchain Technology. Blockchain Technology. https://blockchaintechnologycom.wordpress.com/2016/11/21/advantages-disadvantages/. Accessed 23 Nov 2016
18. Dataflair. Advantages and Disadvantages Of Blockchain Technology. DataFlair, vol. 2018, 01.06.2018
19. Wissenschaftliche Dienste des Deutschen Bundestages, "Daten zu Transaktionen von Zahlungssystemen. vol. 2022 (2022)
20. vom Brocke, J., Simons, A., Niehaves, B., et al.: Reconstructing the giant: On the importance of Rigour in documenting the literature search process. In: ECIS 2009 Proceedings, no. 161 (2009)
21. Cooper, H.M.: Organizing knowledge syntheses: a taxonomy of literature reviews. Knowl. Soc. **1**(1), 104–126 (1988)
22. Gupta, R., Shankar, R.: Managing food security using blockchain-enabled traceability system. Benchmarking: Int. J. **31**(1), 53–74 (2023)
23. Rogerson, M., Parry, G.C.: Blockchain: case studies in food supply chain visibility. Supply Chain Manag. Int. J. **25**(5), 601–614 (2020)
24. Kim, M., Hilton, B., Burks, Z., et al.: Integrating blockchain, smart contract-tokens, and IoT to design a food traceability solution. In: 2018 IEEE 9th Annual Information Technology, Electronics and Mobile Communication Conference: 1–3 November 2018, Vancouver, BC, Canada, pp. 335–340, Institute of Electrical and Electronics Engineers, Piscataway, New Jersey (2018)
25. Sai Radha Krishna, G., Rekha, P.: Food supply chain traceability system using blockchain technology. In: 2022 8th International Conference, pp. 370–375
26. Bumblauskas, D., Mann, A., Dugan, B., et al.: A blockchain use case in food distribution: do you know where your food has been? Int. J. Inf. Manage. **52**, 102008 (2020)
27. Compagnucci, L., Lepore, D., Spigarelli, F., et al.: Uncovering the potential of blockchain in the agri-food supply chain: an interdisciplinary case study. J. Eng. Tech. Manage. **65**, 101700 (2022)

28. Cozzio, C., Viglia, G., Lemarie, L., et al.: Toward an integration of blockchain technology in the food supply chain. J. Bus. Res. **162**, 113909 (2023)
29. Kamilaris, A., Fonts, A., Prenafeta-Boldύ, F.X.: The rise of blockchain technology in agriculture and food supply chains. Trends Food Sci. Technol. **91**, 640–652 (2019)
30. Patel, D., Sinha, A., Bhansali, T., et al.: Blockchain in food supply chain. Procedia Comput. Sci. **215**, 321–330 (2022)
31. Tolentino-Zondervan, F., Ngoc, P.T.A., Roskam, J.L.: Use cases and future prospects of blockchain applications in global fishery and aquaculture value chains. Aquaculture **565**, 739158 (2023)
32. Granillo-Macías, R., González Hernández, I.J., Olivares-Benítez, E.: Logistics 4.0 in the agri-food supply chain with blockchain: a case study. In: International Journal of Logistics Research and Applications, pp. 1–21 (2023)
33. Roy, R., Chekuri, K., Sandhya, G., et al.: Exploring the blockchain for sustainable food supply chain. J. Inf. Optim. Sci. **43**(7), 1835–1847 (2022)
34. Tan, A., Gligor, D., Ngah, A.: Applying Blockchain for Halal food traceability. Int. J. Log. Res. Appl. **25**(6), 947–964 (2022)
35. Xu, Y., Li, X., Zeng, X., et al.: Application of blockchain technology in food safety control: current trends and future prospects. Crit. Rev. Food Sci. Nutr. **62**(10), 2800–2819 (2022)
36. Shew, A.M., Snell, H.A., Nayga, R.M., et al.: Consumer valuation of blockchain traceability for beef in the United States. Appl. Econ. Perspect. Policy **44**(1), 299–323 (2022)
37. Tayal, A., Solanki, A., Kondal, R., et al.: Blockchain-based efficient communication for food supply chain industry: transparency and traceability analysis for sustainable business. Int. J. Commun. Syst. **34**(4), e4696 (2021)
38. Ecocert. Organic agriculture certification and label in the USA. https://www.ecocert.com/en/certification-detail/organic-farming-usa-usda-nop. Accessed 10 Dec 2023
39. Román, S., Sánchez-Siles, L.M., Siegrist, M.: The importance of food naturalness for consumers: results of a systematic review. Trends Food Sci. Technol. **67**, 44–57 (2017)
40. EFSA. Europäer zu den Lebensmittelfragen von heute: neue EU-weite Erhebung erscheint am ersten Internationalen Tag der Lebensmittelsicherheit, 06 July 2019
41. Meyerding, S.G., Trajer, N., Lehberger, M.: What is local food? The case of consumer preferences for local food labeling of tomatoes in Germany. J. Clean. Prod. **207**, 30–43 (2019)
42. Xu, L., Yang, X., Wu, L., et al.: Consumers' willingness to pay for food with information on animal welfare, lean meat essence detection, and traceability. Int. J. Environ. Res. Public Health **16**(19) 3616 (2019)
43. SurveyMonkey. SurveyMonkey: The World's Most Popular Free Online Survey Tool. https://www.surveymonkey.com/. Accessed 10 Dec 2023
44. SurveyCircle. SurveyCircle|Umfrageteilnehmer finden, Forschung unterstützen. 12/10/2023. https://www.surveycircle.com/de/
45. JASP. A Fresh Way to Do Statistics. https://jasp-stats.org/. Accessed 29 Sept 2023

Author Index

F. F.-H. Nah and K. L. Siau (Eds.): HCII 2024, LNCS 14721, pp. 277–279, 2024.
https://doi.org/10.1007/978-3-031-61318-0

Z

Printed in the United States
by Baker & Taylor Publisher Services